P9-CJV-183

# Vibrational Spectroscopy

## for Medical Diagnosis

# Vibrational Spectroscopy
# for Medical Diagnosis

Editors

**Max Diem**
*Northeastern University, Boston, MA, USA*

**Peter R. Griffiths**
*University of Idaho, Moscow, ID, USA*

**John M. Chalmers**
*VS Consulting, Stokesley, UK*

**WILEY**

*Other Wiley Editorial Offices*

John Wiley & Sons Inc., 111 River Street,
Hoboken, NJ 07030, USA

Jossey-Bass, 989 Market Street,
San Francisco, CA 94103-1741, USA

Wiley-VCH Verlag GmbH, Boschstr. 12,
D-69469 Weinheim, Germany

John Wiley & Sons Australia Ltd, 42 McDougall Street,
Milton, Queensland 4064, Australia

John Wiley & Sons (Asia) Pte Ltd, 2 Clementi Loop #02-01,
Jin Xing Distripark, Singapore 129809

John Wiley & Sons Canada Ltd, 6045 Freemont Boulevard
Mississauga, Ontario, Canada L5R 4J3

Wiley also publishes its books in a variety of electronic formats. Some content that appears in
print may not be available in electronic books.

*Library of Congress Cataloging-in-Publication Data*

Vibrational spectroscopy for medical diagnosis / editors, Max Diem, John M. Chalmers,
Peter R. Griffiths.
   p. ; cm.
  Includes bibliographical references and index.
  ISBN 978-0-470-01214-7 (hb : alk. paper)
  1. Vibrational spectra–Diagnostic use. 2. Near infrared reflectance spectroscopy–Diagnostic
use. I. Diem, Max, 1947- II. Chalmers, John M. III. Griffiths, Peter R., 1942-
  [DNLM: 1. Spectrum Analysis, Raman. 2. Diagnostic Imaging–methods.
3. Spectrophotometry, Infrared. 4. Spectroscopy, Near-Infrared. WB 288 V626 2008]
RC78.7.S65V53 2008
616.07'57–dc22
            2007050357

*British Library Cataloguing in Publication Data*

A catalogue record for this book is available from the British Library

ISBN 978-0-470-01214-7 (HB)

Typeset in 10/12.5pt Times by Laserwords Private Limited, Chennai, India.
Printed and bound by Grafos SA, Barcelona, Spain.
This book is printed on acid-free paper responsibly manufactured from sustainable forestry in
which at least two trees are planted for each one used for paper production.

# Preface

Statistics from many countries constantly inform us how prevalent cancer is worldwide, even though a particular variation may be more predominant in one geographical region than another. The traumatic cost on human lives, the dramatic cost of treatment, and the need for rapidly advancing diagnoses and improving diagnostic methods are explicitly seen. In this book, the application of infrared and Raman spectroscopy to the diagnosis of cancer and certain other diseases is analyzed.

Cancer is the second largest cause of death (after heart disease) in North America and Europe. According to a report issued by the National Cancer Institute (NCI) in 2006, the four most common types of cancer in the USA are lung (54.8 deaths per 100 000 in 2002), prostate (28.0 deaths per 100 000), breast (25.4 deaths per 100 000), and colorectal cancer (19.6 deaths per 100 000). The rate of new cases of cancer was 488.6 per 100 000 Americans in 2002, which was approximately the same as a year earlier. During that year, the death rate for all cancers declined by about 1%, to 193.6 per 100 000, down from 195.7 a year earlier and is continuing a steady downward trend.

The projected numbers of new cases of, and deaths from, different types of cancer in the USA in 2007 are given in Table 1.

The numbers in other countries may vary, but the general conclusion that cancer is a major killer and that early diagnosis reduces the risk of mortality is universal. For example, the chapter in this book on malignancy in esophageal and bladder tissue cites that (i) one in three people in the UK will be struck down by cancer in their lifetime and three-quarters of them will succumb to the disease; (ii) the incidence of bladder cancer in the UK is 8% in men and 3% in women, making it the fourth most common cancer in men and the eighth in women; and (iii) in 2002 in the UK 3% of all cancers diagnosed were found in the esophagus (7080) and 5% of all deaths were from esophageal cancer (7008), and that this incidence is rising rapidly in the developed world.

In another chapter it is stated that oral cancer is particularly prevalent in the developing economies, such as India and South-East Asia, but subject to large regional variations. For example, in India the incidence ranges from 7.2/100 000 population in Bhopal to 2.4/100 000 population in Barshi. In the USA, in 2003, 20 424 new cases of oral cavity and pharyngeal, 8981 laryngeal, 1091 nasal, and 5738 thyroid cancers were diagnosed, while in the UK, 7948 new cases, 5504 males and 2444 females, of thyroid cancer were registered in 2000, making it the eighth most common cancer (3% of total cancer diagnoses). In a third chapter it is noted that there are more than 273 000 deaths due to cervical cancer worldwide each year and it accounts for 9% of female cancer deaths, even though in the developed world screening tests are significantly reducing the incidence of cervical cancer.

Lung cancer kills almost 100 000 people per year in the USA, largely as a result of smoking. As smoking habits are changing, these numbers are declining but are still very serious. Oral cancer deaths are smaller, but still significant, in the USA and are frequently

**Table 1.**   Leading cancer types for new cancer cases and deaths in the USA by sex projected for 2007. These data exclude basal and squamous cell skin cancers except urinary bladder cancer.

| Site | New | Percent | New | Percent | Deaths | Percent | Deaths | Percent |
|---|---|---|---|---|---|---|---|---|
| Gender | M | M | F | F | M | M | F | F |
| Prostate | 218 900 | 29 | – | – | 89 500 | 9 | – | – |
| Breast | – | – | 178 500 | 26 | – | – | 40 500 | 15 |
| Lung and bronchus | 114 800 | 15 | 98 600 | 15 | 27 000 | 31 | 70 900 | 26 |
| Colon and rectum | 79 100 | 10 | 74 600 | 11 | 26 000 | 9 | 26 200 | 10 |
| Urinary bladder | 50 000 | 7 | 9600 | – | 9600 | 3 | – | – |
| Uterine corpus | – | – | 39 100 | 6 | – | – | 7400 | 3 |
| Non-Hodgkin lymphoma | 34 200 | 4 | 29 000 | 4 | 9600 | 3 | – | – |
| Melanoma of the skin | 33 900 | 4 | 26 000 | 4 | – | – | – | – |
| Kidney and renal pelvis | 31 600 | 4 | – | – | 8100 | 3 | – | – |
| Thyroid | – | – | 25 500 | 4 | – | – | – | – |
| Ovary | – | – | 22 400 | 3 | – | – | – | – |
| Leukemia | 24 800 | 3 | 19 400 | 3 | 12 300 | 4 | 9500 | 4 |
| Oral cavity and pharynx | 24 200 | 3 | – | – | – | – | – | – |
| Pancreas | 18 800 | 2 | – | – | 16 800 | 6 | 16 500 | 6 |
| Liver and intrahepatic bile duct | – | – | – | – | 11 300 | 4 | 5500 | 2 |
| Esophagus | 10 900 | – | – | – | 10 900 | 4 | – | – |
| Brain and nervous system | – | – | – | – | – | – | 5600 | 2 |
| Total | 766 900 | 100 | 678 100 | 100 | 289 600 | 100 | 270 100 | 100 |

After data from "Cancer Statistics 2007", A. Jemal, R. Seigel, E. Ward, T. Murray, J. Xu and M. J. Thun, *CA Cancer, J. Clin.* 2007; **57**:43–66.

caused by chewing tobacco products. Interestingly, the incidence of oral cancer in tropical countries is largely caused by betel leaf and tobacco chewing.

The rate of cancer cases diagnosed in the USA has stabilized, but the death rate is declining partly because of improved screening. The use of screening tests for breast and cervical cancers is high and remained stable between 2000 and 2003. As of 2003, 69.7% of women over 40 had had a mammogram in the last 2 years, up from just 29% in 1987, and 79.2% had a Papanicolou (Pap) smear test for cervical cancer, up from 73.7%. As a result, deaths from breast cancer are decreasing. However, the NCI report averred that screening for colorectal cancer remains low—just 43.4% of adults over 50 had had at least one endoscopy as of 2003, though that was up from 27.3% in 1987. On the down side, there have been continuing increases in the incidence of prostate cancer and testicular cancer in men, as well as leukemia, non-Hodgkin lymphoma, myeloma, melanoma of skin, and cancers of the thyroid, kidney, and esophagus.

Once the possible presence of cancerous tissue has been detected, by simple screening tests such as Pap smears and mammograms, it usually must be confirmed by a biopsy followed by histopathology of the excised tissue. The most common method involves the use of a mixture of hematoxylin and eosin (H&E), which stains protein-rich regions of the tissue pink and nucleic acid–rich regions blue. This procedure is both time consuming and requires the expertise of an experienced histopathologist to classify the stained tissue.

Even so, it has been said that if you showed a stained tissue to several pathologists, each would classify the tissue somewhat differently. A fast, objective, complementary technique for the classification of tissues that will help confirm the conventional histopathology is clearly required. Also needed is a rapid means of obtaining pathological data while an operation is taking place.

Over the past 10 years, an increasing number of groups around the world has been working towards real-time, noninvasive techniques that utilize light to study abnormalities in tissue, sometimes called *optical diagnosis*. Optical diagnosis relies upon measurement of the interaction of light photons with the constituents of biological tissue. Recent technological developments have made it possible to obtain significant amounts of biochemical data from complex biological tissue in just a few seconds. The resultant data can provide an evaluation of histochemistry or morphology. This information can aid with the deduction of the pathological state of the tissue, and hence lead to a diagnosis.

Ultraviolet, visible, and infrared radiation can interact with tissue in a number of ways, including absorption (sometimes leading to fluorescence or phosphorescence), elastic and inelastic scattering, and reflection from boundary layers. In this book, the approaches designed to investigate the applications of infrared and Raman spectroscopy (known together as vibrational spectroscopy) to medical diagnosis are described. At one stage, it was optimistically believed that the application of molecular-specific spectroscopic techniques would lead to the eventual complete replacement of subjective histopathological diagnoses by objective optical detection and thus provide a real-time, sensitive, and specific measurement of the histological state of the tissue. However, until the efficacy of any of these techniques is proven, it is more likely that optical detection will be used as a complementary technique to improve targeting of biopsy selection and aid the pathologist in defining cancer aggressiveness.

There is an undoubted clinical need for optical diagnosis in a number of important situations. These include instances when sampling limitations severely restrict the effectiveness of excisional biopsy, as exemplified by the high failure rates associated with blind biopsies, whereby the clinician has to randomly select sites for sample collection. There are also situations where conventional excisional biopsy is potentially hazardous; vulnerable regions include the central nervous system, vascular system, and articular cartilage. Furthermore, situations arise where there is a need for immediate diagnosis during an investigative procedure. For example, during surgery for brain cancer, it would be tremendously beneficial if the margin of the tumor could be determined during a single operation, so that all cancerous tissue is removed while as much benign tissue as possible is left intact without the need for a second operation. Finally, when a surgeon has any doubt over a diagnosis, a second diagnosis prior to excision of an organ or lesion, ideally by a completely different technique such as a noninvasive optical probe, would be of enormous benefit to the surgeon.

Advances in both the instrumentation and data analysis software for vibrational spectroscopy that have taken place over the past 15–20 years have allowed mid-infrared and Raman spectroscopy to be applied to this important problem. The first of these was the development of microscopes for use in Fourier transform infrared (FT-IR) spectrometers in the early 1990s. Surprising as it may seem today, the market for FT-IR microscopes was originally thought to be only 10 or 20 per year. Today, however, thousands of these devices are in use worldwide on a plethora of different applications, including the rapid identification of bacteria, a topic that was shown to be possible by the elegant combination of FT-IR microscopy and hierarchical cluster analysis (HCA) that was developed by Dr Dieter Naumann's group at the Robert Koch-Institut, in Berlin, Germany.

Measuring the infrared spectra of tissue and other samples at just a few points with an FT-IR microscope is, however, scarcely sufficient for the purpose of medical diagnosis. This accomplishment was made possible by the development of imaging spectrometers for the simultaneous measurement of thousands of mid-infrared and/or Raman spectra of a single tissue sample. Perhaps not surprisingly, this work was pioneered by the research group of Dr Ira Levin at the National Institutes of Health, Bethesda, Maryland, USA. Several of the first imaging spectrometers for vibrational spectrometry that incorporated array detectors (indium antimonide and mercury cadmium telluride arrays for mid-infrared spectroscopy and charge-coupled device array detectors of Raman spectroscopy) were reported by Levin's group. We are delighted to include chapters by both Dr Naumann and Dr Levin and their coworkers in this book.

In this book, examples of the present state-of-the-art medical diagnosis of tissue and whole cells by mid-infrared and Raman spectroscopy are given. Applications to cancers of the lymph nodes, head and neck cancers, cervix, and brain are described. Other chapters deal with such varied topics as spectroscopic investigation of whole cells for cancer screening purposes and of serum samples for the antemortem identification of diseases such as transmissible spongiform encephalopathy (TSE, also known as *mad cow disease*).

The difference between the vibrational spectra of normal and diseased tissues is sufficiently small that to a classically trained spectroscopist, the measured spectra frequently look the same for all purposes. Experience in this field has shown that diagnosis can almost never be made on the basis of the simple presence or absence of one or two bands and very rarely by using the intensity ratio of two bands. The successful application of vibrational spectroscopy to medical diagnosis has become possible not only because of specific advances in the instrumentation for infrared and Raman spectrometry but, equally critically, through advances in the development of multivariate statistical techniques in analytical chemistry, a field that is known as *chemometrics*. In the past 10 years, chemometric algorithms have been extensively used for quantitative analysis and less commonly for the classification of samples (discriminant analysis). In the field of medical diagnosis, it is the latter application that is particularly important however, as these algorithms must be used to answer general questions such as "Is a particular region of tissue cancerous or normal?" or far more specific questions such as "Can gliomas be subtyped into astrocytomas, oligodendrogliomas, mixed gliomas, and ependymal tumors?" or "Can malignancy grades within the different glioma categories such as grade II (low grade astrocytoma), grade III (anaplastic astrocytoma), and grade IV (glioblastoma multiforme) be assessed?"

Exactly how this is accomplished optimally varies greatly between the research groups working on this topic. Techniques such as linear discriminant analysis (LDA), principal components analysis (PCA), hierarchical cluster analysis (HCA), artificial neural networks (ANN), and soft independent modeling of class analogies (SIMCA), inter alia, have all been successfully applied to this problem. Furthermore, the way in which spectra are pretreated prior to the application of chemometrics also varies. The height and shape of the spectral baseline frequently varies and, therefore, some type of baseline correction is usually applied. Spectra are sometimes normalized to one of the strong bands in the spectrum (often one of the amide bands of the protein that is present in the tissue sample). Some groups have found that they can obtain better results by converting the spectrum to its first or, more commonly, second derivative, while others have found that leaving the spectrum in its original format yielded the better results. In principle, the original spectrum or its first or second derivative should contain the same information, although it is weighted differently. For example, if one calculates the FT of a spectrum, weights the results by a function $x^0$, $x^1$, or $x^2$, and then calculates the inverse transform, the result is the original spectrum, its first derivative or its second derivative, respectively. Converting the spectrum to its derivative has the beneficial effect of resolving small shoulders and the adverse effect of increasing the noise level. It would be a fascinating study to find under what circumstances different ways of pretreating the data (baseline correction, smoothing, conversion to the derivative, etc.) and different algorithms (LDA, PCA, HCA, ANN, SIMCA, etc.) yield the best prediction of a given medical condition.

The opportunity for the spectroscopic and medical communities to meet together is critical. To address this need, several ongoing conferences have been held at which different aspects of medical diagnosis by vibrational spectroscopy have been discussed. The first of these was the (ongoing) biannual workshop, entitled "FT-IR Spectroscopy in Microbiological and Medical Diagnostics", run by Professor Dieter Naumann in Berlin, which was first held in 1998. There are also biannual Biomedical Optics conferences (BiOS 2004, BiOS 2006) held under the auspices of the Society for Photo-Optical Instrumentation Engineers (SPIE) at Photonics West (San Jose, CA).

Another highly successful series of meetings has been the biannual symposia entitled "Shedding Light on Disease: Optical Diagnosis for the New Millennium". The first of these (SPEC 2000) was organized by the Institute for Biodiagnosis in Winnipeg, Canada, which is still the only organization dedicated to the development of spectroscopic approaches to medical diagnosis. The second meeting in this series (SPEC 2002) was held in Rheims, France, the third (SPEC 2004) at Rutgers University, Newark, New Jersey, USA, and the most recent (SPEC 2006) at Heidelberg, Germany. The fifth meeting in this important series will be held in Sao Paulo, Brazil, in 2008. In 2007, one of the coeditors of this volume (Max Diem) ran the International Workshop on Spectral Diagnostics in Boston, MA, USA, which again looks as if it will be held biannually. Also, a workshop focused on tumor diagnosis by FT-IR and Raman spectroscopy, entitled "Tissue Characterization by Optical Spectroscopy" was run recently by Professor Reiner Salzer in Dresden, Germany, in April 2007. The aspect of this workshop that set it apart from others was that it was more oriented toward MDs than PhDs, as it is this community who must ultimately accept diagnoses made on the basis of vibrational spectroscopy. In addition, in the Europe Union (EU), DASIM (diagnostic applications of synchrotron infrared microspectroscopy) has existed over the last 3 years. This EU-funded research collaboration involves about 70 scientists and clinicians from nine European countries. While the primary purpose of this project is to advance diagnostic applications of infrared microspectroscopy and functional imaging using synchrotron light sources, the consortium considers these in relation to both relevant conventional infrared-sourced and Raman applications.

In the opinion of the three editors of this volume, the question to be asked is not whether spectroscopy and hyperspectral imaging will be used for medical diagnosis at some time in the future. Rather, the questions should be "How it will be used?", "How soon?", and "Will there be further advances in the instrumentation for spectroscopic imaging?" We hope that this book helps to shed some light on this important area of medical research.

**Max Diem**
**Peter R. Griffiths**
**John M. Chalmers**
November 2007

# Acknowledgments

Without exception, the first acknowledgment has to go to the authors of the chapters in this book, for their cooperation, timely delivery of manuscripts, and for putting up with the numerous requests that they received from the three of us (as editors) and the team at Wiley; they did this with tolerance, accepting most of our suggested amendments (where justified) speedily and with good grace. Above all, we would like to thank the authors for their excellent chapters; it has been a pleasure to review and edit them. Second, we found once again working with the team from John Wiley & Sons, in particular, Jenny Cossham and Liz Smith, to be a very positive experience. Jenny and Liz have a delightful way of arm-twisting and good-natured cajoling that kept us mostly on track and on schedule, despite some murmurings from ourselves. One of us (PRG) wishes to thank the Alexander von Humboldt Foundation for the award of a senior research fellowship that allowed him to learn more about the subject matter of this book during an enjoyable stay at the Technical University of Dresden. He also wishes to thank Professor Reiner Salzer for hosting this visit.

Finally, and yet again, we would like to thank our wives, Mary Jo Diem, Marie Griffiths, and Shelley Chalmers for putting up with the distraction and temporary loss of their husbands from domestic duties, etc!

**Max Diem**
**Peter R. Griffiths**
**John M. Chalmers**
October 2007

# Contents

# Contributors

**Keith Bambery**  *Monash University, Melbourne, Victoria, Australia*

**Hugh Barr**  *Gloucestershire Hospitals NHS Foundation Trust, Gloucester, UK*

**Michael Beekes**  *Robert Koch-Institut, Berlin, Germany*

**Rohit Bhargava**  *University of Illinois at Urbana-Champaign, Urbana, IL, USA*

**Benjamin Bird**  *Northeastern University, Boston, MA, USA*

**Susie Boydston-White**  *Northeastern University, Boston, MA, USA*

**John M. Chalmers**  *University of Nottingham, Nottingham, UK*

**Tatyana Chernenko**  *Northeastern University, Boston, MA, USA*

**Max Diem**  *Northeastern University, Boston, MA, USA*

**Rina K. Dukor**  *Biotools Inc., Jupiter, FL, USA*

**Heinz Fabian**  *Robert Koch-Institut, Berlin, Germany*

**Sheila E. Fisher**  *University of Leeds, Leeds, UK* and *University of Bradford, Bradford, UK*

**Andrew T. Harris**  *University of Leeds, Leeds, UK*

**Catherine Kendall**  *Gloucestershire Hospitals NHS Foundation Trust, Gloucester, UK*

**Christoph Krafft**  *Dresden University of Technology, Dresden, Germany*

**Peter Lasch**  *Robert Koch-Institut, Berlin, Germany*

**Ira W. Levin**  *National Institutes of Health, Bethesda, MD, USA*

**Christian Matthäus**  *Northeastern University, Boston, MA, USA*

**Don McNaughton**  *Monash University, Melbourne, Victoria, Australia*

**Miloš Miljković**  *Northeastern University, Boston, MA, USA*

**Dieter Naumann**  *Robert Koch-Institut, Berlin, Germany*

**Melissa J. Romeo**  *Northeastern University, Boston, MA, USA*

**Reiner Salzer**  *Dresden University of Technology, Dresden, Germany*

**Siegfried Scherer**  *Technische Universität München, Freising, Germany*

**Nicholas Stone**   *Gloucestershire Hospitals NHS Foundation Trust, Gloucester, UK*

**Mark J. Tobin**   *Australian Synchrotron, Clayton, Victoria, Australia*

**Mareike Wenning**   *Technische Universität München, Freising, Germany*

**Bayden R. Wood**   *Monash University, Melbourne, Victoria, Australia*

# Introduction to Spectral Imaging, and Applications to Diagnosis of Lymph Nodes

## Melissa J. Romeo[1], Rina K. Dukor[2] and Max Diem[1]

[1] Northeastern University, Boston, MA, USA
[2] Biotools Inc., Jupiter, FL, USA

## 1 INTRODUCTION

The field of spectroscopic imaging has expanded explosively in recent years, in particular with relation to biomedical diagnostics. Current instrumentation and computing technology have enabled the rapid analysis of tissue sections, resulting in fast image collection and analysis.

Spectral imaging is the acquisition of individual spectra from adjacent sample pixels, and may be achieved in several ways, discussed in detail in Section 2.4. The traditional method for mid-infrared (IR) spectroscopy is through the use of a focal-plane array (FPA) detector, which is a two-dimensional detector (typically $64 \times 64$, $128 \times 128$, or $256 \times 256$ elements or pixels). Focusing the sample onto an FPA allows spectral signals from corresponding area elements on a sample to be recorded simultaneously either in transmission or reflection mode. In another method, sometimes called *point mapping*, the sample is raster scanned, i.e., the sample is shifted stepwise through the use of an automated stage, and a spectrum is acquired at every step. A third method, sometimes called *linear mapping*, is a combination of these two techniques, in which the sample is raster scanned over a relatively small linear array (typically, $16 \times 1$ or $32 \times 1$) of individual detectors. This method allows the "quilting" of images of nonsquare samples, and offers the better signal quality of independent detector elements. Since all these technically result in an identical hyperspectral data cube (to be discussed later), we refer to all three methods as imaging techniques. Similar arguments hold for the collection of Raman hyperspectral data sets (see Section 2.4).

Compared to standard methods of histology, cytology, and biomedical microscopy, both IR and Raman spectral imaging offer the advantage of probing the biochemical nature of cells and tissues with minimal sample preparation, and without the use of dyes, stains, and fluorescing agents. Furthermore, both techniques are nondestructive in the sense that the samples can be further processed after spectral data acquisition.

The biomedical applications of IR spectral imaging cover a broad spectrum of tissue

---

*Vibrational Spectroscopy for Medical Diagnosis.* Edited by Max Diem, Peter R. Griffiths and John M. Chalmers.

and diseases including cancer, cardiomyopathy,[1] Alzheimer's disease,[2] skin[3,4] and aortic tissue[3] permeability, bone, cartilage, and mineralized tissues,[5–11] and prion diseases (such as bovine spongiform encephalopathy (BSE) and scrapie).[12–16] Within the field of cancer diagnosis, IR spectral imaging and spectral diagnoses have been widely applied, exploring the diagnostic capabilities of this technique for cervical,[17–21] colon,[22–29] pancreatic,[30] prostate,[31–36] lung,[37,38] and breast[39–43] cancers, glioma,[44–50] tumors of the oral cavity,[51–53] lymphoma,[54] melanoma,[26] and metastases in lymph nodes.[55] Raman spectral imaging has also been widely applied for the investigation of biological tissues and includes studies of bone tissue,[56,57] breast,[58–61] skin,[4,62] bladder,[63–65] esophageal cancers,[65–67] and single cells.[68–70] On a gross level, the results of all these studies can be summarized as follows: "normal and malignant tissues can be differentiated spectrally when combined with the use of appropriate statistical methods of analysis."

This introductory chapter explores the capability of vibrational spectroscopy to differentiate not only cellular components but also cell and tissue types. Instrumentation, sampling techniques, and methods of analysis are introduced and, finally, the capability of spectroscopic imaging is shown through the application of this technique for the diagnosis of metastatic cancer in lymph nodes.

## 1.1 Vibrational spectroscopy— sensitivity to structures of biological molecules

The cellular constituents of the human body are composed of water, proteins, nucleic acids, lipids, carbohydrates, and a few other components. Changes in cells or tissues leading to diseases such as cancer are due to some biochemical changes in one or more of these components. As vibrational spectroscopy is sensitive to the structure of these components, changes in the diseased state are reflected in the spectrum of these constituents. Since the pioneering work of

Elliott and Ambrose in 1950 for proteins[71] and Blout and Fields in 1949 for nucleic acids,[72] IR and Raman spectroscopy have been shown to be very sensitive to the conformation of these biological building blocks.[73–79]

### 1.1.1 Nucleic acids

Nucleic acids (ribonucleic acid, RNA, and deoxyribonucleic acid, DNA) carry or transmit within their structure the hereditary information that determines the identity and structure of proteins. Each protein, in turn, participates in the processes that characterize the individuality of the cell. In the 50 years since Blout and Field reported the first IR spectra of nucleic acids,[72] IR spectroscopy has been used in applications such as conformational transitions, identification of base composition, effect of base pairing, and DNA–ligand interaction studies. It has also been used in industry for the quality control of products based on DNA, such as fluorescence probes. IR and Raman spectra of cellular components, among them DNA and RNA, are shown in the Appendix. A few comments about these spectra are appropriate. The spectra shown are those of dried films, although all biomolecules in a cell are in a hydrated state. However, since most of spectral diagnostics is performed on dried samples, we opted for reporting the spectra from dried films. Furthermore, even in live cells, the concentration of biomolecules is so high that standard "solution spectra" offer poor sets of reference data (hemoglobin, e.g., exists in red blood cells at concentrations higher than can be achieved in the laboratory[80]). Finally, the reference spectra shown in the Appendix contain Raman, IR absorption (A), and IR second derivative spectra, $d^2A/d\nu^2$ (henceforth referred to as 2ndD spectra). This follows a suggestion by D. Naumann from the Robert Koch-Institut in Berlin, Germany, who pointed out that the inherent differences in line width between Raman and IR spectra makes a comparison between Raman and 2ndD spectra more meaningful than a comparison between IR and Raman spectra. All 2ndD spectra shown in the Appendix were multiplied by $-1$ to produce "positive" peaks. The reason for emphasizing 2ndD spectra becomes apparent in the discussion of protein spectra below.

For nucleic acids (Figures A1[a] and A2), the bands with strongest intensity are assigned as follows. The $1750-1620\,\text{cm}^{-1}$ region corresponds to in-plane double bond and ring vibrations of the bases. The spectra in this region are sensitive to base-pairing interactions and base-stacking effects, and hydrogen bond formation. The 1230- and 1090-$\text{cm}^{-1}$ bands are assigned to antisymmetric and symmetric phosphate stretching vibrations respectively. A detailed discussion of phosphate vibrations is presented in **Infrared and Raman Microspectroscopic Studies of Individual Human Cells**. Ribose has a strong $C-O$ stretching band at $1120\,\text{cm}^{-1}$, which serves as a marker band for RNA in solution. In the solid, the most significant difference between the two nucleic acids is the ratio of intensity in the triplet of bands around $1055\,\text{cm}^{-1}$.

### 1.1.2 Proteins

The function (or dysfunction) of proteins is related to their structure or structural changes. Both IR and Raman spectroscopy provide information on the secondary structure of proteins, ligand interactions, and folding. As an example of the sensitivity of vibrational spectroscopy to protein conformation, Figures A3 and A4 show IR, 2ndD, and Raman spectra of two proteins with very different conformations: albumin, whose structure is almost all helical and $\gamma$-globulin, a $\beta$-sheet rich protein (no helix and $\sim$40% sheet). In the vibrational spectra, the vibrations of the amide linkage of proteins give rise to nine characteristic bands that are named amide A, amide B, and amides I–VII. Among these bands, the amide I band, which is due mostly to the $C=O$ stretching vibrations of the amide linkage, is by far the best characterized. It gives rise to IR bands in the $1600-1700\,\text{cm}^{-1}$ region and has been used the most for structural studies due to its high sensitivity to small changes in molecular geometry and hydrogen bonding of the peptide group. This

is demonstrated by the 2ndD spectra shown in Figures A3 and A4. The $\alpha$-helical protein shows a strong peak at ca. $1655\,\text{cm}^{-1}$, whereas the sheet protein shows 2ndD peaks at ca. 1635 and $1690\,\text{cm}^{-1}$. The amide II band, due largely to a coupling of $C-N$ stretching and in-plane bending of the $N-H$ group, is extremely weak in Raman spectra. Although it is fairly strong in the IR, giving rise to a band in the $1480-1575\,\text{cm}^{-1}$ region, the amide II band is not often used for structural studies per se because it is less sensitive and is subject to interference from absorption bands of amino acid side-chain vibrations. The amide III band, arising from coupling of $C_{\alpha}-H$ and $N-H$ bending coordinates, gives rise to bands in the $1230-1350\,\text{cm}^{-1}$ region, which is fairly weak in the IR spectrum but quite strong in Raman spectroscopy. This band can also be mixed with vibrations of side chains. In addition to secondary structure information, Raman spectra can also provide insight on dihedral angles of $C-S-S-C$ bonds and their related conformers; it can determine how tyrosines are hydrogen bonded and whether tryptophan residues are in a hydrophobic or hydrophilic environment.

Figure A5 depicts spectral data for collagen, a structural protein found in many tissue types. Collagen exhibits a special secondary structure composed of three independent peptide strands, which form a right-handed triple helix. The spectral characteristics of this triple helix can be detected in many collagen-rich tissue types.

### 1.1.3 Lipids

Lipids are critical to all biological systems as they form the cell membranes that keep biological media organized in their necessary compartments. Lipids of various kinds help regulate the flow of biological molecules from one side of a lipid barrier to the other, often assisted by imbedded proteins that form passages, known as *channels*, for the molecules. The spectrum of a typical lipid is shown in Figure A6. The major absorption bands are at 2960, 2922 (methyl and methylene $C-H$ antisymmetric stretching modes), 2874, 2852 (methyl and methylene $C-H$

symmetric stretching modes), 1738 (ester linkage C=O stretching mode), 1465 (CH deformations), 1255, 1168, 1095, and 1057, and 968 cm$^{-1}$ of the phosphate and phosphodiester vibrations. A phospholipid lipid spectrum is given in Figure A7.

### 1.1.4   Carbohydrates

The most common carbohydrates are sugars, or saccharides. Sugars are present in biological media primarily as hexose sugars, such as glucose, where they are an immediate energy source. Pentose sugars are also present, mainly as the ribose component of the nucleic acid backbone of DNA and RNA, both as the component monomers and the much longer polymers, as well as in the energy-transducing oligomeric species. Polysaccharides in the body are found either in a free state or are bound to proteins known as *glycoproteins*. One polysaccharide in the body that is not bound to a protein is glycogen, a polymeric glucose. Figure A8 shows an IR spectrum of glycogen. The major absorption bands are observed at 1149, 1078, 1043 (weak shoulder), 1016, 996, and 931 cm$^{-1}$.

## 1.2   Cells and tissues

### 1.2.1   Cell biology

The basic unit of life is the cell. The different cell types perform the functions for which they have differentiated from the stem cells originally found in the embryo. In addition to performing their assigned functions, cells divide periodically to produce daughter cells, which are likewise capable of generating new cellular molecules and replicating themselves. Cells were first seen over 300 years ago, shortly after the construction of the first microscope. Cells are very small with diameters less than 0.1 mm, so they are invisible to the naked eye. In all eukaryotic (nucleus-containing) cells, the inner cellular mass is partitioned into a membrane-bound, spherical body called the *nucleus* and an outer surrounding cytoplasm. Cellular DNA is located in the nucleus in varying forms. During the normal function of

a cell, the DNA is distributed fairly uniformly within the nucleus. The DNA is always tightly bound to histone proteins, and only short sections unwind and transcribe. During cell division, the DNA forms rods known as *chromosomes* (23 pairs in human cells). This is the most condensed form of DNA, and individual chromosomes are easily observed microscopically.

There are over 200 different types of cells in the human body,[81] such as heart cells, muscle cells, liver cells, retina cells, red and white blood cells, etc. These cells are assembled into four types of tissue: epithelial, connective, muscular, and nervous tissue. Most tissues contain a mixture of cell types. Epithelial tissues are composed of closely aggregated polyhedral cells with very little intercellular substance. These epithelial cells form coherent cell sheets that line the inner and outer surfaces of the body. Connective tissue is characterized by the abundance of intercellular material produced by its cells. This material is composed mostly of a network of tough protein fibers, collagen, and elastin, embedded in polysaccharide gel. Connective tissue contains several types of cells, each having its own morphological and functional characteristics. The most common cells found in connective tissue are fibroblasts. Nervous tissue is found all over the body as a communication network. Nerve tissue consists of nerve cells or neurons, and several types of glial cells or neuroglia, which support neurons. Muscular tissues can be divided into three main types, skeletal, smooth, and cardiac, each composed of cells with different function and appearance. The four tissues, in association with one another and in variable proportions, form different organs and systems of the body.

### 1.2.2   Cancer

Cancer is the leading cause of death in the western world. It is the number-one cause of death for women in the United States and—if the decline in mortality from heart disease continues at the present rate—cancer will soon be the leading cause of death in the United States and many European countries. Over 1 million cases of cancer occur in the United States every year, excluding curable skin cancers, which add another 700 000 cases annually. The highest mortality rates are seen with lung, female breast, prostate, and colorectal cancers.

Cancer is not a new disease. Hippocrates reported to have distinguished benign from malignant growths. He introduced the term "karkinos", from which the word "carcinoma" is derived. Cancer is a complex family of diseases, and carcinogenesis—the turning of a normal cell into a cancer cell—is a complex multistep process. Figure 1 shows a simplified diagram of cancer progression. In a simple description, a tumor (or neoplasm) begins when a cell within a normal population sustains a mutation that increases its propensity to proliferate when it would normally rest, shown in Figure 1(a). The altered cell and its daughters look normal but they grow and divide excessively. This condition is termed *hyperplasia*, Figure 1(b). After some time, which could be years, one of these cells may suffer another mutation that further loosens control on cell growth. This cell proliferates rapidly and the offspring of this cell appears abnormal in shape and orientation, as shown in Figure 1(c). This state is known as *dysplasia*. Again, after some time, a mutation may occur that alters cell behavior. The affected cells become even more abnormal in growth and appearance. If the tumor is contained within original tissue, it is called in situ cancer, Figure 1(d). This tumor may remain contained indefinitely but often the cells acquire

additional mutations, continue to grow and divide, and invade underlying tissue. At this point, the mass is considered to have become malignant, Figure 1(e). If cells break away from such a tumor, they can travel through the blood stream or the lymph system to other areas of the body and establish new tumors called metastases, which grow in new locations. These tumors may become lethal by disrupting the function of a vital organ. Tumors are of two basic types: benign and malignant. By definition, benign tumors do not invade adjacent tissue borders nor do they metastasize to other sites. The primary descriptor of any tumor is its cell or tissue of origin. Most human malignancies arise from epithelial tissue and are called *carcinomas*.

The "gold standard" in most cancer diagnostics is microscopic evaluation, by a pathologist, of a stained tissue that is obtained from biopsy of a particular organ. Figure 2 shows microscopic images of such biopsies. The top row is comprised of images of normal tissue sections, and the bottom row of images of malignant cells from biopsies of the following organs, all taken under the same magnification: (a) breast, (b) colon, (c) bladder, (d) prostate, (e) cervix, and (f) liver. Although these cells are from different tissues, the differences between "malignant" and "normal" states have many features in common, including the following:

- The morphology and size of cells are different and more variable than those of normal cells (pleomorphism).

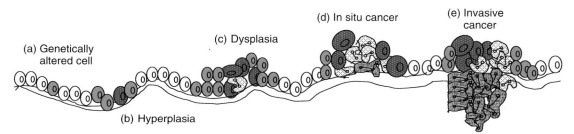

**Figure 1.** Simplified diagram of cancer progression. (a) Normal cells (white) with a single, genetically altered cell (light grey), (b) hyperplasia; over time one of the genetically mutated cells suffers another mutation (dark grey), (c) dysplasia; a mutation that alters cell behavior occurs (dotted white), (d) in situ cancer; cells may acquire additional mutations (blue), and (e) invasive cancer. [Reproduced from Scientific American, September 1996, p. 62–63.]

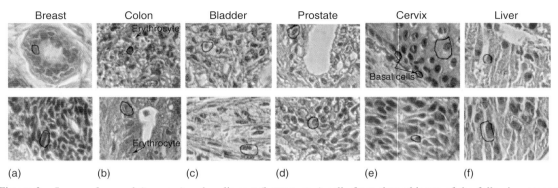

**Figure 2.** Image of normal (top row) and malignant (bottom row) cells from tissue biopsy of the following organs: (a) breast, (b) colon, (c) bladder, (d) prostate, (e) cervix, and (f) liver. Magnification: 100×. Individual cells are circled.

- The nucleus of a malignant cell is larger than the nucleus of a normal cell.
- Large cells with multiple nuclei are observed.
- Normal tissue is invaded by neoplastic cells.

The diagnosis of disease and tumor staging is thus in the hands of a pathologist. This diagnosis will lead to a decision by an oncologist who determines the type of cancer treatment. However, classical pathology is subjective, in that diagnoses are made by comparison with a database stored in a person's head. Therefore, misdiagnoses are common, with false negative rates of up to 20%, and an unknown rate of false positives. Additionally, in some cases (about 10%), a pathologic examination may not produce a firm diagnosis, either because certain tumors are histologically similar or because cells are so poorly differentiated that their tissue of origin cannot be determined. In these cases, other diagnostic procedures might be useful and include, but are not limited to, electron microscopy, immunohistochemistry, cytogenetics, and checking the levels of various tumor markers in the patient's serum or urine.

### 1.2.3 Spectral diagnosis

To alleviate the problems described above, methods of spectral diagnosis were introduced in the late 1990s, and have matured early in this decade. The principle behind spectral diagnosis is the collection of objective, physical data (such as vibrational spectra), which are acquired by instruments that produce reproducible observations, and that are subsequently analyzed by computer algorithms.

It is imperative for the successful development of spectral imaging in the field of biomedicine that the inherent spectral variability in cell and tissue populations is investigated. Exploration of spectral variability and sample heterogeneity will facilitate better understanding of the spectral processes occurring as a consequence of the mechanisms of disease and will lead the field away from the search for a specific marker for cancer, and will prevent inaccurate conclusions about disease processes and the presence of cellular constituents in normal and diseased cells and tissue, as was the case with glycogen and cervical cancer investigations in the 1990s.

## 2 EXPERIMENTAL TECHNIQUES

Tissues are very heterogeneous, comprising several components and different cell types. Different cells have intrinsically different size and shapes, as can be seen from Figure 2, top row. Normal breast cells, for example, are about $10\,\mu m$ in diameter, with a nucleus of about $5\,\mu m$, whereas cervical cells range from 20 to $60\,\mu m$ in diameter, with a nucleus of $5–10\,\mu m$, depending on the stage of maturation of the cell.

## 2.1  Histological samples

Standard histology is carried out on tissue biopsy sections that are prepared as follows. The excised tissue is fixed first, since fixation makes cells permeable to stains and preserves the tissue's physical structure. The procedure usually involves brief immersion of the tissue into formaldehyde solution (formalin) or an alcohol/water mixture. Formaldehyde is known to cross link amino acid side groups containing $-OH$ functions, and renders proteins insoluble and resistant to proteases. After fixation, water is extracted by dipping the tissue in a graded series of ethanol–water mixtures (with decreasing water content), followed by xylene. Subsequently, tissues are embedded in a liquid wax, such as paraffin, that permeates and surrounds the tissue. This embedding medium is then hardened to a solid block by cooling. These tissue blocks are cut into thin slices (3–8 μm thick) with a microtome and mounted on the surface of a glass microscope slide.

Prior to staining the tissue, the embedding medium is removed by dipping the tissue section in a solution of Hemo-De, a mixture of d-limonene and butylated hydroxyanisole, or in xylene followed by ethanol–water mixtures with increasing water content. Appropriate stains are then applied. These stains have preference for a particular part of the cell—the nucleus or membrane. Hematoxylin, the dye of choice for most solid tumors, has an affinity for negatively charged molecules and therefore reveals the distribution of DNA and RNA in a cell.[82] A combination of dyes, hematoxylin, and eosin (H&E), is most commonly used, and stains the nucleus in dark blue, cytoplasm in pale pink, and collagen (in connective tissue) in pink. For any fixation and embedding procedure there is a danger that the treatment may distort the structure of the cell. Immersion of tissues in lipophilic solvents, such as xylene, dissolves the tissue lipids. An alternative method of preparation that lessens this danger is rapid freezing, which precludes either fixation or embedding. The frozen tissue can be cut directly with a cryo-microtome. Although frozen tissues have an advantage that they represent a more native form of the tissue, they are more difficult to prepare, and the presence of ice crystals causes many morphological details to be lost. For spectroscopic analysis, staining is not required, and the microtomed section is placed directly onto the sample substrate and deparaffinized as described above.

## 2.2  Cytological samples

Preparation of cytological samples for spectroscopic analysis is slightly more tedious than preparation of tissue samples. In order to image individual cells, care must be taken to ensure that cells are well separated from one another on the sample substrate. Several commercially available instruments exist, which can be used to prepare samples for spectral cytology and the reader is referred to **Infrared and Raman Microspectroscopic Studies of Individual Human Cells** for an in-depth explanation.

## 2.3  Sample substrates

IR imaging is performed in transmission or transflection modes. For transmission studies, cells and tissues are mounted on an IR transparent substrate such as NaCl, $BaF_2$, $CaF_2$, or KBr. Transflection measurements were traditionally carried out on gold-coated substrates, although now the field is leaning more toward the use of "low e slides" (Kevley Technologies, Chesterland, OH) for this purpose. Made of glass, the Kevley slides are of similar dimensions to traditional glass microscope slides, and are coated with a thin layer of $Ag/SnO_2$. The slides are chemically inert and nearly transparent to visible light; however, mid-IR radiation is almost completely reflected. Thin sections of tissue, 3–8 μm thickness, are placed onto the reflective substrate; the light beam passes through the sample to the substrate, reflects off the silver surface, and passes through the sample a second time. There are several advantages of using "low e slides" over traditional IR transmissive substrates such as $CaF_2$ or $BaF_2$. First of all, "low e slides" cost less than US \$2 apiece, compared to hundreds of dollars for $CaF_2$ or $BaF_2$ windows. Coupled with the low cost of the Kevley slides is the added advantage that they do not need to be used repeatedly, i.e., once the tissue section is mounted onto the slide the sample can be kept. The slides can also be stained after spectral imaging, enabling histopathological evaluation of the same section from which the spectroscopic data were obtained, as opposed to adjacent section evaluation if traditional IR substrates were employed. Normal glass or quartz slides cannot be used in IR microspectroscopy due to the strong $Si-O$ vibrations, which render glass totally opaque over the majority of the mid-IR region. Raman microspectroscopy can be carried out either on fluorescence-free quartz slides or on $CaF_2$ or $BaF_2$ disks.

Reflection measurements can also be undertaken through the use of attenuated total reflection (ATR). The ATR element, also called the internal reflection element (IRE), is a crystal of IR transparent material of high refractive index, for example, zinc selenide (ZnSe) or germanium (Ge), aligned in such a fashion that the probing IR beam undergoes total internal reflection inside the IRE. The sample to be measured is in contact with the IRE. At the interface between the IRE and the sample, a standing wave of radiation called the *evanescent wave* is established that penetrates the medium beyond the surface of the IRE. The sample interacts with the evanescent wave and absorbs IR radiation, resulting in detection of its IR spectrum. The name attenuated total reflection is given because the evanescent wave is attenuated by the sample's absorption.

Single-reflection ATR is particularly suitable for studies of biological samples and has been utilized for the study of arteries and atherosclerotic plaques,[83] wheat stem,[84] and hair.[85] Improved spatial resolution, compared to images collected in transmission or transflection, is achieved due to the high refractive index of the ATR crystal. This results in a higher numerical aperture, which in turn translates into a higher spatial resolution, typically on the order of $3-4\,\mu m$ with a micro-ATR accessory.[86]

## 2.4   Instrumentation for IR and Raman microspectroscopy

Owing to the high water absorption cross section in IR spectroscopy, a significant fraction of IR applications to date has concentrated on in vitro studies of dried tissues and cells. In contrast, Raman spectroscopic applications have pushed toward in vivo diagnostics.

In Raman spectroscopy, excitation of light can be ultraviolet (UV) ($\lambda < 400\,nm$), visible ($400 \leq \lambda \leq 700\,nm$), or near-IR ($\lambda > 700\,nm$). Different tissue types and samples require optimization of excitation wavelengths.[87] For example, some tissues exhibit strong fluorescence, due to the presence of mostly extracellular components, that can obscure the tissue Raman spectra when visible excitation is used. Since fluorescence cannot be excited by low energy photons, near-IR excitation ($785-1064\,nm$) is useful for biomedical studies. Near-IR light penetrates tissues deeply, on the order of millimeters. Alternatively, it is possible to use ultraviolet light at wavelengths shorter than $270\,nm$ to circumvent the fluorescence interference problem. With UV resonance excitation, the tissue penetration depth is on the order of micrometers, so this technique is good for surface layers, although care must be taken to avoid tissue damage when using UV light. Visible excitation Raman spectroscopy appears to be best suited for in situ cell studies because of lower fluorescence of individual cells compared with tissues.

Over the last decade, dramatic technological advances have occurred for many components of vibrational spectroscopic instruments including introduction of microscopes, new multichannel detectors, tunable diode lasers, and optical fibers. The initial Raman spectra of tissues were measured with visible laser excitation, primarily with argon ion lasers, but the spectra were masked by fluorescence.[88] Introduction of Fourier transform (FT-) Raman spectrometers, using 1064-nm excitation (neodymium-doped yttrium aluminum garnet (Nd:YAG) lasers) and cooled InGaAs or Ge detectors allowed collection of fluorescence-free spectra of tissues. However, InGaAs or Ge detectors exhibit substantial noise so the collection time needed to obtain spectra of tissues with good signal-to-noise ratio (SNR) was on the order of 30 min to 1 h.[89–91] Also, FT-Raman spectrometry exhibits certain disadvantages in that laser noise transforms into the high frequency part of the interferogram in a process that has been termed the *multiplexing disadvantage*.[92]

Thus, most of biomedical Raman spectroscopy is now being carried out using dispersive Raman spectrometers and visible to near-IR excitation. Excitation using a diode laser near 800 nm has been found to be the optimum trade-off between loss of Raman intensity as $1/\lambda^4$ for the excitation wavelength and the greatly reduced fluorescence available further into the near-IR region. The development of diode lasers and low-noise cooled silicon charge-coupled device (CCD) cameras sensitive in the near-IR, combined with the use of dispersive systems instead of FT-based spectrometers, has enabled the measurement of fluorescence-free Raman spectra of cells and tissues on a much faster timescale [70], sometimes in less than a second. The use of diode lasers and dispersive systems has provided better sensitivity and the possibility of f-number matching of spectrographs with optical fibers for better throughput.

Microscopy is an integral part of diagnostics, with visible microscopy of stained tissues considered a gold standard. Coupling of a microscope to a Fourier transform IR (FT-IR) or Raman spectrometer provides a very powerful technique of measuring vibrational spectra of individual morphological components of tissues, enabling the creation of pseudocolor maps of tissues, which can be compared directly to tissue histology. In addition to providing the color to which most pathologists are accustomed, these maps also provide fairly detailed biochemical information not available from other techniques. It is important to re-emphasize that these maps are produced from spectra without any prior knowledge of tissue architecture on samples that are not stained.

Spectral mapping allows for combined understanding of morphology and biochemistry that gives rise to observed spectra.

IR microspectroscopy has been described in detail in many excellent reviews, particularly by Messerschmidt[93,94] and Reffner and Martoglio.[95] While it is quite easy to establish the sample area by viewing it with visible radiation, the smallest area that can be analyzed is larger for mid-IR radiation because of the larger diffraction limit in this region since the diffraction limit increases linearly with wavelength. Messerschmidt defines spatial resolution as the "ability to measure the spectrum from an object delineated by the apertures without significant impurity radiation from neighboring objects".[93] Neighboring objects to cells in tissues could be among many possible constituents including other cells, either malignant or normal, connective tissue matrix, or fat lobules. From this example, one might conclude that setting aperture(s) small enough to match precisely the object of interest is all that is needed. Decreasing the aperture dimensions leads to degradation of SNR and, more importantly, as the aperture size approaches the wavelength of light, diffraction becomes a serious limitation. The diffraction-limited spatial resolution or least resolvable separation (LRS) is defined as (0.61 $\lambda$/NA), where $\lambda$ is the wavelength of the radiation and NA is the numerical aperture of the optical system.[93] (The diameter of the diffraction ring, known as an *Airy disc*, is taken as 1.22 $\lambda$/NA; however, the resolving power is the smallest distance between two objects, so the radius of the airy disc is used).[96] The numerical aperture is equal to the product of the refractive index of the medium outside the lens (objective or a condenser) and the sine of half the angle that the marginal ray from the lens makes with the optical axis of the lens and thus ranges theoretically from 0 to 1.0 in air.[96] Most IR microscope objectives have numerical apertures in the range of 0.5–0.7. Using the above formula, the diffraction-limited spatial resolution in the 1700–1000 cm$^{-1}$ region is on the order of 5–12 μm, with higher resolution (lower number) at higher numerical aperture or shorter wavelength. These values represent the absolute minimum spot size achievable with a given objective. But because some radiation falls outside the defined spot, the real sample area is much larger, especially at longer wavelengths, where it could be as high as three times the minimum spot size.

Commercial spectroscopic microscopes fall into two major categories: systems with high numerical apertures with one area-defining aperture and systems with medium numerical apertures with two apertures (confocal microscopes). Both designs lead to improvement in spatial resolution. Numerical aperture also affects throughput, so it should be as large as practical. Magnification of the microscope does not affect throughput or diffraction limit and, in all commercial IR systems, a choice of 15× or 32× magnification is provided, mainly in order to allow convenient visualization and masking of the sample.[93]

### 2.4.1 Synchrotrons

Synchrotron IR light is about 2–3 orders of magnitude brighter than a conventional silicon carbide (Globar®) source used in most commercial mid-IR FT-IR spectrometers. The high brightness (low divergence) of this source allows setting of microscope apertures corresponding to sample geometrical areas equal to the diffraction limit or slightly smaller[97] with resulting spectra of high SNR. The synchrotron beam size is fairly small, for example on the order of $300 \times 600\,\mu m$ at the National Synchrotron Light Source (NSLS) at Brookhaven National Laboratory, but it can vary for different facilities. After demagnification and use of a 32× objective, the geometric size of the beam is $4\,\mu m \times 8\,\mu m$. If no apertures are used, the diffraction-limited spatial resolution depends only on the wavelength of light and numerical aperture of the objective, as discussed in previous section. The numerical aperture of the 32× objective used at NSLS is close to 0.65, therefore the diffraction limit is equal to the wavelength. Synchrotron radiation is ideal for studies of cells and tissues because it allows measurement of low-noise spectra at the diffraction limit,

which matches the size of many individual cells and even nuclei. Measurements with synchrotron radiation are not practical for every day medical use in a hospital or clinical laboratory setting, but high quality spectra obtained in these experiments of individual cell and tissue components can help understand the process of disease and basic cell and cancer biology.

### 2.4.2 Mid-IR focal-plane array detectors

Use of an FPA detector in an IR microscope makes it possible to analyze a fairly large area of sample with very high spatial resolution. In this detection method, spatial resolution is attained from the individual pixels. Each detector pixel serves as an area-defining aperture, thus the sample size examined is determined by the dimension of the array. In 1994, Lewis *et al.*, carried out the first experiments combining an FPA with a step-scanning interferometer coupled to a reflective IR microscope.[98,99] This approach allows for a simultaneous determination of both spatial and spectral information. Modern FPA-based systems collect the hyperspectral data cube in rapid scanning modes for a photovoltaic mercury cadmium telluride (MCT) array up to $128 \times 128$ pixels in size. With the new infinity corrected optics, the field of view is $400\,\mu m \times 400\,\mu m$, with an image on each pixel of $6.25\,\mu m \times 6.25\,\mu m$. According to the previous discussion of the diffraction limit, a spatial resolution of $6.25\,\mu m$ can only be achieved in the short wavelength range of the spectrum. A detailed discussion of theoretical and observed spatial resolution has been presented by Carr[97] and Lasch and Naumann.[27]

FPA-based IR microscopes yield spectra of slightly inferior quality when compared to the spectra collected on linear array IR microspectrometers, due to the superior noise characteristics of photoconductive single element detectors (see Figure 3). The new PerkinElmer Spotlight PE400 (PerkinElmer, Inc.) allows spectral data acquisition of up to $200\,\text{pixels}\,s^{-1}$ at superior SNR, compared to the FPA-based IR instruments under comparable sampling conditions.

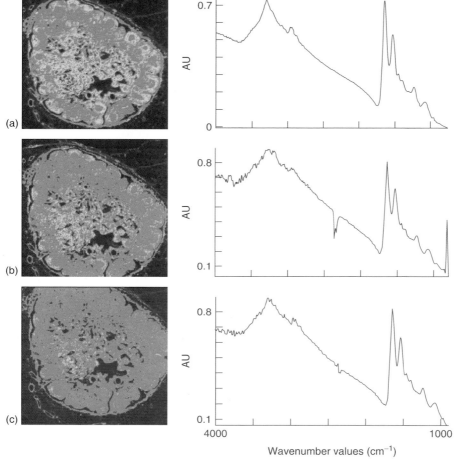

**Figure 3.** Comparison of IR spectral images and spectra generated from FPA and linear array detectors. All comparison spectra were taken from the same sample position. (a) Linear array detector, 1 scan, $8\,cm^{-1}$, $25\,\mu m$ spatial resolution, raw spectra. (b) FPA, 1 scan, $8\,cm^{-1}$, pixel aggregation equivalent to $25\,\mu m$ spatial resolution, raw spectra. (c) FPA, 4 scans, $8\,cm^{-1}$, pixel aggregation equivalent to $25\,\mu m$ spatial resolution, raw spectra.

### 2.4.3 CCD detection in Raman microspectroscopy

Three basic strategies have evolved for the measurement of microscopic Raman images. The first methods is a point-by-point collection carried out by translating the sample beneath a fixed small laser focus, and detecting the inelastically scattered light simultaneously via a high sensitivity CCD detector mounted in the focus of a single monochromator. For this purpose, only the center few elements of a $1024 \times 128$ element CCD are illuminated by the dispersed light. Although this method appears to be inefficient, since only one point is sampled at a given time, it provides the best illumination geometry, and concentrates the laser beam into a very small volume. This focal volume is given by the same diffraction limits discussed above, and measures about 300 nm in diameter and about 1200 nm in height for mid-visible excitation and a lens with a numerical aperture of one.

The second Raman imaging method is known as *line illumination*, where the laser beam is projected as a line on the sample. The image of the exciting line is projected through a dispersive spectrometer onto a CCD detector such that one dimension of the CCD detects the dispersed Raman spectrum, and the other dimension each point along the line simultaneously. The entire image can be obtained by sweeping the laser line image across the sample in discrete steps.

The third imaging technique involves global illumination of the sample, and to project the Raman scattered light of the entire sample area through a tunable optical filter onto the surface of a CCD detector. The Raman images are then collected at discrete wavelengths. Although this method appears at first to be much more efficient than single point illumination, the fact that the laser beam needs to be diffused over a relatively large area (ca. $80 \mu m \times 80 \mu m$) reduces the power density enormously; consequently, best spectra are still obtained in single-mode mapping experiments.

The use of a tunable filter in the global imaging setup leads to some loss of spectral resolution. The spatial resolution of this method is constrained by the size of the CCD pixels, which are typically $25$-$\mu m$ squares. Higher spatial resolution can be achieved by magnification of the image on the CCD up to the diffraction limit. Near-IR Raman has a larger diffraction limit size ($0.7 \mu m$ and lower) than the corresponding limit in the visible ($0.4$–$0.7 \mu m$) by the ratio of the wavelengths used in these two regions.

## 2.5 Practical considerations

Sample preparation, lateral spatial resolution, data preprocessing, and multivariate methods of analysis are all important considerations in spectral imaging. The experimental approaches are as varied as the applications of spectral imaging in the biomedical field. The lateral spatial resolution chosen for spectral imaging is a trade-off between many factors. The most important consideration is the size of the samples to be analyzed. Cell size varies with the cell type and the tissue where it grows. If a spatial resolution is selected that is too low, optical dilution and spectral contamination arise.[27] However, a high spatial resolution, which ensures that specific areas of the sample are adequately represented, results in longer acquisition times, larger data files, and slower processing times.[27,100]

Sample preparation is also a concern. Most researchers utilizing tissue sections for spectral analysis choose to use paraffin-embedded thin tissue sections over frozen sections. Archived, paraffin-embedded tissues are advantageous in that they are readily available and have associated histopathological results. For cell investigations, the choice of fixation methods varies and includes dry, alcohol, and formalin fixation. Formalin fixation has been found to preserve cytoplasm components better than alcohol fixation, which tends to cause leaching of lipids.[34] In our laboratory, we have found no significant spectral differences between formalin fixation and dry fixation. However, formalin fixation is preferred as the samples are better preserved over time.

The purpose of preprocessing spectral data is to remove, as much as possible, spectral artifacts and spectral variance that is unrelated to the system under investigation. In spectral imaging, the first step of preprocessing is usually the removal of spectra from regions of the image where there is no sample and/or the removal of spectra that do not meet a required signal-to-noise threshold. For IR data, if the instrument is not purged for sample collection, an additional step may be required to remove the effects of water vapor in the spectra, as absorptions from water vapor overlap the amide I band. The presence of atmospheric $CO_2$ is not an issue as it occurs in a spectral region where there is no biologically important spectral information. Baseline correction, often through the use of second derivatives, is performed in order to remove baseline effects and scattering artifacts.[101] The choice of baseline correction methods depends on the quality of the spectrum in terms of SNR. Second derivatives decrease the SNR of the spectra and are generally not utilized for noisy IR data, or for Raman imaging.

For IR data, the Savitzky–Golay algorithm for calculating derivatives offers the added advantage of smoothing the data, although care must be taken not to choose a large smoothing window, as this results in band broadening. The final step in data preprocessing of IR spectral mapping data sets is normalization of the data, which serves to remove spectral intensity differences arising from sample thickness variations (for Raman maps, this step is not crucial since the intensity does not depend on sample thickness in confocal measurements). Vector normalization is ideal for this purpose as it does not result in the loss of spectral information, which occurs if min–max normalization is employed, i.e., spectral information in the form of intensity values from the ratioed band, usually the amide I band, is lost.

Ideally, the field of spectral diagnostics should work toward a set of standards with which to carry out spectral imaging experiments within a given tissue type or disease state. The standardization of the field would greatly advance translation into clinical applications as it would allow for easier collaborations between research groups and the development of global data sets containing hundreds, if not thousands, of images, a necessary step for acceptance by the medical community and one not practical with laboratories working in isolation.

# 3 MULTIVARIATE METHODS OF DATA ANALYSIS

Hyperspectral images and data hypercubes contain an enormous amount of data and as a consequence multivariate statistical methods must be employed in order to objectively and efficiently analyze the data.

The term multivariate data analysis incorporates many different methods and techniques and can be divided into two different types: clustering methods and ordinal methods. Clustering methods, such as hierarchical cluster analysis (HCA), are algorithmic approaches that aim to divide or collect samples into groups, depending on similarities. Ordinal methods, such

as principal component analysis (PCA), use mathematical properties to decompose data matrices. These methods work by obtaining new coordinates describing the variance, or covariance, in the data set. This enables complex data consisting of many variables to be reduced to a lower dimensionality.[102]

Pattern recognition refers to the ability to assign an object to one of several possible categories according to the values of measured parameters or variables. In chemometrics, pattern recognition can be further divided into two groups: unsupervised and supervised. Unsupervised pattern recognition includes cluster analysis and hierarchical techniques; the interpretation of the number of clusters and populations can often be subjective, as this information is not known prior to analysis. In supervised pattern recognition techniques, such as the classification techniques used in artificial neural networks (ANNs), the number of groups is already known and there are representative samples of each group. Classification uses information from known samples to identify and categorize future samples.

The concept of distance is very important in classification procedures and follows from the assumption that proximity in multivariate space is indicative of similarity between samples. Therefore, samples that are near in variable space are considered to have the same characteristics, whereas a large separation is suggestive of different characteristics. The most commonly used method for determining similarity between samples is to measure the Euclidean distance, $d$, of samples, $x$, in variable space:[103]

$$d_{1,2} = \left[ \sum_{j=1}^{M} (x_{1j} - x_{2j})^2 \right]^{\frac{1}{2}} \qquad (1)$$

where $M$ is the number of variables.

## 3.1 Hierarchical cluster analysis

Following data preprocessing, the hypercubes representing the IR or Raman spectral information can be classified via HCA. The necessity of this step is obviated by the size of the spectral images collected, usually in the order of 40 000 IR spectra per image. It is impossible and undesirable to analyze these spectra visually, as this process would impart a level of subjectivity. The spectral differences between normal and diseased tissues are often so subtle that they cannot be differentiated by eye, even if second derivatives are employed to enhance the differences. HCA is sensitive to these subtle differences and the results discussed in Section 4.1.3 indicate that there is a strong correlation of clusters based on tissue type and disease state, as confirmed by histopathology. This provides strong evidence that the observed spectral changes are the result of biochemical differences between tissue types and disease states. The strength of spectral imaging coupled with HCA has been previously demonstrated in the diagnosis of many disease states including differentiation of normal from malignant tissues of the colon,[23,25–27] cervix,[20,104] breast,[40,43] brain,[45] and skin.[26,105]

HCA is an unsupervised multivariate statistical method. Unsupervised analysis methods require no initial input as to the nature of the data set. The first step in HCA is the calculation of the correlation matrix from the preprocessed data set, resulting in correlation coefficients, $C_{LM}$, between all pairs of mean centered $(S - \bar{S})$ spectra L and M.

$$C_{LM} = \frac{\sum_{i=1}^{N}(S_i^L - \bar{S}^L)(S_i^M - \bar{S}^M)}{\sqrt{\sum_{i=1}^{N}(S_i^L - \bar{S}^L)^2}\sqrt{\sum_{i=1}^{N}(S_i^M - \bar{S}^M)^2}} \quad (2)$$

The correlation coefficients are a measure of the similarity of the spectra. A correlation coefficient of 1 means the spectra are identical, whereas a correlation coefficient of 0 indicates the spectra are not related. The correlation matrix is searched to find the two spectra with the highest coefficient, i.e., the two spectra that are the most similar to each other. These two spectra are grouped together to form one object or cluster and the matrix is recalculated, with the dimensionality reduced by 1. This process is repeated until all the spectra have been clustered together. There are many different methods that are used to perform the merging of clusters; Ward's algorithm is preferred as tighter clusters with high homogeneity are produced.[106] In the initial stages of this process, it is usually individual spectra that are clustered together, but as the process continues objects containing multiple clusters are merged to form new objects. The results of this clustering process are presented schematically as a dendrogram, a treelike structure that displays the similarity of each cluster from another. The reader is referred to Figures 2 and 3 in **Infrared Spectroscopy in the Identification of Microorganisms** for examples of dendrograms. The dendrogram is used to determine how many clusters are required, but for spectroscopic analysis of tissue the number of clusters is determined on the basis of the histopathology of the tissue section.

In spectral imaging, the outcome of HCA is a pseudocolor map, where all pixels in a cluster are automatically assigned a color. Thus, pixels of the same color in the pseudocolor map exhibit the most similarities. Although HCA is an excellent method for preliminary identification of abnormalities in cells and tissues, it is an unsuitable method for analyzing and diagnosing an unknown data set. HCA is based on the similarity of spectra within a given data set, and is difficult to apply absolutely on unknown data. Therefore, HCA is employed for determining the clusters of spectra to be used to train a supervised diagnostic algorithm, such as an ANN. The development of successful ANNs in the pharmaceutical and biospectroscopy fields, based on analyses of IR spectra, have been reported.[26,43,53,107–109]

## 3.2 Artificial neural networks

ANNs are mathematical models and algorithms that have been designed to mimic the information processing and knowledge acquisition methods of the human brain. Neural networks have significant advantages over standard pattern recognition algorithms in that they do not use rules but rather "learn" from training sets in a similar way to human learning. ANNs have the ability to learn directly from data, and can process data that only broadly resembles that on which they were trained.[110] After learning the patterns of inputs and outputs, they are able to classify patterns and make predictions based upon new patterns of inputs.[111]

The basic unit in an ANN is the neuron or node, which has many input paths each modified by a weight. The generation of an output from a node for a given input involves two steps. The first step is the evaluation of the net input, followed by a nonlinear transformation, designed to be flexible to different learning situations. An ANN consists of many nodes organized into groups called layers. The first layer is the spectral inputs, which—when using spectral data—are intensities at certain wavenumbers. The neural network software utilizes measures of covariance to decide the best-input wavenumbers, based on similarities within a class and differences between the classes. Neural networks often employ a hidden layer of nodes, which serve to assess which spectral information is most valuable and characteristic of the desired output.

During training, the weights are continually adjusted until a correspondence between the inputs and the expected output (diagnosis) is achieved. Networks are typically trained on tens of thousands of IR spectra and the process of network learning is quite time consuming. Once the algorithm is developed, however, analysis of 10 000 IR spectra can be performed in seconds.

# 4 APPLICATIONS

## 4.1 Metastatic cancer in lymph nodes

Lymph nodes are small rounded or bean-shaped masses of lymphatic tissue surrounded by a capsule of connective tissue. Lymph nodes filter the lymphatic fluid, trapping cancer cells and bacteria from the lymph fluid. Lymph nodes are critical for the body's immune response and are the principal site for the initiation of many immune responses. Cancer can arise in lymph nodes in one of two ways. Cancer cells can grow directly in the lymph nodes, such as with Hodgkin's and non-Hodgkin's lymphoma. The second way is through the spread of cancer cells into the lymphatic fluid, where trapped cells in the lymph nodes begin to proliferate and form what is referred to as a *secondary* or *metastatic cancer*. Secondary cancer cells resemble cells from the primary tumor, i.e., metastatic breast cancer cells growing in a lymph node resemble the characteristic morphology and architecture of breast cells.

Lymph node analysis plays a vital role in the staging and typing of tumors, and the detection of secondary tumors within the lymph nodes is often the first sign that cancer exists, as many tumors may be asymptomatic in the organ of origin. It is especially important then to be able to develop a rapid, automated, and objective diagnostic method which can not only identify that a secondary tumor exists within a lymph node but also determine the site of the primary tumor.

IR imaging plays a key role in the development of such a diagnostic technique and has been successfully employed in the diagnosis of breast, colon,[55] and thyroid metastatic tumors in lymph node sections.

### 4.1.1 Instrumentation

Tissue samples were imaged via a PerkinElmer (PE) Spectrum One/Spotlight 300 IR spectrometer (PerkinElmer Corp., Sheldon, Connecticut). This system is a totally integrated IR microspectrometer, and incorporates a $16 \times 1$ element ($400\,\mu m \times 25\,\mu m$) MCT array detector for imaging applications. The instrument provides either 1:1 or 4:1 imaging on the MCT detectors, resulting in a nominal resolution of 25 or $6.25\,\mu m$, respectively. Visual image collection via a CCD camera is completely integrated with the microscope stage motion and IR spectral data acquisition. The visible images are collected under white light LED illumination, and are "quilted" together to give pictures of arbitrary size and aspect ratio. The desired regions for the IR images are selected from these visual images. Spectral images presented in this section were recorded at 25 μm spatial resolution and 8 cm$^{-1}$ spectral resolution.

### 4.1.2  Data processing

Spectral hypercubes were imported into Cytospec™ software (http://www.cytospec.com) as a binary .FSM file (the proprietary file format of PerkinElmer Spotlight 300 imaging IR spectrometer). Spectra were converted from transmittance into absorbance, followed by a quality test to remove spectra from the image where there was no sample. In order to baseline correct and enhance the subtle differences between the spectra of tissues, a Savitzky–Golay second derivative with a 13-point smoothing window was applied to the hypercube. The spectral region was reduced to 1800–950 cm$^{-1}$ and the data was vector normalized to eradicate spectral differences due to sample thickness.

D-values clustering algorithm and Ward's linkage algorithm enabled a uniform basis of HCA[24,112] for all maps. The decision on how many clusters were necessary to accurately describe the tissue section was determined such that the HCA pseudocolor map matched the histopathological diagnosis of the stained slide. The output of HCA is a pseudocolor map, where each pixel represents one IR spectrum. Pixels of the same color indicate spectra that exhibit the most similarity to one another on the basis of correlation.

Single data files from each cluster were exported in ASCII file format and served as inputs for subsequent ANN analysis. The inputs for the neural network discussed in this section were derived from a random selection of IR spectra from clusters identified by histopathological correlation as being representative of breast or thyroid metastatic carcinoma, as well as representative spectra from normal tissue types present in lymph nodes. The outputs of the network were a desired diagnosis and tissue type, corresponding to the different clusters or classes used for training.

ANN classification and feature selection were performed with NeuroDeveloper 2.5 (Synthon GmbH, Heidelberg, Germany). Data for each lymph node tissue type were split into three sets. The training data set was used to establish network parameters that provided the best possible classification. The validation data set was used to optimize performance of the network. The test data set served to confirm that a given network had sufficiently broad generalization power to serve reliably as a diagnostic algorithm. Depending on the number of available spectra, the general scheme was to have one-fifth of the data pool used for training and validation (split 80 and 20%, respectively), with the remaining spectra used for testing purposes.

The application of a spectral feature selection algorithm reduced the complexity and dimensionality of the classifier. This also improved the quality of the classification model and removed the high redundancy of information across the entire spectral range. Spectral feature selection was based on the calculation of the covariance of the spectral data points. Ranked in descending order according to the covariance measurement, the best-selected features were used as inputs for the ANN classifier. Generally, the 120 data points with the highest covariance were made available for the neural network.

Three-layer, feed-forward networks with 5–120 input neurons, 4–20 hidden nodes, and 4–7 output nodes were tested. Resilient back-propagation (Rprop)[113] was used as the learning algorithm. Testing Rprop parameters were in the following range: $\Delta_0 = 0.075{-}0.1$, $\Delta_{max} = 30{-}50$, and $\alpha = 4{-}5$. The training process was stopped when errors of training and validation data sets converged.

### 4.1.3  Results and discussion

The sensitivity of IR imaging, coupled with multivariate statistics, is illustrated in Figure 4 with accurate reproduction of tissue architecture and morphology. Panel (a) illustrates an H&E stained

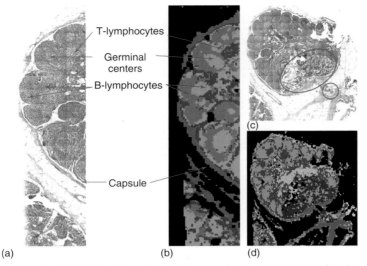

**Figure 4.** (a) Visual image of an H&E stained lymph node tissue section (without metastasis), (b) resulting pseudocolor map, (c) H&E stained tissue section of an entire lymph node, cancerous areas are indicated in green, and (d) resulting pseudocolor map.

section of a lymph node. The secondary follicles containing germinal centers are clearly seen. Surrounding the node is the fibrous capsule and some fatty tissue. Panel (b) is the pseudocolor map generated from the IR spectral image of the same tissue section in panel (a) (the section was stained after IR image collection). The spectral clusters clearly show the architecture of the lymph node, reproducing the morphology of the secondary follicles containing the germinal centers. In fact, IR imaging is such a sensitive technique that it is able to differentiate proliferating B-lymphocytes (germinal center, light green) from nonactivated B-lymphocytes (dark green). The node capsule is shown in brown and the T-lymphocytes in blue. Panel (c) shows a stained visual image of a lymph node containing metastatic colon carcinoma, circled in green. The corresponding eight-cluster pseudocolor map (panel d) shows the differentiation of the colon metastases (red). The node capsule and connective tissue are shown in brown and B-lymphocytes are represented by the green cluster.

Panels (b) and (d) of Figure 5 illustrate another example of the detection of metastatic colon cancer in lymph nodes. Panel (a) shows the

stained visual image of a lymph node containing metastatic colon adenocarcinoma, circled in green. The resulting pseudocolor IR data generated map (panel b) clearly differentiates the adenocarcinoma (red and burnt orange). The differentiation of this metastases into two distinct clusters again demonstrates the sensitivity of this technique, with discrimination observed between the glandular (red) and stromal (burnt orange) components of the cancer.

Panels (c) and (d) of Figure 5 give an example of an axillary lymph node containing a large metastatic breast infiltration, seen as pink staining in the upper half of the image in panel (c). The corresponding pseudocolor map generated from IR spectroscopic data (panel d) reproduces the node architecture and discriminates the breast metastasis from the normal lymph node tissue. The breast metastatic tumor is shown, and again differentiation of the tumor into the stromal (burnt orange) and glandular (red) components is observed.

Thyroid metastatic cancer is also easily detected through the combination of spectral imaging and multivariate statistics. Of the many types of thyroid carcinoma, only two have the possibility

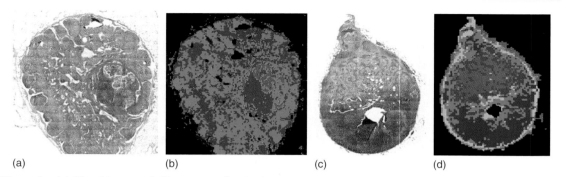

**Figure 5.**   (a) Visual image and (b) corresponding IR data generated pseudocolor map of a lymph node with metastatic colon adenocarcinoma. (c) Visual image and (d) corresponding IR data generated pseudocolor map of an axillary lymph node with metastatic breast cancer.

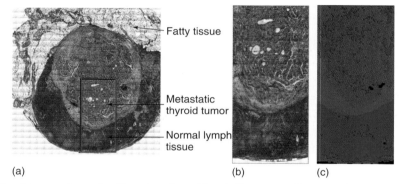

**Figure 6.**   (a) Visual image of unstained lymph node; red box indicates sampling area. (b) Expanded visual image of red box. (c) Pseudocolor map of resulting IR image, three clusters.

of infiltrating the lymphatic system and forming secondary tumors in the lymph nodes: papillary thyroid carcinoma (PTC) and its follicular variant (FVPTC, follicular variant of papillary thyriod carcinoma).

Figure 6 illustrates a section of a lymph node containing metastatic papillary carcinoma cells. The lymph node is surrounded by fatty tissue and the interior of the lymph node has been infiltrated with metastatic thyroid tumor cells. An IR spectral image was collected from the red highlighted area of panel (a), a magnified image of which is shown in panel (b). Panel (c) displays the resultant IR spectral image pseudocolor map from HCA of the image shown in panel (b). Three clusters are presented and clear differentiation of the metastatic tumor (red) within the lymph node

is observed. The blue cluster indicates regions of normal lymph tissue and the purple cluster represents fibrosis, a reactive change to the presence of the tumor.

As mentioned previously, secondary tumors reproduce the morphology of the primary site, and, in PTC and FVPTC, this can result in the formation of what are known as *cystic cavities*. The next example presents an important finding for the advantages of spectral imaging for optical diagnosis and eventual implementation into the medical community. Figure 7 illustrates a lymph node containing an infiltration of metastatic thyroid tumor, indicated by the cystic cavity seen in the right half of the red box. The red box in panel (a) indicates the area that was imaged. Panel (b) illustrates the enlarged section of this region.

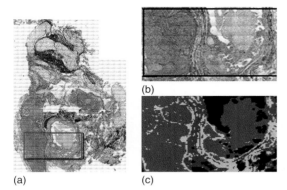

(a)  (c)

**Figure 7.** (a) Visual image of unstained lymph node; red box indicates sampling area. (b) Expanded visual image of red box. (c) Pseudocolor map of resulting IR image, three clusters.

The IR spectroscopic pseudocolor map in panel (c) shows the result of three clusters from HCA.

Interestingly, the initial diagnosis rendered by a pathologist, and the spectral diagnosis represented by the pseudocolor map in panel (c), were in contradiction: the spectral results indicated that the metastatic thyroid tumor on the right of Figure 7(b) and the lymph node tissue on the left (panel b) were both cancerous, whereas the original histopathological inspection of this slide had diagnosed the lymph node tissue as normal. Re-examination of the H&E stained tissue section of this node revealed that the left side of panel (b) was actually metastatic thyroid tumor. The lymph node exhibited two patterns of metastatic tumor, aggregated type on the right (where it attempts to mimic thyroid structure) and disaggregated type with individual tumor cells on the left. With this in mind, the cluster assignments for the pseudocolor map in panel (c) are as follows: the red cluster represents metastatic thyroid tumor; the purple cluster represents fibrosis, and the orange cluster represents secretions from the cystic cavity. Since IR spectral imaging detects chemical variations between normal and cancerous cells, even without architectural changes being exhibited, it was able to differentiate cancer even though the histopathologist had initially overlooked it.

HCA is an important first step in the development of a diagnostic methodology for cancer as pseudocolor maps serve as the basis of comparison with standard histopathology. However, due to its computational limitations for large spectral images and inability to transfer clustering information between data sets, it is necessary to employ ANNs for the development of the final diagnostic algorithm.

ANNs have been successfully applied as diagnostic algorithms for the identification of both breast and thyroid metastases in lymph nodes. The first ANN was trained on spectra from the spectral image displayed in Figure 5, panel (d). The spectral data were separated into two classes of spectra: normal and abnormal. The resulting network was then applied to two unknown spectral images of lymph nodes heavily infiltrated with breast carcinoma (Figure 8). Areas in red represent malignant cells, and normal lymph node tissues are shown in blue.

Figure 9 represents the diagnostic images resulting from a neural network trained on the spectra from the spectral image displayed in Figure 6.

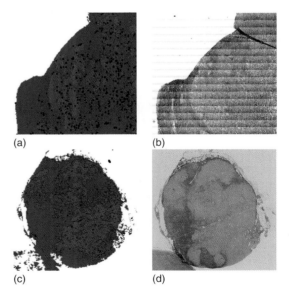

(a)  (b)

(c)  (d)

**Figure 8.** (a) and (c) ANN images from IR data of lymph node tissue containing metastatic breast carcinoma (red), (b) and (d) corresponding H&E stained tissue sections. Panels (a) and (c) generated by Miloš Miljković.

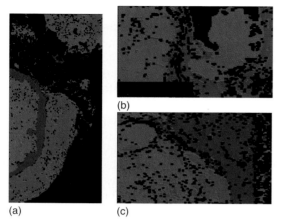

(a)          (c)

**Figure 9.** ANN images of lymph nodes containing metastatic thyroid carcinoma. Panels (a) through (c) were produced from lymph node sections from three individual patients.

Red regions show areas of metastatic thyroid carcinoma, blue regions show normal lymph node tissue, and purple regions indicate fibrosis. The image in panel (a) was generated from a different spectral image taken from the same lymph node section from which the network was trained. As expected, the metastatic regions are clearly differentiated from the normal and fibrotic tissue. The true test of the diagnostic capability, however, is the application of the network to unknown data, i.e., spectral images that were not involved in network training. Panels (b) and (c) of Figure 9 show the diagnostic images of two lymph node sections taken from different patients and reflect the diagnostic capability of the network. The metastatic cancerous regions are clearly defined in red, normal tissue in blue, and fibrosis in purple. These results highlight the possibilities of using spectral imaging as an objective diagnostic for the detection of metastatic cancer.

### 4.1.4   Remaining problems: correcting the dispersion artifact

IR spectra are sometimes contaminated by a dispersion artifact. The causes of this artifact and our attempts to address the problem have been detailed previously.[114,115] The pseudocolor maps in Figures 4(d) and 5(b) were generated from the spectral region 1580–950 cm$^{-1}$. By removing the amide I band from the spectral image, the correlation matrix is not dominated by the amide I band shift characteristic of the artifact and therefore the clusters formed result from biochemical changes within the cells rather than being a reflection of the degree of dispersion present in the spectral image.

The presence of this artifact in IR spectra is undesirable due to the peak shifts and changing spectral band ratios overriding the subtle spectral changes that occur within biological samples such as tissue and cells. In spectral images heavily dominated by the presence of dispersion, the shift of the amide I band can be between 10–30 cm$^{-1}$. The cluster analysis, which bases the assignment of spectra into a given class based on the covariance between the individual spectra within the hypercube, is dominated by the shift in the amide I band. Clusters are likely based on this shift, and to a lesser extent the amide I/II ratio, rather than on the subtle differences between normal and diseased cells that arise due the biochemical processes of cancer mechanisms and proliferation.

Presently, we are collaborating with researchers from CenSSIS (Center for Subsurface Sensing and Imaging Systems, Northeastern University, Boston, USA) to investigate a technique known as *unsupervised spectral unmixing*.[116] Spectral unmixing searches for the pure component spectra in a spectral image or hypercube, the pure component spectrum representing the dispersive component is then subtracted from all the spectra in the hypercube, thus removing the artifact. Preliminary results of this unmixing are given in Figure 10, which gives a representative example of uncorrected and corrected spectra. IR spectra A and C show a strong dispersive component. This is reflected in the position of the amide I band (1633 and 1627 cm$^{-1}$ respectively), the amide I/II band ratio, and the sharp dip around 1750 cm$^{-1}$. The IR spectra B and D resulted from spectral unmixing. Note the position of the amide I band (1646 cm$^{-1}$ in both traces) and the corrected amide I/II band ratio. The spectral unmixing

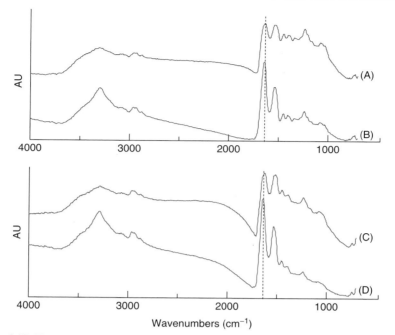

**Figure 10.** (A) and (C) IR spectra exhibiting a strong dispersion artifact. (B) and (D) Corrected IR spectra.

algorithm was tested on a small section of an IR spectral image, and the resulting cluster maps (not shown) gave more distinct clusters and better correlation with the architectural morphology of the tissue. It is believed that this algorithm can be routinely applied to IR spectral images as a means of overcoming the dispersion artifact.

## 5 CONCLUSIONS

Spectral imaging for the detection of cancer, as a technique to be used and accepted by the medical community, has made many advances over the last 5 years. Although still in its early stages, spectral imaging promises to serve as an important adjunct to routine cytological and histopathological screening. It is inexpensive, can be automated, measurements are fast, staining is not required, and diagnosis is objective. However, before these methods can be routinely applied in a clinical setting, it is imperative that the spectral information and variability is understood on a cellular level. These considerations

are discussed in greater detail in **Infrared and Raman Microspectroscopic Studies of Individual Human Cells**.

## END NOTE

[a.] The prefix A in a figure number indicates that the figure appears in the Appendix of this book.

## ABBREVIATIONS AND ACRONYMS

| | |
|---|---|
| ANNs | Artificial Neural Networks |
| ATR | Attenuated Total Reflection |
| BSE | Bovine Spongiform Encephalopathy |
| CCD | Charge-Coupled Device |
| FPA | Focal-Plane Array |
| FT- | Fourier Transform |
| FT-IR | Fourier Transform IR |
| FVPTC | Follicular Variant of Papillary Thyroid Carcinoma |
| HCA | Hierarchical Cluster Analysis |

| | |
|---|---|
| H&E | Hematoxylin and Eosin |
| IR | Infrared |
| IRE | Internal Reflection Element |
| LRS | Least Resolvable Separation |
| MCT | Mercury Cadmium Telluride |
| Nd:YAG | Neodymium-doped Yttrium Aluminum Garnet |
| NSLS | National Synchrotron Light Source |
| PCA | Principal Component Analysis |
| PE | PerkinElmer |
| PTC | Papillary Thyroid Carcinoma |
| SNR | Signal-to-Noise Ratio |
| UV | Ultraviolet |

## REFERENCES

1. Q. Wang, W. Sanad, L.M. Miller, A. Voigt, K. Klingel, R. Kandolf, K. Stangl and G. Baumann, *Vib. Spectrosc.*, **38**, 217–222 (2005).

2. K. Gough, M. Rak, A. Bookatz, M.D. Bigio, S. Mai and D. Westaway, *Vib. Spectrosc.*, **38**, 133–141 (2005).

3. S.G. Kazarian and K.L.A. Chan, *Biochim. Biophys. Acta*, **1758**, 858–867 (2006).

4. C. Xiao, D.J. Moore, C.F. Flach and R. Mendelsohn, *Vib. Spectrosc.*, **38**, 151–158 (2005).

5. X. Bi, G. Li, S.B. Doty and N.P. Camacho, *Osteoarthr. Cartil.*, **13**, 1050–1058 (2005).

6. X. Bi, X. Yang, M.P.G. Bostrom and N.P. Camacho, *Biochim. Biophys. Acta (BBA) – Biomembr.*, **1758**, 934–941 (2006).

7. A.L. Boskey and N.P. Camacho, *Biomaterials*, **28**, 2465–2478 (2007).

8. A.L. Boskey and R. Mendelsohn, *Vib. Spectrosc.*, **38**, 107–114 (2005).

9. D. Faibish, A. Gomes, G. Boivin, I. Binderman and A.L. Boskey, *Bone*, **36**, 6–12 (2005).

10. R. Mendelsohn, C.R. Flach and D.J. Moore, *Biochim. Biophys. Acta (BBA) – Biomembr.*, **1758**, 923–933 (2006).

11. R. Mendelsohn, E.P. Paschalis and A.L. Boskey, *J. Biomed. Opt.*, **4**, 14–21 (1999).

12. J. Dubois, R. Baydack, E. McKenzie, T. Booth and M. Jackson, *Vib. Spectrosc.*, **32**, 95–105 (2003).

13. J. Kneipp, M. Beekes, P. Lasch and D. Naumann, *J. Neurosci.*, **22**, 2989–2997 (2002).

14. J. Kneipp, L.M. Miller, S. Spassov, F. Sokolowski, P. Lasch, M. Beekes and D. Naumann, 'Biomedical Vibrational Spectroscopy and Biohazard Detection Technologies', SPIE, San Jose, 17–25 (2004).

15. A. Kretlow, Q. Wang, J. Kneipp, P. Lasch, M. Beekes, L. Miller and D. Naumann, *Biochim. Biophys. Acta (BBA) - Biomembr.*, **1758**, 948–959 (2006).

16. Q. Wang, A. Kretlow, M. Beekes, D. Naumann and L. Miller, *Vib. Spectrosc.*, **38**, 61–69 (2005).

17. K.R. Bambery, B.R. Wood, M.A. Quinn and D. McNaughton, *Aust. J. Chem*, **57**, 1139–1143 (2004).

18. J.-I. Chang, Y.-B. Huang, P.-C. Wu, C.-C. Chen, S.-C. Huang and Y.-H. Tsai, *Gynecol. Oncol.*, **91**, 577–583 (2003).

19. S.R. Lowry, *Cell. Mol. Biol.*, **44**, 169–177 (1998).

20. W. Steller, J. Einenkel, L.-C. Horn, U.-D. Braumann, H. Binder, R. Salzer and C. Krafft, *Anal. Bioanal. Chem.*, **384**, 145–154 (2006).

21. B.R. Wood, L. Chiriboga, H. Yee, M.A. Quinn, D. McNaughton and M. Diem, *Gynecol. Oncol.*, **93**, 59–68 (2004).

22. P. Lasch, L. Chiriboga, H. Yee and M. Diem, *Technol. Cancer Res. Treat.*, **1**, 1–7 (2002).

23. P. Lasch, W. Haensch, E.N. Lewis, L.H. Kidder and D. Naumann, *Appl. Spectrosc.*, **56**, 1–9 (2002).

24. P. Lasch, W. Haensch, E.N. Lewis, L.H. Kidder and D. Naumann, *Appl. Spectrosc.*, **56**, 1–9 (2002).

25. P. Lasch, W. Haensch, D. Naumann and M. Diem, *Biochim. Biophys. Acta*, **1688**, 176–186 (2004).

26. P. Lasch and D. Naumann, *Cell. Mol. Biol.*, **44**, 189–202 (1998).

27. P. Lasch and D. Naumann, *Biochim. Biophys. Acta*, **1758**, 814–829 (2006).

28. P. Lasch, A. Pacifico and M. Diem, *Biopolymers*, **67**, 335–338 (2002).

29. T. Richter, G. Steiner, M.H. Abu-Id, R. Salzer, R. Bergmann, H. Rodig and B. Johannsen, *Vib. Spectrosc.*, **28**, 103–110 (2002).

30. Y.-J. Chen, Y.-D. Cheng, H.-Y. Liu, P.-Y. Lin and C.-S. Wang, *Chang Gung Med. J.*, **29**, 518–527 (2006).

31. R. Bhargava, D.C. Fernandez, S.M. Hewitt and I.W. Levin, *Biochim. Biophys. Acta (BBA) – Biomembr.*, **1758**, 830–845 (2006).

32. D.C. Fernandez, R. Bhargava, S.M. Hewitt and I.W. Levin, *Nat. Biotech.*, **23**, 469–474 (2005).

33. E. Gazi, J. Dwyer, P. Gardner, A. Ghanbari-Siahkali, A.P. Wade, J. Miyan, N.P. Lockyer, J.C. Vickerman, N.W. Clarke, J.H. Shanks, L.J. Scott, C.A. Hart and M. Brown, *J. Pathol.*, **201**, 99–108 (2003).

34. E. Gazi, J. Dwyer, N.P. Lockyer, J. Miyan, P. Gardner, C. Hart, M. Brown and N.W. Clarke, *Biopolymers*, **77**, 18–30 (2005).

35. A.S. Haka, L.H. Kidder and E.N. Lewis, 'Proceedings of SPIE-The International Society for Optical Engineering', SPIE-The International Society for Optical Engineering, 47–55 (2001).

36. C. Paluszkiewicz, W.M. Kwiatek, A. Banas, A. Kisiel, A. Marcelli and M. Piccinini, *Vib. Spectrosc.*, **43**, 237–242 (2007).

37. Y. Yang, J. Sule-Suso, G.D. Sockalingum, G. Kegelaer, M. Manfait and A.J. El Haj, *Biopolymers*, **78**, 311–317 (2005).

38. K. Yano, S. Ohoshima, Y. Gotou, K. Kumaido, T. Moriguchi and H. Katayama, *Anal. Biochem.*, **287**, 218–225 (2000).

39. H. Fabian, P. Lasch, M. Boese and W. Haensch, *J. Mol. Struct.*, **661–662**, 411–417 (2003).

40. H. Fabian, N.A.N. Thi, M. Eiden, P. Lasch, J. Schmitt and D. Naumann, *Biochim. Biophys. Acta (BBA) – Biomembr.*, **1758**, 874–882 (2006).

41. H. Fabian, R. Wessel, M. Jackson, A. Schwartz, P. Lasch, I. Fichtner, H.H. Mantsch and D. Naumann, 'Proceedings of SPIE-The International Society for Optical Engineering', SPIE-The International Society for Optical Engineering, 13–23 (1998).

42. L.H. Kidder, A.S. Haka, P.J. Faustino, D.S. Lester, I.W. Levin and E.N. Lewis, 'Proceedings of SPIE-The International Society for Optical Engineering', SPIE-The International Society, 178–186 (1998).

43. L. Zhang, G.W. Small, A.S. Haka, L.H. Kidder and E.N. Lewis, *Appl. Spectrosc.*, **57**, 14–22 (2003).

44. N. Amharref, A. Beljebbar, S. Dukic, L. Venteo, L. Schneider, M. Pluot, R. Vistelle and M. Manfait, *Biochim. Biophys. Acta (BBA) – Biomembr.*, **1758**, 892–899 (2006).

45. K.R. Bambery, E. Schultke, B.R. Wood, S.T.R. MacDonald, K. Ataelmannan, R.W. Griebel, B.H.J. Juurlink and D. McNaughton, *Biochim. Biophys. Acta (BBA) – Biomembr.*, **1758**, 900–907 (2006).

46. C. Beleites, G. Steiner, M.G. Sowa, R. Baumgartner, S. Sobottka, G. Schackert and R. Salzer, *Vib. Spectrosc.*, **38**, 143–149 (2005).

47. C. Krafft, L. Shapoval, S.B. Sobottka, K.D. Geiger, G. Schackert and R. Salzer, *Biochim. Biophys. Acta (BBA) – Biomembr.*, **1758**, 883–891 (2006).

48. C. Krafft, S.B. Sobottka, K.D. Geiger, G. Schackert and R. Salzer, *Anal. Bioanal. Chem.*, **387**, 1669–1677 (2007).

49. C. Krafft, S.B. Sobottka, G. Schackert and R. Salzer, *Analyst*, **129**, 921–925 (2004).

50. C. Krafft, S.B. Sobottka, G. Schackert and R. Salzer, *J. Raman Spectrosc.*, **37**, 367–375 (2006).

51. C. Conti, P. Ferraris, E. Giorgini, T. Pieramici, L. Possati, R. Rocchetti, C. Rubini, S. Sabbatini, G. Tosi, M.A. Mariggio and L.L. Musio, *J. Mol. Struct.*, **834–836**, 86–94 (2007).

52. C. Conti, E. Giorgini, T. Pieramici, C. Rubini and G. Tosi, *J. Mol. Struct.*, **744–747**, 187–193 (2005).

53. C.P. Schultz, *Technol. Cancer Res. Treat.*, **1**, 95–104 (2002).

54. E. Burattini, F. Malvezi-Campeggi, M. Chilosi, C. Conti, P. Ferraris, F. Monti, S. Sabbatini, G. Tosi and A. Zamo, *J. Mol. Struct.*, **834–836**, 170–175 (2007).

55. M.J. Romeo and M. Diem, *Vib. Spectrosc.*, **38**, 115–119 (2005).

56. K. Golcuk, G.S. Mandair, A.F. Callender, S. Nadder, D.H. Kohn and M.D. Morris, *Biochim. Biophys. Acta*, **1758**, 868–873 (2006).

57. M. Kazanci, P. Roschger, E.P. Paschalis, K. Klaushofer and P. Fratzl, *J. Struct. Biol.*, **156**, 489–496 (2006).

58. J. Kneipp, T. Bakker Schut, M. Kliffen, M. Menke-Pluijmers and G. Puppels, *Vib. Spectrosc.*, **32**, 67–74 (2003).

59. R. Manoharan, K. Shafer, L. Perelman, J. Wu, K. Chen, G. Deinum, M. Fitzmaurice, J. Myles, J. Crowe, R.R. Dasari and M.S. Feld, *Photochem. Photobiol.*, **67**, 15–22 (1998).

60. K.E. Shafer-Peltier, A.S. Haka, M. Fitzmaurice, J. Crowe, J. Myles, R.R. Dasari and M.S. Feld, *J. Raman Spectrosc.*, **33**, 552–563 (2002).

61. C. Yu, E. Gestl, K. Eckert, D. Allara and J. Irudayaraj, *Cancer Detect. Prev.*, **30**, 515–522 (2006).

62. Z. Huang, H. Lui, X.K. Chen, A. Alajlan, D.I. McLean and H. Zeng, *J. Biomed. Opt.*, **9**, 1198–1205 (2004).

63. B.W.D. de Jong, T.C.B. Schut, J. Coppens, K.P. Wolffenbuttel, D.J. Kok and G.J. Puppels, *Vib. Spectrosc.*, **32**, 57–65 (2003).

64. B.W.D. de Jong, T.C. BakkerSchut, K. Maquelin, T. vander Kwast, C.H. Bangma, D.J. Kok and G.J. Puppels, *Anal. Chem.*, **78**, 7761–7769 (2006).

65. N. Stone, M.C.H. Prieto, C.A. Kendall, G. Shetty and H. Barr, 'Biomedical Vibrational Spectroscopy III: Advances in Research and Industry', SPIE, San Jose, 60930U–609309 (2006).

66. C. Kendall, N. Stone, N. Shepherd and H. Barr, 'Diagnostic Optical Spectroscopy in Biomedicine II', SPIE, Munich, 237–248 (2003).

67. G. Shetty, C. Kendall, N. Shepherd, N. Stone and H. Barr, *Br. J. Cancer*, **94**, 1460–1464 (2006).

68. C. Eliasson, A. Loren, J. Engelbrektsson, M. Josefson, J. Abrahamsson and K. Abrahamsson, *Spectrochim. Acta, Part A*, **61**, 755–760 (2005).

69. C. Krafft, T. Knetschke, A. Siegner, R.H.W. Funk and R. Salzer, *Vib. Spectrosc.*, **32**, 75–83 (2003).

70. C. Matthäus, S. Boydston-White, M. Miljković, M. Romeo and M. Diem, *Appl. Spectrosc.*, **60**, 1–8 (2006).

71. A. Elliot and E.J. Ambrose, *Nature*, **165**, 921 (1950).

72. E.R. Blout and M. Fields, *Biol. Chem.*, **178**, 335–43 (1949).

73. K. Brandenburg and U. Seydel, 'Fourier Transform Infrared Spectroscopy of Cell Surface Polysaccharides', in "Infrared Spectroscopy of Biomolecules", eds H.H. Mantsch and D. Chapman, Wiley-Liss, New York, 203 (1996).

74. R.N.A.H. Lewis and R.N. Mcelhany, 'Fourier Transform Infrared Spectroscopy in the Study of Hydrated Lipids and Lipid Bilayer Membranes', in "Infrared Spectroscopy of Biomolecules", eds

H.H. Mantsch and D. Chapman, Wiley-Liss, New York, 159 (1996).

75. J. Liquier and E. Taillandier, 'Infrared Spectroscopy of Nucleic Acids', in "Infrared Spectroscopy of Biomolecules", eds H.H. Mantsch and D. Chapman, Wiley-Liss, New York, 131 (1996).

76. P. Pancoska, J. Kubelka and T.A. Keiderling, *Appl. Spectrosc.*, **53**, 655 (1999).

77. W.K. Surewicz and H.H. Mantsch, 'Infrared Absorption Methods for Examining Protein Structure', in "Spectroscopic Methods for Determining Protein Structure in Solution", ed H.A. Havel, VCH Publishers, New York, 135 (1996).

78. L.G. Tensmeyer and E.W. Kauffman II, 'Protein Structure Revealed by Nonresonance Raman Spectroscopy', in "Spectroscopic Methods for Determining Protein Structure in Solution", ed H.A. Havel, VCH Publishers, New York, 69 (1996).

79. M. Tsuboi, *Appl. Spectrosc. Rev.*, **3**, 45 (1969).

80. B.R. Wood, B. Tait and D. McNaughton, *Biochim. Biophys. Acta (BBA) – Biomembr.*, **1539**, 58–70 (2001).

81. B. Alberts, D. Bray, J. Lewis, M. Raff, K. Roberts and J.D. Watson, 'Molecular Biology of the Cell', Garland Publishing, New York (1983).

82. D. Sheehan, 'Theory and Practice of Histotechnology', Mosby, St Louis (1980).

83. C.S. Colley, S.G. Kazarian, P.D. Weinberg and M.J. Lever, *Biopolymers*, **74**, 328–335 (2004).

84. M. Kansiz and E. Miseo, *Spectroscopy (Suppl.)*, **27**, 414 (2007).

85. K.L.A. Chan, S.G. Kazarian, A. Mavraki and D.R. Williams, *Appl. Spectrosc.*, **59**, 149–155 (2005).

86. K.L.A. Chan and S.G. Kazarian, *Appl. Spectrosc.*, **57**, 381–389 (2003).

87. E.B. Hanlon, R. Manoharan, T.-W. Koo, K.E. Shafer, J.T. Motz, M. Fitzmaurica, J.R. Kramer, R.R. Dasari and M.S. Feld, *Phys. Med. Biol.*, **45**, R1 (2000).

88. R.H. Clarke, E.B. Hanlon, J.M. Isner and H. Brody, *Appl. Opt.*, **26**, 3175 (1987).

89. R.R. Alfano, C.H. Liu, W.L. Sha, H.R. Zhu, D.L. Atkins, J. Cleary, R. RPrudente and E. Clemer, *Lasers Life Sci.*, **42**, 23 (1991).

90. A. Mizuno, T. Hayashi, K. Tashibu, S. Maraishi, K. Kawauchi and Y. Ozaki, *Neurosci. Lett.*, **141**, 47 (1992).

91. R.P. Rava, J.J. Baraga and M.S. Feld, *Spectrochim. Acta, Part A*, **47**, 509 (1991).

92. S. Fendel, R. Freis and B. Schrader, *J. Mol. Struct.*, **410–411**, 531–534 (1997).

93. R.G. Messerschmidt, 'Minimizing Optical Nonlinearities in Infrared Microspectroscopy', in "Practical Guide to Infrared Microspectroscopy", ed H.J. Humecki, Marcel Dekker, Inc, New York, 3 (1995).

94. R.G. Messerschmidt and M.A. Harthcock, 'Infrared Microspectroscopy: Theory and Applications', Marcel Dekker, Inc, New York (1988).

95. J.A. Reffner and P.A. Martoglio, 'Uniting Microscopy and Spectroscopy', in "Practical Guide to Infrared Microspectroscopy", ed H.J. Humecki, Marcel Dekker, Inc, New York (1995).

96. B.K. Johnson, 'Optics and Optical Instruments', Dover Publications Inc., New York (1960).

97 G.L., Carr, *Rev. Sci. Instrum.*, **72**, 1613–1619 (2001).

98. E.N. Lewis, A.M. Gorbach, C. Marcott and I.W. Levin, *Appl. Spectrosc.*, **50**, 263 (1996).

99. E.N. Lewis, P.J. Treado, R.C. Reeder, G.M. Story, A.E. Dowrey, C. Marcott and I.W. Levin, *Anal. Chem.*, **67**, 3377 (1995).

100. R. Bhargava, D.C. Fernandez, S.M. Hewitt and I.W. Levin, *Biochim. Biophys. Acta*, **1758**, 830–845 (2006).

101. B. Mohlenhoff, M. Romeo, M. Diem and B.R. Wood, *Biophys. J.*, **88**, 3635–3640 (2005).

102. N. Vogt, *Chemom. Intell. Lab. Syst.*, **1**, 213–231 (1987).

103. O.M. Kvalheim and T.V. Karstang, 'SIMCA – Classification by Means of Disjoint Cross Validated Principal Components Models', in "Multivariate Pattern Recognition in Chemometrices",

"Volume 9: Data Handling in Science and Technology", ed R. Brereton, Elsevier, Netherlands, 209–245 (1992).

104. B.R. Wood, L. Chiriboga, H. Yee, M.A. Quinn, D. McNaughton and M. Diem, *Gynecol. Oncol.*, **93**, 59–68 (2004).

105. A. Tfayli, O. Piot, A. Durlach, P. Bernard and M. Manfait, *Biochim. Biophys. Acta*, **1724**, 262–269 (2005).

106. M. Adams, 'Chemometrics in Anaytical Chemistry', The Royal Society of Chemistry, Cambridge (1995).

107. M. Romeo, F. Burden, M. Quinn, B. Wood and D. McNaughton, *Cell. Mol. Biol.*, **44**, 179–187 (1998).

108. G. Schiffer, T. Udelhoven, H. Labischininski and J. Schmitt, 'Workshop on FTIR Spectroscopy in Microbiological and Medical Diagnostic', Robert Koch-Institut, Berlin (2002).

109. J. Schmitt, M. Beekes, A. Brauer, T. Udelhoven and D. Naumann, 'Workshop on FTIR Spectroscopy in Microbiological and Medical Diagnostic', Robert Koch-Institut, Berlin (2002).

110. D. Hammerstrom, *IEEE Spectr.*, **30** (6), 26–32 (1993).

111. D. Maddalena, *Chem. Aust.*, **60**, 218–221 (1993).

112. J.R. Mansfield, L.M. McIntosh, A.N. Crowson, H.H. Mantsch and M. Jackson, *Appl. Spectrosc.*, **53**, 1323–1330 (1999).

113. M. Riedmiller and H. Braun, 'IEEE International Conference on Neural Networks', IEEE, New York, San Francisco, 586–591 (1993).

114. M. Romeo and M. Diem, *Vib. Spectrosc.*, **38**, 129–132 (2005).

115. M.J. Romeo and M. Diem, *Vib. Spectrosc.*, **38**, 115–119 (2005).

116. M. Berman, A. Phatak, R. Lagerstrom and B.R. Wood, *J. Chemom.* (2006) Accepted for Publication.

# Infrared and Raman Microspectroscopic Studies of Individual Human Cells

**Melissa J. Romeo[1], Susie Boydston-White[1], Christian Matthäus[1], Miloš Miljković[1], Benjamin Bird[1], Tatyana Chernenko[1], Peter Lasch[2] and Max Diem[1]**

[1] Northeastern University[a], Boston, MA, USA
[2] Robert Koch-Institut, Berlin, Germany

## 1   INTRODUCTION

The subject of this chapter, vibrational spectroscopy of individual human cells, has not yet been reviewed in detail in any book or monograph, since the amount of data available was insufficient to warrant a review. This has changed recently, since several groups have initiated research programs[1-6] to investigate vibrational spectroscopic properties of individual cells with the specific aim of detecting disease at the level of one single cell.

The development of infrared microspectroscopy (IR-MSP) is closely tied to spectroscopic investigations of cells. The original work of detecting and identifying pathogens by mid-IR spectroscopy was performed in the early 1990s by Naumann's research group at the Robert Koch-Institut in Berlin, who pioneered the use of infrared (IR) spectroscopy and IR microspectroscopy, coupled to multivariate statistics such as cluster analysis, for the identification of bacteria.[7] However, because of the extremely small size

(ca. 1 μm) of prokaryotes, these studies were not carried out at the level of single cells, but small colonies of cells. Spurred by these highly successful studies, researchers at the National Research Council of Canada (NRCC) in Ottawa began to utilize IR spectroscopy to study exfoliated human cells.[8,9] The rational for these studies was to improve on the low accuracy of standard cytological methods, such as the Papanicolaou[10] ("Pap", for short) test used in screening for cervical disease. However, exfoliated cervical cells were certainly not an easy sample system to be used in such pioneering and novel studies, since samples of cervical cells are often contaminated by red blood cells (erythrocytes), lymphatic cells such as polymorphonuclear lymphocytes (PMNs), cervical mucous, and bacteria.[11-14] Furthermore, even the cervical cells themselves are mixtures of squamous and glandular (columnar) epithelial cells. This heterogeneous mixture of cells was analyzed macroscopically, i.e., not at a cell-by-cell level,

*Vibrational Spectroscopy for Medical Diagnosis*. Edited by Max Diem, Peter R. Griffiths and John M. Chalmers.
© 2008 John Wiley & Sons, Ltd. ISBN 978-0-470-01214-7.

but averaged over 1000–100 000 of cells. Thus, it is not surprising that these early studies yielded only very qualitative correlation between cytology and spectroscopy.

A gross correlation between certain spectral features and disease was discovered during these early efforts: dysplastic (precancerous) cells generally exhibited a significantly lower glycogen IR spectroscopic signal as compared to normal (superficial) cells.[9,15,16] Although this finding was certainly correct, we demonstrated that the reverse statement, namely, that low cellular glycogen implies precancerous cervical disease, does not hold.[16] This aspect is elaborated upon in the next paragraph, and indicates how carefully one has to be in correlating spectral results with medical diagnoses.

Squamous cervical cells originate from a layer of cells known as the *basal layer*. In this basal layer, cells divide actively, and the daughter cells migrate to the surface of the epithelium over a time span of about 2 weeks. In this process, they mature and accumulate glycogen. In precancerous and cancerous cervical tissue, this maturation and other aspects of cell development

are perturbed, and immature, precancerous cells without glycogen may be found in the superficial layer. Thus, the observation of reduced glycogen concentration in cervical dysplasia was certainly correct. However, there exist other conditions where reduced or zero glycogen levels are observed among superficial cells. One of these conditions, metaplasia, refers to the presence of immature, but noncancerous cells at the surface.[17] Metaplastic cells resemble normal basal cells spectrally, and are completely devoid of glycogen signals. Thus, the absence or presence of glycogen signals cannot be used as a reliable indicator of disease.

Furthermore, superficial cells lose their glycogen toward the end of their normal life, when they are particularly susceptible to exfoliation. Thus, exfoliated cells can have widely varying glycogen content, as shown by the spectra shown in Figure 1, along with the photomicrographs (a and b) of two mature cervical cells. One of them exhibits significant glycogen contribution as seen by the strong absorption profile between 1200 and 950 cm$^{-1}$ (spectrum A), whereas cell (b) shows no observable glycogen (spectrum B).

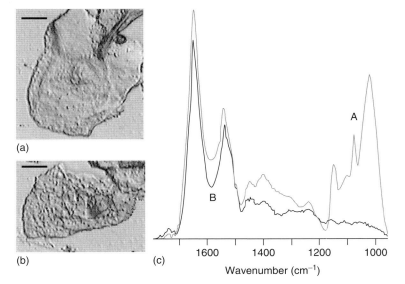

(a)

(b)

(c)

**Figure 1.** Photomicrographs of unstained superficial squamous cervical cells. (a) Glycogen-rich cell and (b) glycogen-free cell. The length of the scale bar corresponds to 20 μm. (c) Infrared absorption spectra of glycogen-rich cell a (trace A) and glycogen-free cell b (trace B).

This behavior was found in other glycogen-accumulated epithelial cells, and is discussed in detail later. Although the spectra displayed in Figure 1 are by no means up to the quality of data that can be recorded with modern IR-MSP spectrometers, and took enormous efforts to collect, they demonstrate that the glycogen signals cannot be used as an indicator for disease in cervical cells. This is particularly true if spectral data are acquired for pellets of cells, rather than for individual cells.

Efforts to circumvent these difficulties started in the late 1990s, when IR microspectrometers (also known as *IR microscopes*) became available that permitted routine acquisition of spectra from individual cells. However, rather than simplifying the interpretation of spectra from single cell ensembles, the spectra collected from individual cells showed enormous heterogeneity. Thus, multivariate statistical methods that emphasize common features and de-emphasize irrelevant variations in the spectra needed to be employed for the analysis of spectra of individual cells.[18]

Thus, vibrational spectroscopy of cells is very different from normal "chemical" spectroscopy: when collecting the IR spectrum of an analyte at different laboratories and at different times, even when using different instruments, the spectral results are generally highly reproducible. Any differences in results can be attributed to physical factors such as temperature, purity, and instrumental response, and a "true" spectrum of the analyte exists. For cells, such a true spectrum does not exist per se, but has to be viewed as a most probable spectrum from a class of spectra with large variance. This aspect is new in the field of spectroscopy, and may have contributed to the negative attitude shown by some researchers toward this new enterprise. Furthermore, the fact that two near-identical cells may present quite different spectral patterns points to the fact that sophisticated methods of multivariate statistics need to be employed, and aspects of cell biology need to be understood, for this research to yield useful results.

## 2 METHODS

This section starts with a discussion of methods for sample preparation for exfoliated and cultured cells. Certain aspects, in particular, the choices for substrates and methods of fixation, are still under investigation in order to find the most cost-effective, reproducible methods, which are also acceptable to cellular biologists and cytologists. This aspect is highly important since the information sought by biochemists and spectroscopists (biochemical composition and variations therein) may differ from that of cytologists, who are most interested in cell morphology.

### 2.1 Exfoliated cells

One of the funded projects of the author's research group is the development of methodologies to screen exfoliated cells for disease. Any sample of exfoliated cells, however, will exhibit significant heterogeneity, and the correlation with cytology is by no means straightforward. Thus, model systems for exfoliated cells had to be developed from samples that were readily available, and could be collected under a local Institutional Review Board (IRB) protocol. In particular, we were interested for the original studies in a large number of "normal" samples in order to establish spectral variance. Thus, three sources of epithelial cells were identified from which samples were readily accessible: oral mucosa cells, cells isolated from urine and containing both squamous and urothelial cells, and canine cervical cells. Oral mucosa (buccal) cells were harvested from volunteer graduate and undergraduate students under a local IRB protocol. To collect cells, the inside of the cheek was gently swiped with a sterile polyester swab. We estimate that a single swiping of the oral cavity results in about $10^5 - 10^6$ exfoliated cells. Visual microscopic inspection of the cell samples produced indicates that over 99% of the cells are large squamous cells with small, well-delineated nuclei and a large cytoplasmic area. Details of sample preparation are discussed in Section 2.3.

Urine-borne cells were isolated by filtration and centrifugation from urine. Two major cell types are contained in urine: squamous cells from the distal end of the urethra and urothelial cells from the lining of the bladder. This work was carried out to develop a screening method for bladder cancer. The preparation of spectroscopic samples is presented in Section 2.3

Canine cervical cells were collected in collaboration with a local veterinary clinic. During spaying of dogs, the ovaries and the uterus, including the cervix, are surgically removed. Cervical cells were exfoliated from the exposed cervices using miniature dental brushes. These experiments were performed with permission from the Animal Welfare Committee of Hunter College, City University of New York. The exfoliated canine cervical cells were further processed like all other squamous cells.

## 2.2   Cultured cells

In contrast to exfoliated cells, which were collected from live donors, cultured cells originate from cancerous or otherwise transformed cells, taken originally from live donors, but subsequently kept alive in cell culture using a medium in which the cells can survive. When this medium is supplemented, for example, with fetal calf serum that contains biochemical signals that causes cells to divide, these cells can be kept in an "exponential growth" phase in which they divide approximately every 24 h, and duplicate the number of cells.

### 2.2.1   Cell division and the cell cycle

Most of the time, normal human cells are not divisionally active, but perform the tasks determined by their level of differentiation. This stage of cellular activity, which may last for days or weeks, is known as the *G0 phase*. Upon receiving a biochemical signal to divide, the cell ceases its normal metabolic activity to prepare for cell division, which takes about 24 h.[19] During the first divisional stage, known as the *Gap* 1 (*G*1)

*phase*, which lasts about 8 h, cells produce the proteins necessary for cell division, among them the histone core proteins around which the newly produced DNA is wound. During the next phase, known as the *S phase*, cells reproduce their entire genome. This implies that a duplicate set of genes, about $3 \times 10^9$ bp, is constructed. At this point, the cell contains twice the DNA of a cell in G0 or G1, and is said to be *tetraploid*. This phase also lasts about 6–8 h. After the S phase, the DNA-replicating proteins are degraded, and proteins and energy for the actual cell division are accumulated in the Gap 2 (G2) phase, which lasts about 6 h. Finally, during mitosis (M, which can be further subdivided into specific stages), two new daughter cells are produced within about 1 h. Cancerous or transvected cells often process back into this cell division cycle after mitosis; i.e., the mechanisms to control cell divisions may be damaged. In cell cultures, the number of cells in each of the stages of cell cycle (G1, S, G2, and M) is determined statistically by the time requirement of each stage. For certain experiments, it is advantageous to "synchronize" a cell culture. In a synchronized culture, ideally all cells are at the same stage of the cell division cycle. This can be accomplished by several procedures, one of which is discussed next.

During mitosis, adherent cells tend to round up and lose their firm attachment to the surface on which they are growing. During this phase, cells can be shaken off, washed, and reseeded onto sample substrates. The success of this procedure can be gauged by the occurrence of cell pairs after a few minutes. In this way, one can pinpoint the "age" of cells to within about 1 h in the 24-h cell cycle.

In this context, it is advantageous to discuss some aspects of DNA/chromatin structure within the cell's nucleus during cell division. In the nucleus of a cell, DNA exists as a tight complex with proteins known as *histones*, and is enormously condensed to reduce its apparent length. In the first level of condensation, DNA winds around histone octamers to form nucleosome fibers, which results in a packing ratio[b] of about six. The next level of organization results in a solenoid structure with a 30-nm diameter and represents the "average" degree of condensation of an interphase (nonmitotic) cell and corresponds to a packing ratio of 40. During mitosis, the solenoid further condenses to form the metaphase chromosomes in which the packing ratio ranges between 7000 and 10 000.

### 2.2.2  Cell culture protocols

Several cancerous or transformed cell lines were cultured for different aspects of this work. For adherent cell cultures, the following protocol for HeLa cells may serve as an example. Cervical adenocarcinoma (HeLa) cells, purchased from the American Tissue and Cell Culture Corporation (ATCC, Manhasset, VA), cell line CCL-2, were seeded in 75-cm$^3$ sterile plastic cell culture flasks (Fisher Scientific) at a concentration of about $2 \times 10^4$ cells per cm$^2$. Typically, the growth medium consisted of 20 mL Dulbecco's Modified Eagle's Medium (DMEM, ATCC), supplemented with 10% Fetal Bovine Serum (FBS, ATCC). To prevent bacterial contamination, 2.5 $\mu$g mL$^{-1}$ of Amphotericin B (ATCC) and 100 IU mL$^{-1}$ penicillin/streptomycin (ATCC) were added to the medium. Cells were incubated at 37 °C in an atmosphere containing 5% CO$_2$.

Different protocols were followed from this point on, depending on the experiments planned. Adherent cells can be grown directly onto substrates, or removed from the cell culture flasks and deposited onto substrates. The method of growing cells on a substrate is discussed in Section 2.3.1. Removal of cells from the cell culture flasks is performed by partial digestion of the surface proteins that are responsible for the strong cell adhesion. This can be accomplished by treating the cells in the culture flasks by the protease trypsin, which is added as a commercial trypsin-ethylene diamine tetraacetate (EDTA, ATCC) mixture to the flask. Trypsinized cells present a totally different morphology than cells grown on the substrate; they are very nearly spherical, and are generally much smaller. Also, because of their spherical shape, the nuclei cannot be detected easily, and staining with standard

hematoxylin/eosin (H&E) does not produce satisfactory results. Spectrally, these near-spherical cells are much more likely to present artifacts that are discussed later in this chapter. However, the trypsinized cells can be refrozen, or seeded onto substrates for subsequent experiments (see Section 2.3.1).

## 2.3  Sample preparation

In principle, it would appear quite straightforward to pipette a cell suspension onto a substrate and evaporate the solvent under mild vacuum or a stream of dry air. This method, in general, produces poor samples: cells do not spread evenly, but tend to clump. Furthermore, if isotonic saline is used to prevent cell lysis (cell death by bursting), salt crystals form around the cells, which may produce substantial scattering of light. If pure water is used as a solvent, cells may lyse in the time required for the solvent to evaporate. Thus, we use one of three methods discussed below to produce samples of cell for spectroscopic studies.

### 2.3.1  Cells grown on substrates

For cells grown directly onto substrates, carefully cleaned and sterilized windows—either CaF$_2$ or "low-e" slides (vide infra)—are deposited into the cell culture flasks, and a few milliliters of a suspension of trypsinized cells are pipetted onto the substrates. After about 2 days, cells can be seen growing on the substrate immersed in the medium. After a few division cycles, these cells have a totally different morphology than the trypsinized cells, as shown later. The substrates are removed from the medium, and cells are fixed, washed, and air dried (vide infra). The last washing step should be in deionized water to prevent salt crystals (from the balanced saline solution (BSS)) forming on the cells during drying. To prevent cell lysis, the final drying step should be performed very quickly under a stream of dry, compressed air.

Cells prepared in such a way generally show a distinct nucleus containing nucleoli, and a large cytoplasmic region. The cytoplasm of nonmitotic cells often exhibits distinct "pseudo-pods", or spindle-like protrusions, which offer the cells some motility, and provide the firm attachment of the cells to the surface. Their attachment to the substrate is so strong that they can be treated chemically on the substrate without dislodging them. Cells grown on a substrate show reasonably reproducible spectra, although the overall amplitude of spectra from different cells may vary by about an order of magnitude. The spectra resemble those of exfoliated cells quite closely, which also show intensity variations by an order of magnitude. Cells grown on a substrate show slightly different spectra for the nucleus and the cytoplasm.

### 2.3.2  CytoSpin® methodology

Exfoliated or trypsinized cultured cells may be deposited onto spectral substrates by one of two different methods developed originally for cytology. The aim of the deposition techniques is to produce sparse monolayers of cells such that all cells are well separated from their nearest neighbors to provide unobstructed view (or measurement apertures) of each cell. This is quite different from the "smears" produced by physicians from cervical brushes, which contain thick clumps of cells that are not suitable for visual or automatic analysis.

The CytoSpin® (Thermo Shandon, Pittsburgh, PA) and the ThinPrep® methods described in the next section are referred to as *liquid-based* methods since the cells are deposited from a cell suspension. In the CytoSpec method, ca. 0.5 mL of cell suspension, with a concentration of about $5 \times 10^4$ cells per mL, is placed into a special, conical funnel shown schematically in Figure 2(a). This funnel is clamped against the microscope slide. Between the funnel and the slide is a layer of wicking paper with a 5-mm-diameter hole. The entire assembly, shown in Figure 2(b), is placed into a special rotor of a centrifuge, and spun between 800 and 1200 rpm

for 20–300 s. The centrifugal force presses the liquid onto the slides, where it is wicked away by the absorbent layer of paper, leaving the cells on the substrate. A typical exfoliated cell sample prepared by this method is shown in Figure 2(c). The sampling area, 5 mm in diameter, contains about $10^3$ cells with more than a cell's size distance between cells.

### 2.3.3  ThinPrep® methodology

The ThinPrep® method, developed and commercialized by Cytyc Corporation (Marlborough, MA), uses a cup with a filter bottom that is immersed into the cell suspension. This suspension uses a specialized solvent, which contains methanol as a fixative. The suspension is suctioned through the porous membrane, the pores of which transmit erythrocytes and cellular debris in the sample. The large, cervical, or buccal cells are deposited on the filter membrane. When a sufficient number of large cells have been deposited, the cup is inverted and stamped against a microscope slide. This step is aided by a short pulse of compressed air to dislodge the cells from the membrane. Optionally, the slide can be stained automatically.

This process produces sample slides with very homogeneous cell deposits. The sample spot is much larger (ca. 20 mm diameter) than in the CytoSpin method and, consequently, contains many more cells. Unfortunately, the high concentration of methanol used in the ThinPrep method changes the spectral properties and morphology of the cells, as compared to unfixed or formalin-fixed cells. However, we have used the ThinPrep method with standard buffered, aqueous solutions, and obtained cell deposits with cells indistinguishable from the CytoSpin method. Fixation aspects are discussed in the next section.

## 2.4  Substrate and fixation issues

The choice of substrates for data acquisition is a crucial issue in IR-MSP. As compared to Raman microspectroscopy (RA-MSP), for which

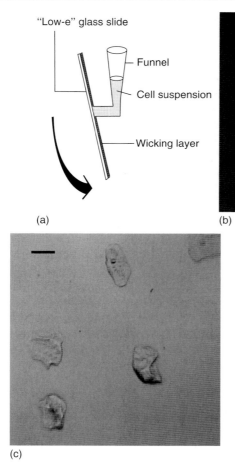

(a)

(b)

(c)

**Figure 2.** Sample preparation using the Thermo-CytoSpin methods (see text for details). (a) and (b) Schematic diagram and view of the CytoSpin funnel. (c) Sparse monolayer of oral mucosa cells produced by CytoSpin method. The length of the scale bar corresponds to 50 μm.

fluorescence-free quartz ($SiO_2$) substrates may be used, $CaF_2$, $BaF_2$, or other nonabsorbing materials must be used for IR-MSP in transmission mode. ($CaF_2$ is suitable for RA-MSP as well). However, specially coated glass slides can be used for transflection IR-MSP. These slides, known as *low-e* (low emissivity) slides (Kevley Technologies, Chesterland, OH), were originally developed for skyscraper windows. They consist of Ag-coated glass; the coating is sufficiently thin to be transparent in the visible, but highly reflective in the IR region. Thus, any cell on these slides can be inspected by visual microscopy, and IR spectral data can be collected, in transflection, from these cells.

Fixation is another major issue for spectral cytology. Exfoliated cells, such as the oral mucosa cells, discussed in detail below, can be used unfixed after quick drying. Once dry (about 1 min), the cells were found to be stable for many days, which was established by measuring spectra for several days, or even weeks, apart. Neither morphologic nor spectroscopic changes could be observed as a function of time. It is well known that dry fixation is a particularly mild form of fixation that preserves structure and composition of biological samples such as cells and tissues. The process of precipitating and drying proteins appears to have a similar effect to fixation in that it renders the protein matrix insoluble and relatively inert. Formalin fixation, using 5–10% buffered aqueous solution of formaldehyde, for about 10 min, produces spectra that are indistinguishable from unfixed cells. This was demonstrated by analyzing data sets of hundreds of fixed and unfixed spectra by multivariate data analysis methods. These results are discussed in Section 3.

When cells are grown onto a substrate and allowed to dry, treatment with ethanol produces relatively minor spectral changes, mostly due to the removal of phospholipids, and only minor changes are observed in the protein amide I and amide II (ca. 1650 and 1550 cm$^{-1}$, respectively) bands of such cells. However, when wet or live cells are treated with ethanol or methanol, significant spectral changes may be observed in the amide I/amide II region.[20] This explains the significant spectral changes in the cells produced and fixed by the ThinPrep method. The spectral changes are, of course, of no consequence to standard cytology, which is concerned with nuclear and cellular morphology, and not with the conservation of the biochemical identity of cellular components.

## 2.5 Special sample preparation

### 2.5.1 Component digestion

*Ethanol treatment.* Dry preparations of cells were exposed for 5 min to 96% ethanol (Fisher Scientific) to remove most phospholipids. The effect of ethanol treatment causes dehydration of proteins and nucleic acids, aggregation of proteins, and removal of phospholipids. Although this is expected to perturb the spectra somewhat, we find that the spectra are relatively unchanged if cells are dry before ethanol treatment (see Section 2.4). Thus, the cells showed minimal spectral changes aside from those expected due to the removal of the phospholipids. This is further discussed at a later stage in this chapter.

*RNase digestion.* Cells attached to substrates were incubated twice with a solution of RNase A (1 mg mL$^{-1}$, type III, from bakers yeast, Sigma–Aldrich Co.) for 15 min at 37 °C to remove cytoplasmic and nuclear RNA.[21,22] Subsequently, the slides were rinsed twice with distilled water and allowed to air-dry. Normally, RNase treatment is preceded by washing with detergents to render a cell's interior accessible to the enzyme. Since the ethanol treatment removes most of the cell membrane, we found that RNase digestion works without the detergent treatment, reducing the cytoplasmic RNA concentration significantly.

### 2.5.2 Cell culture synchronization, and special staining for G1, S, and G2 phases

In order to study cellular effects that may depend on the stage of the cell in its division cycle, synchronization experiments were carried out in which the cells in culture all go through the phases in the cell cycle in unison. Synchronization was accomplished by a procedure known as *mitotic shake-off* (see Section 2.2.1).

Mitotic shake-off was carried out as follows.[23] After removing the culture medium and adding 20 mL of phosphate buffered saline (PBS), the flasks were shaken on a laboratory vortex mixer at a medium–high setting for 45 s to remove debris, poorly adherent and dead cells. The PBS was removed and replaced with 15 mL of fresh, warmed DMEM supplemented with 10% FBS. The flasks were returned to the incubator and allowed to incubate for 1 h. After incubation, the flasks were each again shaken for 45 s on the vortex mixer. The medium was immediately centrifuged for 6 min at 80 g to collect the cells, which were resuspended in warmed DMEM culture medium. Subsequently, 90 lowe slides were immersed in sets of three in 30 Petri dishes in 25-mL DMEM, supplemented with 100-IU mL$^{-1}$ penicillin/streptomycin and 10% FBS. Fifty microliters of cell suspension were pipetted onto each slide, and the slides placed in the incubator at 37 °C and 5% CO$_2$.

At 1-h intervals, one Petri dish was removed from the incubator. All cells on the three slides removed at a given time point should be in approximately the same stage of the cell cycle (or at the same "biological age"), since they all went through mitosis at the same time. To further determine the cells' biological age, we used common biochemical markers that indicate the presence of cell cycle–specific proteins, cyclin E for G1 phase, cyclin B1 for G2 phase, and 5-bromo-2′-deoxyuridine (BrdU) incorporation[24] for S phase. BrdU incorporation needs to be carried out before IR data acquisition, whereas the cyclin stains could be added to the cells afterwards.

BrdU is a DNA-base analog that is incorporated into the replicating DNA without affecting the division of cells. Slides selected for S-phase staining were incubated for 30 min under standard conditions with medium containing 50 mM BrdU (EMD Biosciences, Madison, WI, USA).[24] Subsequently, the BrdU labeled and the other slides selected for spectral data acquisition were washed, and allowed to dry in a dry air stream. Spectral and imaging data were collected as described in Section 2.6.2.

Following data collection, cells were subjected to immunohistochemical staining to determine their exact biochemical age. This staining affords a more accurate determination of a cell's phase in the cell cycle than the elapsed time since the last mitosis, since the duration of the cell cycle phases may vary from cell to cell. Utilizing the combination of time since last mitosis along with staining for phase-specific cellular events affords the best determination of a cell's biochemical age.

Cyclin B1 and cyclin E were visualized by treating the cells on the slides with mouse monoclonal anticyclin B1 (Abcam plc, Cambridge, UK) and rabbit polyclonal anticyclin E (Abcam)[25] and two secondary antibodies, goat polyclonal antimouse conjugated with Alexa Fluor 360® (Molecular Probes) and goat polyclonal antirabbit IgG conjugated with Texas Red® (Calbiochem). The fluorophores attached to the antibodies were visualized via a Nikon Optiphot-2 microscope (see below). Rat monoclonal BrdU, conjugated with fluorescein isothiocyanate (FITC, Abcam) was used to detect BrdU incorporated into the cell's

nuclear DNA as a consequence of DNA synthesis during the S phase of the cell cycle.

### 2.5.3 DAPI staining for chromatin during mitosis

To visualize chromatin and allow a comparison with Raman images of condensed DNA, cells were stained after spectral data acquisition with 4′,6-diamidino-2-phenylindole (DAPI) at a concentration of $0.2 \, \mu g \, mL^{-1}$ at room temperature for 3 min. After staining, the slides were washed in BSS, water, and quickly dried in a dry air stream. All fluorescent stains were visualized via a Nikon Optiphot-2 microscope equipped with a mercury lamp and an episcopic-fluorescence attachment EFD-3, utilizing the appropriate excitation and emission filters and the Plan Apo DIC 60× oil objective, and a Sony camera for image capture.

## 2.6 Data acquisition

In the next section, the instruments used for the acquisition of the spectral data presented in this chapter are presented. The PerkinElmer Spectrum One/Spotlight 300 (PerkinElmer, Shelton, CT) IR-MSP instrument in the authors' laboratory was purchased for tissue and cell studies in 2003, and represented the most advanced and versatile IR-MSP instrument at that time. Synchrotron-based Fourier-transform infrared (FT-IR) measurements were carried out at the National Synchrotron Light Source (NSLS) at Brookhaven National Laboratory on beam line U10B. Two different Raman microspectrometers were used in the course of this study: the original measurements were carried out using a LabRam Microspectrometer (Jobin Yvon Inc., Edison, NJ), whereas the later observations were made using a CRM 200 (WITec, Inc, Ulm, Germany).

### 2.6.1 Raman instrumentation

The JY-LabRam-Microscope (Jobin Yvon Inc., Edison, NJ) is constructed around an Olympus

BX30 microscope. Excitation is provided by a HeNe laser, which delivers around 10 mW of laser power at 632.8 nm at the sample. The instrument uses the same 100× objective for focusing the laser to a spot size approximately 1 μm in diameter, and for collecting the Raman back-scattered radiation.[c] After passing a notch filter, the scattered Raman radiation is dispersed by a 30-cm-focal-length monochromator, employing either a 1800- or a 600-mm$^{-1}$ grating. Detection of the Raman scattered photons is provided by a cooled charge-coupled device (CCD) detector. Individual Raman spectra of subcellular regions were recorded at ca. 1 cm$^{-1}$ spectral resolution. A method first proposed by Deckert and Kiefer[26] was used, in which the read-out pattern commonly observed in Raman spectra collected via CCD detectors is cancelled by collecting the same spectral range with a slightly different grating position. Raman images were recorded at 60 s exposure time per pixel. The step size of the computer-interfaced mechanical microscope stage was 1 μm. The data acquisition time for a single cell image was 15 h on average.

Recent Raman data were collected using a Confocal Raman Microscope, Model CRM 2000 (WITec, Inc., Ulm, Germany). Excitation (ca. 30 mW each at 488, 514.5, or 632.8 nm) was provided by an air-cooled Ar ion laser or HeNe (Melles Griot, Models 05-LHP-928 and 532, respectively). Exciting laser radiation was coupled into the Zeiss microscope through a wavelength-specific single mode optical fiber. The incident laser beam was collimated via an achromatic lens and passed a holographic band-pass filter, before it was focused onto the sample through the microscope objective. A Nikon Fluor (60×/1.00 NA, WD = 2.0 mm) water immersion or a Nikon Plan (100×/0.90 NA, WD = 0.26 mm) objective was used in the studies reported here (NA: numerical aperture, WD: working distance).

The sample is located on a piezo-electrically driven microscope scan stage with $X-Y$ resolution of ca. 3 nm and a repeatability of ±5 nm, and $Z$ resolution of ca. 0.3 nm and ±2 nm repeatability. Samples for Raman studies are grown or deposited onto optically flat CaF$_2$ windows, which

are immersed into water, buffer, or cell culture medium if the water immersion objective is used.

Raman back-scattered radiation is collected through the microscope objective, and passes a holographic edge filter to block Rayleigh scattering and reflected laser light, before being focused into a multimode optical fiber. The single mode input fiber (with a diameter of typically 50 μm) and the multimode output fiber (with a diameter of 50 μm as well) provide the optical apertures for the confocal measurement. The light emerging from the output optical fiber is dispersed by a 30-cm-focal-length, f/4 Czerny–Turner monochromator, incorporating a turret with three interchangeable gratings (1800 mm$^{-1}$, blazed at 500 nm, 600 mm$^{-1}$, blazed at 500 nm, and 600 mm$^{-1}$, blazed at 750 nm). The light is detected by a back-illuminated deep-depletion, $1024 \times 128$ pixel CCD camera operating at $-82\,°C$.

Lateral spatial resolution $\delta_{lat}$ of the acquired confocal sampling area is determined by the diffraction limit, and is given by[27]

$$\delta_{lat} = 0.61\lambda/NA \qquad (1)$$

Depending on the laser wavelength and objective used, a lateral resolution between ca. 300 and 435 nm can be achieved. The axial (depth) resolution is given by

$$\delta_{ax} = 2\lambda n/(NA)^2 \qquad (2)$$

($n$, refractive index; $\lambda$, wavelength) resulting in a theoretical depth resolution for the water ($n = 1.33$) immersion objective ($NA = 1$) between ca. 1300 and 1700 nm, for 488 and 633 nm excitation respectively.

Everall et al.[28] have recently shown that the axial resolution is actually compromised in many-layer systems by the different refractive indices of the sample, and that a substantial error is encountered at depth exceeding 20–25 μm. The cells studies here never exceeded 10 μm in thickness; consequently, we believe that our $X-Z$ and $Y-Z$ optical sections are reasonably accurate in terms of the depth sampled. We calibrated the

WITec CRM 2000 using a multilayer plastic film and found that, to a depth of about $15\,\mu m$, our observed depth profiles agreed well with the film's manufacturer specification.

Spectral resolution depends on the excitation wavelength and grating groove density; furthermore, the spectral resolution varies significantly over the spectral range projected onto the CCD detector. For 488-nm excitation and the $600\text{-mm}^{-1}$ grating, the spectral band pass varies from about $5.5\,cm^{-1}$ per pixel at about $200\,cm^{-1}$ to about $3.3\,cm^{-1}$ per pixel at $4500\,cm^{-1}$. The $1800\,mm^{-1}$ grating compresses the wavenumber coverage to about $1400\,cm^{-1}$.

### 2.6.2  Infrared instrumentation

The PerkinElmer system consists of a Spectrum One FT-IR spectrometer bench coupled to a Spectrum Spotlight 300 IR microscope, henceforth referred to as the PE300. This totally integrated imaging IR microspectrometer incorporates a $16 \times 1$ element ($400\,\mu m \times 25\,\mu m$) HgCdTe (mercury-cadmium-telluride, MCT) array detector and a single point, $100\,\mu m \times 100\,\mu m$ MCT detector mounted on the same Dewar. Both detectors operate in photoconductive mode at liquid nitrogen temperature. The $D^*$ of each element in the array detector exceeds $4.5 \times 10^{10}\,cm\,Hz^{1/2}\,W^{-1}$. The detectors were designed for use with 1300 K sources typically used in IR spectroscopy, and cover the spectral range down to $720\,cm^{-1}$. The single-point MCT detector even allows data collection down to $650\,cm^{-1}$.

The symmetrically arranged objective and condenser provide an image magnification of $6\times$ (at 1:1 imaging, see below), and each has a numerical aperture of 0.58. For imaging applications, the use of specifically designed optics permit 1:1 or 4:1 imaging of the sampled area on the detector elements, resulting in $25\,\mu m \times 25\,\mu m$ or $\sim6\,\mu m \times 6\,\mu m$ pixel size. Visual image collection via a CCD camera is completely integrated with the microscope stage motion and IR spectra data acquisition. The visible images are collected under white light illumination, and are "quilted" together to give pictures of arbitrary size

and aspect ratio. The desired regions for the IR maps are selected on the visual images, and are restricted in size only by available memory. Spectral maps are collected in rapid scan mode at a maximum rate of about 80 pixels per second. Single cell maps collected via the PE300 were collected at a slower rate, due to the weak signals observed in the cellular cytoplasm. Typically, between 32 and 64 interferograms, collected at $4\,cm^{-1}$ spectral resolution, were added together, requiring 0.6 or 1.2 s per pixel, respectively. Using the 16 MCT detector linear array, these collection parameters result in a very acceptable acquisition rate of about 10–20 pixels per second.

For single-point measurements, individual cells are selected from the visually acquired sample image as seen on the screen. For each cell position on the sample substrate, the aperture is selected to straddle the cell, typically $50\,\mu m \times 50\,\mu m$. Cell position and apertures are stored for each cell. Data acquisition of all stored positions proceeds automatically. The microscope and the optical bench are continuously purged with purified, dry air. In addition, the sample area in the focal plane of the microscope is enclosed in a purged sample chamber.

### 2.6.3  Synchrotron-based infrared instrumentation

Synchrotron radiation IR-MSP data were collected at beam line U10B of the vacuum UV ring at NSLS, using a Magna 860 FT-IR instrument coupled to an infinity corrected, confocal Spectra Tech Continuum IR microscope (Thermo Nicolet, Madison, WI), equipped with $32\times$ Cassegrain objective and condenser. Owing to the highly collimated IR beam of the synchrotron source, apertures as small as $8\,\mu m \times 8\,\mu m$ could be used with adequate signal quality. Spectral maps of single cells were collected in transmission mode at $8\,cm^{-1}$ spectral resolution. One hundred and twenty-eight interferograms, collected at an aperture of $10\,\mu m \times 10\,\mu m$, were added together. At comparable signal-to-noise ratio (SNR), the synchrotron data took longer to acquire than those using the PE300; however, the spatial resolution was higher for the synchrotron-based system.

## 2.7  Methods of data analysis

The earliest efforts to use spectroscopic methods to diagnose disease used mostly visual inspection of spectra and simple band intensity ratios to correlate spectral features and histopathology. In contrast, the results presented here, and in other chapters of this volume, utilize supervised and unsupervised methods of multivariate statistics to maximize the spectral information used in the diagnostic process.

### 2.7.1  Data preprocessing

Although the software packages available with modern IR and Raman microspectrometers are quite sophisticated, we prefer to utilize our own software packages, written in MATLAB® (TheMathWorks, Inc.), for the analyses presented in this chapter. With the amount of data being collected ($>10^4$ spectra per hour), it is essential for the analysis software to read the native file formats, which differ widely between instruments and are not necessarily easily available. The PerkinElmer instrument, utilized in most of the data reported herein, reports data as linear vectors of intensities, spaced at $2.00\,cm^{-1}$ intervals, and appropriate file headers. Other instrument software packages use a constant data-point spacing of about $1.93\,cm^{-1}$. This number arises by dividing the total spectral range covered by the interferogram, $0-15\,802.8\,cm^{-1}$ (the frequency of a typically used HeNe reference laser), by 8192 for an 8k Fourier transform. To convert from one to the other data-point spacing, the MATLAB spline interpolation function can be used.

Once data are imported into the data processing program, they are truncated to a common wavenumber range, typically $800-1800\,cm^{-1}$, and corrected for a constant offset. For automatic processing, we generally do not correct for a sloping baseline, since this procedure may produce artifacts. However, for a simple visual

inspection and presentation of data, a straight line may be subtracted from the spectra.

Data may be smoothed by Savitzky and Golay[29] odd-point, sliding smoothing window. If this procedure is employed (mostly for visualization of underivatized spectra), it is employed uniformly for all spectra. Individual spectra may be normalized (either min–max or vector normalization) to account for variations in sample thickness (see below). For multivariate statistical analyses (hierarchical cluster analysis (HCA), principal component analysis (PCA) or artificial neural networks) data were generally converted to second-derivative spectra, $(d^2I/d\nu^2)$ via a Savitzky–Golay sliding smoothing/derivative window.[29] Although this procedure decreases the SNR of the spectra (see below), derivatization offers a number of advantages. First and foremost, sloping baselines are removed and, therefore, second-derivative spectra appear mostly with a totally flat baseline (see Section 3.2.1). Furthermore, second-derivative spectra exhibit collapsed bandwidth with seemingly "higher spectral resolution". Indeed, Fourier self-deconvolution (FSD) and second-derivative methods are offered in standard software packages as "resolution-enhancing techniques".

At a fairly recent spectroscopy conference (Spec2004, Rutgers University, Newark, NJ), a lively debate among participants arose about the virtue of employing second-derivative spectroscopy as a resolution and spectral contrast enhancing method. Next, arguments will be produced that strongly suggest that derivatization is, indeed, a useful technique. Figure 3(A) shows two computed spectra, one consisting of a single, 50:50 Gaussian/Lorentzian peak (black trace), with a center wavenumber $\nu_0 = 1100\,cm^{-1}$, a full width at half height (FWHH) of $\Delta\nu_0 = 30.0\,cm^{-1}$, and an intensity $I_0 = 1.0$. The other spectrum (gray trace) is a superposition of two 50:50 Gaussian/Lorentzian profiles with $\nu_1 = 1099\,cm^{-1}$, $\Delta\nu_1 = 23.0\,cm^{-1}$, $I_1 = 0.8$, and $\nu_2 = 1113\,cm^{-1}$, $\Delta\nu_2 = 23.0\,cm^{-1}$ $I_2 = 0.5$. The band parameters were selected such that the resulting superposition produces a computed spectrum that barely shows a shoulder and a shift in peak

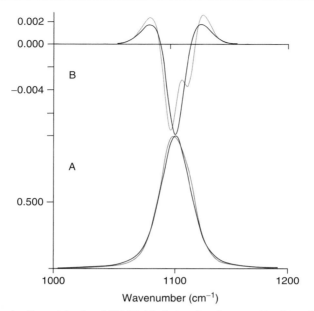

**Figure 3.** (A) Single Gaussian/Lorentzian band (50:50, black trace) and composite of two Gaussian/Lorentzian bands (gray, peak shift: $14\,cm^{-1}$, FWHH: $23\,cm^{-1}$). See text for details. (B) Second-derivative spectra of bands shown in (A).

position of only $2\,cm^{-1}$. Clearly, these computed spectra look very similar to the human eye. Similarly, the correlation coefficient $C^{LM}$, defined by

$$C^{LM} = \frac{\sum (X_i^L - \underline{X}^L)(X_i^M - \underline{X}^M)}{\sqrt{\sum (X_i^L - \underline{X}^L)^2}\sqrt{\sum (X_i^M - \underline{X}^M)^2}}$$

(3)

between spectrum L (the single band) and M (the superposition) is very close to unity (0.999). In Equation (3), the summation is carried out over all data points (200) in the spectrum, and $\underline{X}^L$ and $\underline{X}^M$ represent the mean values of each spectrum (i.e., the spectra are mean centered and vector normalized).

When both spectra shown in Figure 3 are converted to their second derivative, using a 13-point Savitzky–Golay smoothing/derivative window, the spectra shown in Figure 3(B) are obtained. Notice that the overall amplitudes of the spectra are reduced from unity to about 0.01 in Figure 3(B) panel. Thus, noise is amplified relative to the spectral amplitude unless smoothing

is applied during the computation of the derivatives. In the derivative of the composite spectrum (gray trace), the peaks are split by $4\,cm^{-1}$, and the weaker band appears as a distinct peak. Visually, these peaks are much better resolved, which is also borne out in the correlation coefficient between the spectra, which is 0.902. HCA, which uses the similarity of spectra based on the correlation coefficient, certainly discriminates second-derivative spectra much more readily than the untreated original spectra. Thus, we maintain that spectral differentiation brings an advantage to data analysis by multivariate spectral methods.

### 2.7.2 Principal component analysis (PCA)

The analysis of spectra from individual cells was carried out by PCA, which is a well-established multivariate data analysis method ideally suited to distinguish small, reoccurring spectral variations in large data sets containing uncorrelated variations. There are, of course, other methods of data analysis. However, we prefer PCA since it is, at this level, a completely unsupervised method to establish whether or not spectra of cells group into classes due to cell type, donor identity, or disease, among other factors. For a more detailed description of PCA, the reader is referred to books on chemometrics, for example, Adams.[30]

For PCA, the entire spectral data set, containing $n$ spectra, is written as one matrix **S**. In this matrix, each column represents one spectrum $S(\nu)$ of $m$ intensity data points. We assume that the spacing between data points is constant over the entire spectrum; therefore, we need to deal with intensity values only. The spectral vectors are normally preprocessed in some way: for example, smoothing (Savitzky–Golay, see above), normalization, or computation of first- or second-derivative may be performed. We prefer to use second-derivative, vector-normalized data, since these procedures provide the best discrimination against irrelevant variations in the spectra (such as sloping baseline, background, etc.). This is demonstrated in Section 3.

The *intensity* correlation matrix is constructed from the spectral matrix **S** according to

$$\mathbf{C} = \mathbf{S}\mathbf{S}^{\mathrm{T}} \qquad (4)$$

or

$$C_{kl} = \sum S^i(\nu_k)S^i(\nu_l), \quad i = 1, \ldots, n \qquad (5)$$

**C** is a $(m \cdot m)$ matrix, in which the off-diagonal terms $C_{kl}$ are the correlation between intensity values at wavelengths $\nu_k$ and $\nu_l$, summed over all spectra. If vector normalized, mean-centered spectral data are used, the correlation matrix equals the covariance matrix of the data set. Diagonalization of the intensity correlation matrix, according to

$$\mathbf{P}^{\mathrm{T}}\mathbf{C}\mathbf{P} = \mathbf{\Lambda} \qquad (6)$$

yields the eigenvector matrix **P**, from which "principal components" **Z** are calculated according to

$$\mathbf{Z} = \mathbf{S}\mathbf{P} \qquad (7)$$

The eigenvalues **Λ** express the variance contained in each of the principal components. Thus, from the viewpoint of linear algebra, the principal components are the original spectra expressed in a rotated coordinate system, based on the maximum variance of the original spectra. Subsequently, we may express each of the original spectra $S(\nu)$ in terms of the new principal components (also known as *loading vectors*)

$$\mathbf{S} = \boldsymbol{\alpha}\mathbf{Z} \qquad (8)$$

or

$$S_i(\nu) = \sum_{\mathrm{p}} \alpha_{ij} Z_j(\nu) \quad j = 1, \ldots, p \qquad (9)$$

where the "scores", $\alpha$, are given by

$$\boldsymbol{\alpha} = \mathbf{P}^{\mathrm{T}} \qquad (10)$$

For spectral data sets of individual cells, one finds that a large fraction of the total spectral variance is contained in the first few "loading vectors". Typically, 5–8 loading vectors contain more than 99% of the variance. Thus, the spectral expansion given by summation in Equation (9) can be truncated after the $p$th term, where $p$ is the number of relevant loading vectors or principal components. Thus, the score matrix $\boldsymbol{\alpha}$, which determines how much each principal component contributes to each spectrum, has the dimension $(p \cdot n)$. This method results in a much-reduced size of the data set, since all spectra are expressed in terms of a few (typically <10) basis functions and a "score vector" of about $p$ entries.

Similar spectra exhibit similar scores $\alpha$, which may be used to discriminate, or group, spectra. This is accomplished by plotting the values $\alpha_i$ and $\alpha_j$ (that is, the contribution of $\mathrm{PC}_i$ and $\mathrm{PC}_j$ to each spectrum) against each other, where each data point represents one spectrum. (The significance of these results is discussed in Section 3.) If grouping is observed, there are quantifiable and significant variations in the spectra, which can be used to construct discriminant algorithms for distinguishing cell types, state of differentiation and maturation of cells, and disease. At this point, we introduce PCA as an unsupervised, preliminary test to determine whether or not there are significant spectral differences.

In the figures using PCA for spectral discrimination, PC2 and PC3 are used most frequently as the components of choice. PC1 contains the mean value of all spectra in the data set, and generally is of low diagnostic value. Below PC4, the variance contained in a principal component is quite small; thus, most distinguishing features are contained in PC2 through PC4.

On the basis of the covariance of data revealed by PCA, discriminant algorithms can be developed for a fast, supervised analysis of unknown spectra. We have utilized artificial neural networks (ANNs) for discrimination between spectra, which show distinct groupings. However, at this point, the ANN-based results of individual cells constitute only a small part of our efforts, and are not discussed here any further.

### 2.7.3   Hierarchical cluster analysis (HCA)

HCA is a powerful method for data segmentation based on local decision criteria. These criteria are based on finding the smallest "distances" between items such as spectra, where the term *distances* may imply Euclidean or Mahalanobis distances, or correlation coefficients (vide infra). HCA is, per se, not an imaging method, but can be used to construct pseudocolor maps from hyperspectral data sets from cells or tissue sections.[31] In Section 3.1.3, high spatial resolution maps of individual cells that were based on hierarchical clustering are presented.

HCA starts by computation of the *spectral* correlation matrix, $\mathbf{C}'$ according to

$$\mathbf{C}' = \mathbf{S}^\mathrm{T}\mathbf{S} \qquad (11)$$

or

$$C_{ij} = \sum S^i(\nu_N)S^j(\nu_N), \quad N = 1, \ldots, n \quad (12)$$

This matrix is of dimension $(n \cdot n)$ ($n$ is the number of spectra in the data set) in which the off-diagonal terms $C_{ij}$ are the correlation between spectra $i$ and $j$, summed over all data points of the spectral vector $S(\nu_N)$. The correlation matrix is subsequently searched for the two most similar spectra coefficients, i.e., two spectra $i$ and $j$ for which the correlation coefficient $C_{ij}$ is closest to unity. Subsequently, these two spectra are merged into a new object, and the correlation coefficient of this new object and all other spectra is recalculated. The process of merging is repeated, but the items to be merged may be spectra, or spectra and merged objects. In the merging process, a membership list is kept that accounts for all individual spectra that are eventually merged into a cluster. We have used Ward's algorithm[32] for the process of merging of spectra, and the repeated calculating of the correlation matrix. Once all spectra are merged into a few clusters, color codes are assigned to each cluster, and the coordinates from which a spectrum was collected is indicated in this color. In this way, pseudocolor maps are obtained that are based strictly on spectral similarities. Mean cluster spectra may be calculated that represent the chemical composition of all spectra in a cluster.

## 3   RESULTS AND DISCUSSION

IR and RA-MSP both monitor the total chemical composition contained in the sampling volume. Since a cell is composed of thousands of structural and metabolic proteins, nucleic acids, many different phospholipids, sugars and carbohydrates, and small molecules that may be involved in signaling pathways, the observed spectrum of a cell is a superposition of the spectra of thousands of cell constituents. In addition, it appears that this superposition may be nonlinear, i.e., not obeying the Lambert–Beer law (see below). The interior of a cell is not a homogeneous solution of the above-mentioned components, but a highly compartmentalized and organized structure. Only the largest of these organizations can be detected in IR-MSP (i.e., nucleus and cytoplasm), whereas organelles as small as $1\,\mu\mathrm{m}$ in diameter can be observed in RA-MSP. This is, of course, due to the diffraction limit for spatial resolution, which is proportional to the wavelength, which determines the smallest object that can be resolved in microscopy.

## 3.1 Spatially resolved results for individual cells

We begin the discussion by presenting single-point, high lateral spatial resolution IR measurements of cells, with the main aim of identifying biochemical composition, and variations therein. These results were collected both on synchrotron based and conventional IR microspectrometers. At a spatial resolution approaching the diffraction limit, some subcellular organization, and the effect of a cell's level of activity on the spectra of the nucleus and the cytoplasm can be evaluated separately.

### 3.1.1 Mid-infrared spectra and spectral maps of cultured and exfoliated human cells

The original and pioneering studies in this field were reported in 1998 by the NSLS group at Brookhaven National Laboratory.[33] This study demonstrated that spectral differences exist for single cells between the cytosol and the cell nucleus. However, in this study, the experimental design was chosen to maximize the SNR and the spatial resolution in the short wavelength (C−H stretching) region. Consequently, the signal quality in the long wavelength ($>6.7\,\mu$m, or below $1500\,\mathrm{cm}^{-1}$) region was compromised. Here, we emphasize the longer wavelength region because of its superior sensitivity to distinguish biomolecular composition.

In Figure 4, we reproduce some of the earliest spatially resolved spectral results collected in the authors' laboratory of an exfoliated oral mucosa cell.[21] These data were collected by single-point mapping using an instrument that incorporated a conventional, rather than the synchrotron source. Although these results were obtained at low spatial resolution, later repetition of these

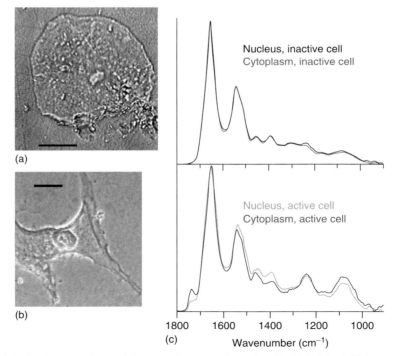

**Figure 4.** (a) Bright-field image of an exfoliated oral mucosa (buccal) cell. (b) Bright-field image of a cultured skin fibroblast cell. The lengths of the scale bars correspond to $30\,\mu$m. (c) Top: IR spectra of nucleus and cytoplasm of buccal cell shown in (a). Bottom: IR spectra of nucleus and cytoplasm of fibroblast shown in (b).

experiments at a spatial resolution approaching the diffraction limit produced exactly the same results.

A photomicrograph of an oral mucosa cell is shown in Figure 4(a). The cell is typical for a superficial squamous cell: it is a large flat cell with large cytoplasm, and a well-delineated, very small nucleus. The small size of the nucleus is due to the fact that it is pyknotic. Pyknosis is a stage in the cell's life, shortly before its death, when most metabolic activity of the cell ceases, and its main function is to protect underlying tissue from oxidation and other environmental stresses. Since the cell need no longer divide and since its metabolism has slowed down, neither transcription nor replication of genes takes place, and the nucleus shrinks to about 5 $\mu$m diameter or less. However, the pyknotic nucleus still contains the entire genome.

Averaged IR spectra from the nucleus and the cytoplasm are shown in Figure 4(c), top traces. The spectra, which are averages from individual measurements, are totally dominated by protein features, with only the smallest peaks discernible at the frequencies of DNA/RNA (at ca. 1235 and 1085 cm$^{-1}$, see below). Given the fact that all DNA is still contained in the nucleus, yet cannot be observed, led us to the formulation of "optically dense" DNA, which is discussed at the end of this section.

Very different spectral patterns, shown in Figure 4(c), lower traces, are observed for an actively growing fibroblast cell, shown in Figure 4(b). The spectra show distinct phosphate peaks at ca. 1235 and 1085 cm$^{-1}$, which were not observed in the oral mucosa cell. In the context of this discussion, the biochemical nomenclature is used, where "phosphate" refers to the phosphodiester linkage

$$-O-PO_2^--O-$$

The central phosphorus atom is surrounded approximately tetrahedrally by four oxygen atoms. The central phosphorus atom bears a negative charge that is countered in DNA by Na$^+$ ions. The central PO$_2^-$ group exhibits multiple bond

character. The terms *symmetric* and *antisymmetric* phosphate stretching vibration refer to the vibrations of the central PO$_2^-$ group, and are normal modes observed at ca. 1085 and 1235 cm$^{-1}$, respectively. These vibrations are conserved between many species containing this group, i.e., DNA, RNA, and phospholipids. The vibrations of the $-O-P-O-$ moiety are referred to as the phosphodiester vibrations, which are less intense in the IR.

The fibroblast shown in Figure 4(b) is a rapidly dividing cell with a high level of transcription and protein synthesis. Contributions of the phosphate vibrations at 1235 and 1085 cm$^{-1}$, due to DNA, RNA, and phospholipids, are observed in the spectra of both the cytoplasm and the nucleus (see below).

Furthermore, we found that the spectra of actively growing cells are quite similar, regardless of whether the cells were cancerous or not.[21] Figure 5 shows a comparison between the spectra of a sarcoma cell and a cultured, benign human fibroblast. Inspection of the spectra of the nuclei (traces 1) and cytoplasm (traces 4) of the sarcoma cells (Figure 5a) and the fibroblasts (Figure 5b) reveals that the spectra are quite similar. Thus, the large differences observed between spectra of the oral mucosa cell, and the ones of the fibroblast and sarcoma cells, are due to different levels of metabolic activity, rather than disease.

The spectra shown in Figure 5 were collected at 10 $\mu$m $\times$ 10 $\mu$m aperture at the NSLS at Brookhaven. These spectra are averages of one spectrum each of the nuclei of 15 individual cells, and two spectra of the cytoplasm, where the signal is much weaker, for each of the two cell types. After each of the sample treatments (ethanol, RNase digestion), the 15 cells were reinvestigated. In addition to improving the SNR of the observed spectra, this averaging is also expected to minimize cell-to-cell variations in these data. The spectra shown in Figure 5 were amplitude normalized at the amide I band (ca. 1650 cm$^{-1}$). Since the cells are not synchronized, and their stage in the cell cycle is not known, the standard deviation between spectra is larger than for synchronized

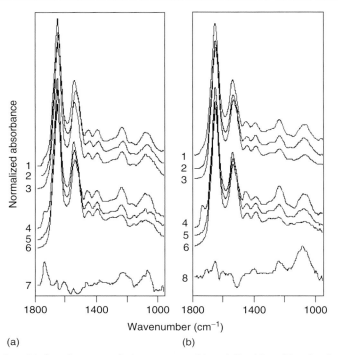

**Figure 5.** Synchrotron-based infrared spectra of giant sarcoma (a) and fibroblast (b) cells. See text for details.

cell cultures, but still does not exceed 15% of the total absorbance value observed.

In Figure 5, traces 1 and 4 represent nuclear and cytoplasmic regions, respectively, of untreated cells. Traces 2 and 5 represent nuclear and cytoplasmic regions after ethanol extraction, and traces 3 and 6 after RNase digestion. These spectra reveal an enormous amount of information on the contributions to the observed spectra of the classes of biomolecules found in a cell. The spectra of nuclei and the cytoplasm for both fibroblasts and sarcoma cells differ mostly in the lipid band (at $1738 \, \mathrm{cm}^{-1}$ due to $C{=}O_{\mathrm{ester}}$, see traces 1 and 4 in Figure 5a and b). Furthermore, we find that the spectra of the two cell lines are very similar, although one cell line represents cancerous, and the other the benign cells.

Upon treatment with ethanol (traces 2 and 5 in Figure 5a and b), the $C{=}O_{\mathrm{ester}}$ peak at $1738 \, \mathrm{cm}^{-1}$ disappears, indicating nearly complete removal of phospholipids from the samples. After ethanol treatment, the intensity profiles of the symmetric

and antisymmetric phosphate stretching bands at 1085 and $1235 \, \mathrm{cm}^{-1}$, respectively, display small differences between the nucleus and the cytoplasm. This indicates that the concentration of the remaining $-PO_2{}^- -$ groups from DNA and RNA is higher in the nuclei than in the cytosol.

A difference spectrum for the cytosol, before and after ethanol treatment (trace 4–trace 5), is shown in trace 7 of Figure 5. This trace, amplified four times, represents the spectra of phospholipids reasonably well, and we may conclude that the spectral components removed from the cells were, indeed, phospholipids. The observed difference spectra were larger for the cytosol than for the nucleus.

In both Figure 5(a) and (b), traces 3 and 6 depict the effect of RNase digestion. This step removes cytoplasmic and nuclear RNA, resulting in nearly complete removal of RNA contribution from the spectra of cytosol. These results demonstrated, for the first time, that the cytoplasmic RNA is detectable by IR-MSP. This RNA is most likely in the form of ribosomes involved in protein synthesis. Nucleic acid features persist in the nuclear region, due to DNA and possibly undigested RNA in the nucleus. Further digestion with DNA removed nearly all nucleic acid contributions (data not shown, see Reference 21). Thus, we suggest that successive treatment of the cells with ethanol and RNase removes most of the lipids and major parts of the RNA, while DNA is still present.

The difference spectrum shown in trace 8 shows the effect of RNAse digestion in the nucleus. The difference spectrum is that of pure RNA,[17] with the distinct structure of the $1080 \, cm^{-1}$ peak and an intensity ratio of nearly 2:1 for the $1080/1235 \, cm^{-1}$ vibrations. The small derivative line shape in difference spectra under the amide I peak ($1650 \, cm^{-1}$) is most likely due to a small shift of this peak.

The remaining spectra of the cytoplasm (traces 6 in Figure 5a and b) are dominated by protein features, and resemble the spectra of the cytoplasm of the oral mucosa cell. However, the spectra of the nuclei, after treatment with ethanol and RNase (traces 3 in Figure 5a and b), still show nucleic acid contributions at 1085 and $1235 \, cm^{-1}$, which are stronger than those observed in the oral mucosa cell. Furthermore, the DNA peaks appear a bit more intense in the cancerous (Figure 5a) than in the benign cells (Figure 5b).

The treatment of the cells with ethanol to remove phospholipids, followed by RNase digestion to remove cytoplasmic and some of the nuclear RNA, has allowed a detailed assessment of the DNA spectral contributions in the nuclear regions. We find small DNA contributions in the nuclear regions of the cultured cells (Figures 4 and 5) that are clearly absent in the cytoplasmic regions. Furthermore, we do not detect these DNA contributions in the pyknotic nucleus of the oral mucosa cells. This confirms that DNA is not observed when it is completely condensed (see Section 2.2.1).

The spatially resolved data reported here indicate that most of the observed phosphate signals in an active cell are due to RNA and phospholipids, which are found in high concentrations in the

cytoplasm of active cells. Thus, one major conclusion from our results is that an interpretation of the $-PO_2^- -$ bands is possible only if one accounts for the spectral contributions of phospholipids. One of the most fascinating aspects of the spatially resolved data is the detection of the sheer abundance of the phospholipids in the cytoplasm outside the nucleus. This aspect is discussed in more detail in Section 3.1.2. We believe that the intensity of the sharp peak at $1738 \, cm^{-1}$ (due to the ester linkage between the fatty acid and the glycerol unit) results mostly from membranes of subcellular organelles such as the endoplasmic reticulum and Golgi apparatus that contain large amounts of phospholipids. It is unlikely that the cell outer membrane can be observed in transmission microspectroscopy, since its thickness is insufficient for a detectable IR signal.

Figure 6 depicts IR spectral imaging data for a cultured HeLa cell grown on a $CaF_2$ substrate, collected using the PE300 instrument at $6.25 \, \mu m$ pixel size. The nucleus and the nucleoli in the nucleus are discernible in the visual image (Figure 6a). Figure 6(b) shows the distribution of the integrated intensity due to the protein amide I vibration over the cellular position. This image shows very large protein spectral intensities from the region of the nucleus. These intensities are nearly an order of magnitude higher than those observed for the cytoplasm. This observation confirms similar results, discussed later, that the large protein vibrations are due to two factors: the total protein content of the cell is largest in the nucleus, since histone and many other nuclear proteins are localized there. Second, the shape of the nucleus presents a longer path to the IR beam than the surrounding cytoplasm, which is often only a few micrometers thick.

The result of cluster analysis (see Section 2.7.3) applied to the data shown in Figure 6(b) is displayed in Figure 6(c). Cluster analysis produces images based on spectral similarities. Thus, Figure 6(c) represents an image in which the nuclear region (black), the cytoplasm (dark gray), and the region of the edges of the cell (light gray) can be distinguished. For cluster analysis, the input data were vector normalized; thus,

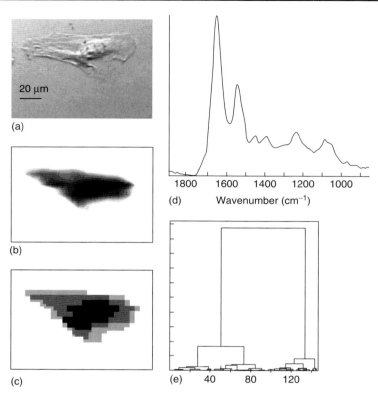

**Figure 6.** (a) Bright-field image of HeLa cell grown on $CaF_2$ substrate. (b) Amide I intensity map for cell shown in (a). (c) HCA map for cell shown in (a). (d) Mean cluster spectrum of nuclear region. (e) Dendrogram depicting clustering process for HCA map shown in (c).

cluster analysis detects spectral differences that are independent of the cell thickness. Figure 6(e) shows the dendrogram, which is a visualization of the clustering process. The cluster on the right side (abscissa values 100–150) corresponds to the nucleus (black region in Figure 6c), whereas the center cluster (40–100) corresponds to the dark gray area and the left cluster (0–40) to the light gray area of the cytoplasm. The dark and light gray areas are closely related, and are due to different regions of the cytoplasm. A representative spectrum from the nuclear cluster is shown in Figure 6(d). For the nucleus, the intensity in the amide I band is on the order of 0.2 absorbance units (AU). Distinct spectral features due to the symmetric and antisymmetric stretching vibrations of the $PO_2^-$ are manifested as peaks at ~1080 and 1235 cm$^{-1}$, respectively.

The absence of a distinct peak due to phospholipids at 1740 cm$^{-1}$ indicates that the $PO_2^-$ vibrations are most likely due to the DNA and RNA. Since the cell studied here is an actively growing cell, the spectrum of the cytoplasm close to the nucleus contains nucleic acid (RNA) features, in accordance with the discussion of the actively growing fibroblast and sarcoma cell spectra discussed above.

There are enormous spectral differences between the proliferative fibroblasts, giant sarcoma, and HeLa cells on the one hand, and the terminally differentiated and inactive cells, such as oral mucosa and other squamous epithelial cells, on the other. These latter cells occupy the outermost layer of squamous epithelium and are terminally differentiated. Compared to the cell size, these nuclei are small ($<10 \mu$m), and the DNA in such nuclei is nearly completely invisible in IR microspectroscopy. This observation is discussed next.

The presence of signals attributed to DNA in nuclei of human cells are manifested by bands at ca. 1235 and 1085 cm$^{-1}$ due to the antisymmetric and symmetric stretching modes of the PO$_2^{-1}$ moieties of DNA, respectively,[21,33,34] as discussed above. An unambiguous assignment of these peaks to DNA in human tissue and cells[21,22] was established by selectively removing the DNA from the sample by digestion via the enzyme DNase. In active cells, the PO$_2^-$ stretching peaks of DNA may exhibit an amplitude of 0.05 AU; the equally strong DNA peaks in the C=C-stretching regions cannot be observed directly since they overlap the amide I protein band.

In spite of the presence of DNA, the IR spectra of nuclei are still dominated by the contributions of proteins, which are nearly 10 times stronger than those collected outside the nucleus in a typical cell. This is expected, since the nucleus in a dried cell offers a longer path length to the probing IR radiation; furthermore, the nucleus contains proteins at a very high concentration. Therefore, typical protein amide I signals (at ca. 1650 cm$^{-1}$) from a nucleus are between 0.2 and 0.4 AU. DNA signals, if observed, are much smaller, and vary considerably in magnitude. Their small spectral amplitude is due to the fact that DNA concentration in the nucleus is only about 1%. The variation of the DNA intensities appears to depend on the degree of condensation: most highly condensed DNA, such as that found in terminally differentiated and metabolically inactive cells cannot be observed in transmission or transflection measurements. This was first observed for superficial cervical cells[34] and verified for oral mucosa cells[21] whose main task is to protect the underlying tissue; consequently, they become metabolically inactive. These inactive cells can no longer divide, and their protein synthesis is at a minimal level. Therefore, the nuclei shrink to about 5–8 μm in diameter (pyknosis).

DNA is not distributed uniformly throughout the nucleus, but rather, is tightly wrapped around histone proteins to form chromatin. These chromatin structures may have a diameter between 30 and 800 nm, depending on the cell's biological activity. These small and highly absorbing "chromatin particles" are much smaller than the diffraction limit; consequently, their spectral contributions (albeit strong) are averaged over the entire area of the sampling beam, and are therefore not discernible. Thus, very little or no information on the condensed chromatin is contained in the IR spectrum of a compact (pyknotic) nucleus. However, such highly condensed particles exhibit large light scattering cross sections. We have shown recently for mitotic cells that the strongest Raman signals are, indeed, observed for the most highly condensed chromatin, for example, in the metaphase and anaphase of mitosis.[3] Raman signals of the DNA of metabolically active, but nondividing cells are quite weak. This is discussed in Section 3.1.2.

In active or dividing cells, however, the DNA is less compact, and therefore detectable in IR microspectroscopy.[35] Further evidence for the dependence of IR signals of DNA on the degree of chromatin condensation was reported in several studies. It was found that avian erythrocytes (which contain inactive, condensed nuclei) and human erythrocytes, which are anucleated, exhibit virtually identical IR spectra.[35] Furthermore, Jamin *et al.*[36] reported a large increase in the DNA signals of apoptotic cells. In apoptosis, DNA is degraded into oligonucleotides with a few hundred base pairs that subsequently diffuse out of the nucleus. The accumulation of these oligonucleotides was observed in high spatial resolution, high quality synchrotron FT-IR microspectroscopic data. This observation suggests that the DNA, while highly condensed in the nucleus, is unobservable, yet its degradation products are readily detected by IR microspectroscopy.

### 3.1.2 High spatial resolution Raman maps of cultured and exfoliated cells

We present in Figure 7 confocal Raman images of a hydrated exfoliated oral mucosa cell. Except

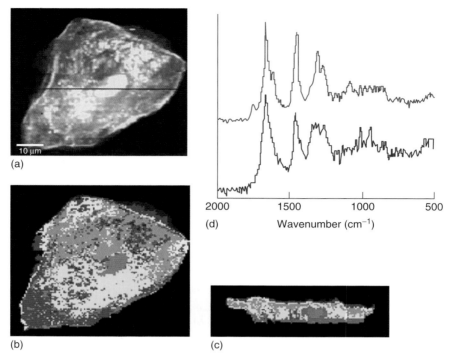

**Figure 7.** (a) Map of integrated C-H stretching intensities of an oral mucosa cell. Brighter yellow hues correspond to higher intensities. (b) Pseudo-color X–Y map of the cell shown in (a), based on HCA of the Raman imaging data set. (c) Corresponding X–Z map, taken as a confocal section along the black line indicated in (a). In panels b and c, cluster colors are assigned arbitrarily. (d) Normalized mean cluster Raman spectra of phospholipid-rich (top) and protein-rich (bottom) regions of the cytoplasm.

that this particular cell is in an aqueous environment, it represents exactly the same cell type shown in Figure 4(a). Figure 7(b) is constructed from HCA of 14 400 individual spectra collected using a $60\times$ water immersion objective ($NA = 1$), and about 15 mW laser power at 514.5 nm excitation. The pixel size in this image is determined by the diffraction limit (see Equations (1) and (2)); given the excitation wavelength and the numerical aperture, we arrive at a diffraction-limited spatial resolution of about 320 nm in the X–Y plane, and about 1400 nm in the Z-direction. Figure 7(c) shows an X–Z section of the cell; this scan was taken along the direction of the black line in Figure 7(a). These images demonstrate the excellent image quality and biochemical information that is available from Raman spectral imaging, followed by multivariate data

analysis methods. The red spots in Figure 7(b), for example, differ greatly in chemical composition from the surrounding cytoplasm (blue trace in Figure 7d), and are due to phospholipids (red trace in Figure 7d). From this image, the small size of the pyknotic nucleus in a terminally differentiated squamous cell can be estimated to be less than 10 μm in diameter. Furthermore, one finds that the distribution of the phospholipids, which were not observable in the IR spectra (Figure 4c), is very inhomogeneous.

Figure 8 shows a Raman map of a HeLa cell grown on $CaF_2$ a substrate, obtained from HCA of the data set. Although the bright-field image reveals low contrast (since it was taken of a cell in aqueous surrounding), the Raman image reveals the much larger nucleus in this actively growing cell. Furthermore, we can discern the nucleoli,

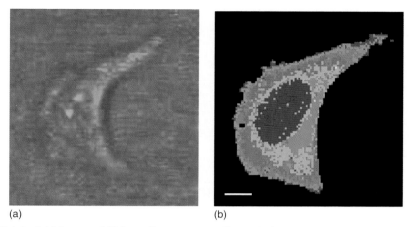

(a)                                        (b)

**Figure 8.** (a) Bright-field image of HeLa cell grown on a $CaF_2$ window, acquired in aqueous medium through a $60\times$ water immersion objective. (b) Raman cluster image of cell shown in (a), revealing nucleus, nucleoli, and various regions of the cytoplasm. The length of the scale bar corresponds to $20\,\mu m$.

and various fractions of the cytoplasm. In fact, we demonstrated that some of the cytoplasmic clusters can be associated with mitochondria-rich areas in the perinuclear region of the cell.[37] Depending on the number of clusters utilized in HCA, details such as the nucleoli in the nucleus can be visualized. Given the similarity in chemical composition of nucleus (protein, DNA and RNA) and the nucleolus (protein and RNA), it is astounding that Raman spectroscopy, coupled with HCA, can distinguish the corresponding spectral patterns. These maps provide images of the biochemical composition of cells unsurpassed by any other method that does not use contrast agents or stains.

### 3.1.3 High spatial resolution spectral maps of mitotic cells

Here, we report Raman and IR microspectro-scopic images of human cells at different stages of mitosis by following the spectral signatures of condensed chromatin, and other biochem-ical components, utilizing inherent protein and DNA spectral markers. Most of the Raman spec-tral images presented here are univariate maps obtained by plotting DNA or protein intensities versus the coordinates from which spectra were recorded.

We have reported infrared microspectroscopic studies that have concentrated on the detection of variations of a cell's biomolecular composi-tion during progression through the cell cycle (see References 5 and 23, and Section 3.2.4), which precedes mitosis. These studies represent the necessary background information required for an eventual application of vibrational spec-troscopic imaging methods to screen cells for the presence of disease that may manifest itself in increased cell proliferation.

Here, however, we present IR and Raman spectral maps for the events that take place during mitosis, when the cell physically divides into two daughter cells. The overall aim of this study is to develop spectroscopic methods to detect cell proliferation in samples of cultured or exfoliated human cells, and to complete the study of temporal variation of spectral signa-tures during the entire cell division cycle. The exact sequence of events during mitosis is, of course, well understood from light and fluo-rescence microscopy studies. Spectral imaging methodology, however, affords the advantage of monitoring the distribution of biochemical components, using only their inherent vibrational fingerprints without employing any stains or other commonly used probes.

We monitor mitosis by mapping the IR or Raman scattering light intensities of the chromatin during mitosis. Mitosis typically lasts less than 1 h for most cell types grown in culture, and can be divided into several subphases: (i) during prophase, the chromosomes replicated during synthesis (S) phase undergo extensive condensation but are still contained within the nuclear envelope; (ii) later in prophase, the mitotic spindle begins to grow and the nuclear envelope breaks down; (iii) in metaphase, the microtubules begin to interact with the chromosomes, which leads to the alignment of the two complete sets of diploid chromosomes lining up along the "equator" of the parent cell; anaphase commences with separation of the daughter chromosomes away from the equatorial metaphase plate and toward one of the two spindle pole regions; (iv) in telophase, two separate groups of chromosomes have formed at each pole, and a nuclear envelope begins to appear around each set of chromosomes to form two nuclei that are temporarily in one cell; furrowing and division of the cell takes place during cytokinesis to yield two distinct daughter cells.

Mitotic cells can be distinguished from interphase cells by microscopic inspection of live cells growing on a substrate in culture. Cells exhibiting mitotic figures were selected, fixed, and spectral maps acquired. The spectral maps reported below for different stages of mitosis (i.e., prophase, metaphase, anaphase, and telophase) do not represent a time course of mitosis of one individual cell, but time points in mitosis of several different cells. Each Raman spectral data set required 12–24 h of data acquisition time.

Figure 9 shows HeLa cells undergoing the typical stages of mitosis: prophase, metaphase, anaphase, and telophase. The left column of images shows the cells as seen in cell culture under an inverted phase contrast microscope. The second column shows Raman maps of the nucleic acid band intensity at 785 cm$^{-1}$, normalized with respect to the protein peak, whereas the third column depicts the protein amide I intensity (1655 cm$^{-1}$). The protein intensity during mitosis is contributed largely by the microtubules and the dense histone-packed chromatin. The far right column shows, for comparison, the fluorescence images of the DAPI-stained cells demonstrating the density and distribution of the condensed chromatin within the cell at each of the phases of mitosis.

The chromosomal condensation during metaphase and anaphase, shown in the second and third rows respectively, is manifested by a large intensity increase of the DNA-related peaks. This can be seen clearly in Figure 10, which shows Raman spectra collected from the chromatin (trace B) and cytoplasm (trace A) of a cell during anaphase. The bottom row of images in Figure 9 was taken for a cell in telophase. In contrast to the images of the cells in anaphase and metaphase, the DNA signal is reduced, since the DNA distribution has already become relatively diffuse during this stage of mitosis. However, the protein signals clearly indicate the presence of two newly formed nuclei.

In Figure 9 (right column), the corresponding fluorescence images, using DAPI stain, are depicted. The Raman chromatin (DNA) intensities correlate very well with the DAPI fluorescence of chromatin. However, the Raman maps are constructed from the Raman scattering signals inherent in the cell, such as the protein/DNA distribution, rather than stains.

The Raman images reveal the chromatin distribution for the prophase, metaphase, and anaphase cells in exquisite detail. However, the chromatin exhibits low signals in interphase cells, as well as for the late-telophase cells. In an interphase cell, the protein–DNA chromatin complexes are packed much less densely, with a packing ratio of 40, as compared to a metaphase chromosome with a packing ratio of 10 000. Thus, the DNA or chromatin may be too diffuse to yield a detectable Raman signal in the volume probed in RA-MSP, which is on the order of 1 $\mu$m$^3$ or about 10$^{-12}$ g. It appears that there exists a complementary relationship of Raman and IR microspectroscopy with regard to the detectability of DNA. We have discussed before that highly condensed DNA, such as in a pyknotic nucleus, cannot be observed in IR microspectroscopy.[35] By contrast, it appears

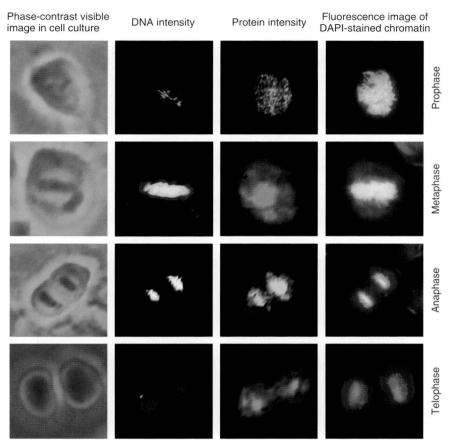

**Figure 9.** Phase contrast, Raman and fluorescence images of HeLa cells during various phases of mitosis: (top row) prophase, (second row) metaphase, (third row) anaphase, (bottom row) late telophase. All images are collected at 40× magnification. The cell in the prophase is about 25 μm in diameter. (Column 1) Phase-contrast images of live cells in culture. (Column 2) Raman scattering intensity plots for the DNA scattering intensities. The colors range from black (low intensity) to red to white (high intensity). (Column 3) Raman scattering intensity plots for the protein scattering intensities. The colors range from black (low intensity) to green to yellow (high intensity). (Column 4) Fluorescence images of the DAPI-stained cells.

that Raman signals are more prominent for most highly condensed DNA.

We now turn to IR imaging of mitotic cells. The cell shown in Figure 11 is a nearly spherical cell in metaphase. Although the SNR of the IR spectra is more than an order of magnitude higher than that of the Raman spectra, and the data acquisition times nearly 100 times faster, the IR images reveal very little detail about the distribution of condensed chromatin structures in the important region of the mid-IR spectrum between 800 and

$1800\,cm^{-1}$ ($12.5-5.6\,\mu m$). In Figure 11(b), just a hint of the chromatin features in the nucleus can be detected. This is, of course, due to the fact that the spatial resolution of the IR spectral images is diffraction limited. This limit is about $10\,\mu m$ at $1000\,cm^{-1}$, $6\,\mu m$ at the protein amide I vibration ($1650\,cm^{-1}$), and about $3\,\mu m$ in the C−H stretching vibrations at $\sim 2950\,cm^{-1}$. The spatial resolution of IR spectral images in the C−H stretching region, therefore, is twice as good as in the amide I region. Thus, spatially resolved

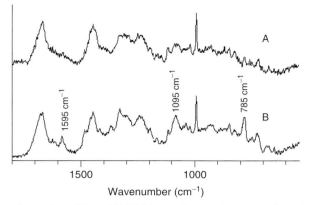

**Figure 10.** Raman spectra of cytoplasm (A) and DNA/chromatin complex (B), collected during anaphase of mitosis. Major bands due to DNA are indicated by their wavenumber.

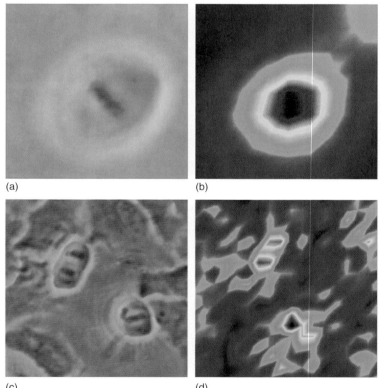

**Figure 11.** Phase-contrast images of HeLa cells in (a) metaphase and (c) anaphase. (b) Infrared spectral map based on the amide I absorption intensities of the cell shown in (a). In the center of the nucleus, the largest intensity correlates with the position of the chromatin. Dark blue indicates low intensity, and yellow to red higher and highest intensity, respectively. (d) Resolution enhanced infrared spectral map based on the 2925 cm$^{-1}$ absorption region of the cell shown in (c). The two chromatin regions of the cell in anaphase are detectable. Color scheme is as in (b). The cell in (a) measures about 25 μm in diameter.

images in the IR are best collected in the short wavelength region, provided useful compositional information can be obtained from this region.

Using standard resolution targets, the spatial resolution of the PE300 instrument was found to be about twice the diffraction limit at all mid-IR wavelengths. Thus, we needed to employ FSD to improve the *spatial* resolution to about the diffraction limit. One-dimensional FSD has been utilized to improve *spectral* resolution.[38] It is based on the principle that spectral resolution is contained in the high frequency part of the Fourier spectrum. FSD is usually carried out by transforming a FT-IR spectrum back into "mirror-displacement" domain, and multiplying the resulting interferogram with a resolution-enhancing function. Such a resolution-enhancing function is generally an exponential function that amplifies the high frequency components, albeit at an increase in the noise level, reducing the SNR.

Similarly, *spatial* FSD can be achieved by reverse Fourier transformation of an $X-Y$-frequency data set, and applying a resolution-enhancing function in the $X$ and $Y$ dimensions. After transforming the data set back into spectral and $XY$ space, an image with enhanced spatial resolution is obtained. This method works best if data are collected at a pixel and step size smaller than the diffraction limit (oversampling), and can improve the spatial resolution by about a factor of 2.

Figure 11(d) shows the map of an anaphase cell after FSD. This map displays the intensity at $2924\,cm^{-1}$. The picture reveals the newly separated chromosomes, and represents the best spatially resolved single cell spectra available from nonsynchrotron-based measurements.

### 3.1.4 Mie scattering of cellular nuclei

In Figure 12, we present results that confirm that the nuclear DNA is not observed in absorption,

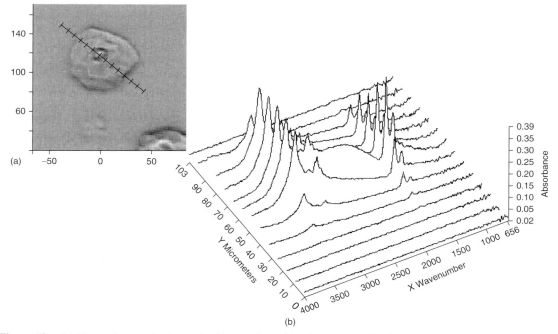

(b)

**Figure 12.** (a) Photomicrograph of unstained human buccal (oral mucosa) cell, with positions indicating the locations of spectral data acquisition. Axes labeled in $\mu$m. (b) 3-D plot of IR spectra obtained at the positions indicated in (a). Notice the absence of DNA features for the spectrum of the nucleus, and its broad scattering background, centered at about $2400\,cm^{-1}$.

yet the light scattering due to the nucleus is observable. The photomicrograph in Figure 12(a) shows a cell with the nucleus clearly visible. The aperture of the microspectrometer was set to straddle the nucleus, $10\,\mu\text{m}$ on edge. Individual spectra were subsequently acquired at positions indicated by the crosses along the black line in Figure 12(a). A sequence of consecutive spectra is shown in Figure 12(b). This plot shows that the intensity of the spectral amplitude increases from the cytoplasm to the nucleus, and demonstrates again that, in the strongest spectrum due to the nucleus, no DNA absorption signals are observed. However, we observe a broad spectral feature, centered around $2400\,\text{cm}^{-1}$ between the amide I and the C−H stretching region, which occurs only in the nuclear spectrum. This broad, spectral feature is observed at quite different wavelengths for different cells (see Figure 13a and b). We attribute these spectral features to Mie-type scattering of the nucleus, which is discussed next.

Dielectric spheres are known to scatter electromagnetic radiation if the wavelength of the light is comparable to the size of the dielectric sphere. The theory of this scattering process was first described theoretically by Mie.[39] We recently published a summary of the theoretical background for Mie scattering computations.[40]

Mie theory assumes a spherical scattering particle in the field of a plane electromagnetic wave. Therefore, it is best to express the wave in terms of spherical polar coordinates, resulting in complicated expressions of the electric and magnetic vectors in terms of Bessel (cylindrical coordinate) functions. Therefore, the scattering cross section, $Q_{\text{sca}}$, of a dielectric sphere interacting with a plane electromagnetic wave is given by a series expansion of the size parameter $\rho$, and complicated expressions in the half-integer order Bessel (Ricatti−Bessel) functions of the first kind and their complex equivalents (Hankel functions). The size parameter $\rho$ is defined by

$$\rho = 2\pi r m_{\text{o}}/\lambda \qquad (13)$$

where $r$ is the radius of the sphere, $\lambda$ the wavelength of the light, and $m_{\text{o}}$ the ratio of refractive indices between the sphere and the surroundings.

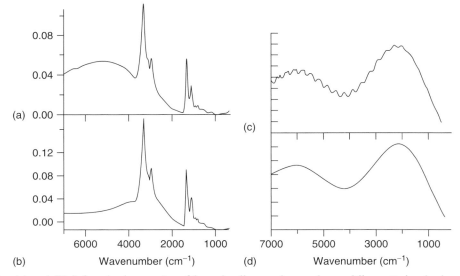

**Figure 13.** (a) and (b) Infrared microspectra of buccal cells superimposed on a Mie scattering background which may have its first peak at different wavelengths. Notice the compressed scale above $2000\,\text{cm}^{-1}$. (c) Mie scattering calculated for a dielectric sphere with $m_{\text{o}} = 1.33$ and $r = 5\,\mu\text{m}$. (d) Mie scattering using the approximate formula given in Equation (14).

The Bessel functions account for the undulating wavelength dependence of Mie scattering, which is shown in Figure 13(c). The angle dependence of Mie scattering is given by series expansions in the Legendre polynomials and their derivatives with respect to the scattering angle.

We have calculated the Mie scattering from first principle, using software available as shareware.[41] In Figure 13(c), we show the Mie scattering cross section for a sphere with refractive index of 1.33 and a radius of 5 μm, using the rigorous calculations. We have also used an approximate formula, which was reported in the literature:[42]

$$Q_{\text{sca}} = 2 - (4/\rho)\sin\rho + (4/\rho^2)(1 - \cos\rho) \tag{14}$$

with the size factor $\rho$ defined previously. This approximation (which is the truncated form of the summation in the complete scattering formula) reproduces the rigorous calculation to within 1% (cf. Figure 13d), but fails to reproduce the high frequency ripples on the Mie-scattering curve shown in Figure 13(c). Using the approximate calculations, we were able to model the Mie-scattering background satisfactorily, and were able to correct the distorted spectra by subtracting the modeled Mie background.

### 3.1.5 Mie scattering from entire cells

In the following section, we shall introduce an effect that we believe is due to entire cells undergoing Mie scattering. During the mitosis stage of cell division (see Section 2.2), cells become nearly spherical, the nucleus of the cell disappears, and the chromatin may be found distributed throughout the cell (before it is separated into the new nuclei of the forming daughter cells). Cells with similarly spherical morphology may also be created when adherent cell cultures are trypsinized to detach cells from a substrate (see Section 2.2). Since the spherical cells created during mitosis or trypsination are much larger (~20 μm diameter) than the nuclei discussed before (8–10 μm diameter), the first Mie-scattering maximum is expected to occur at much lower wavenumber, around 1200 cm$^{-1}$. This may give rise to very distorted spectral patterns between 1000 and 1800 cm$^{-1}$.

These very distorted spectral patterns, possibly due to interference of Mie scattering and the cellular absorption, were first reported by Holman et al.[43] for selected cells grown on a substrate. These authors suggested that the cells with this unusual spectral pattern were undergoing mitosis. We observed very similar, distorted spectral patterns for many of the trypsinized cells,[44] which are nearly spherical in shape, and their size can vary widely (see Figure 14b). Depending on their size, Mie scattering was observed with the first Mie-scattering maximum occurring as low as 1200 cm$^{-1}$. This is shown in Figure 14(a), trace A. The Mie-scattering contribution could be modeled (trace B), as discussed before, using the approximate cell size, and subtracted to reveal spectral traces that appeared quite normal (trace C).

The discussion presented in the last two sections demonstrates that spectroscopy of individual cells is a difficult subject that requires a careful analysis of the observed spectra. Given the difficulties described so far, it comes as no surprise that some of the earlier efforts to analyze cellular spectra were unsuccessful.

## 3.2 Infrared spectroscopy of whole cells

In this section, we present spectral results collected for entire individual cells with a specific aim, namely, to detect abnormalities in samples of cells. Section 3.1 laid the groundwork for understanding cellular spectra in terms of chemical composition, distribution of subcellular organelles, and possible interference due to cellular morphology. Here, we are poised to utilize this knowledge to detect disease in cells that were exfoliated (e.g., cervical cells obtained during the "Pap" cervical cancer screening procedure), or obtained by any other means (lavages, thin needle aspiration, etc.) If such cell samples were highly homogeneous, one could devise measurement

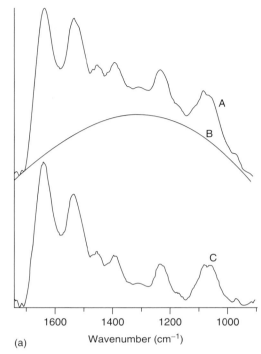
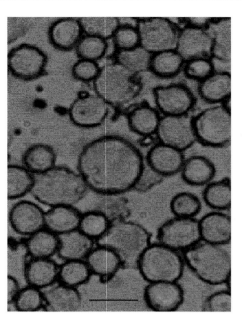

(a)          Wavenumber (cm⁻¹)                    (b)

**Figure 14.** (a) Infrared spectrum of nearly spherical, trypsinized cell (trace A), and Mie contribution calculated for a spherical cell with $r = 12 \mu m$. Trace C: corrected spectrum after Mie background subtraction. (b) View of trypsinized HeLa cells. The length of the scale bar corresponds to $30 \mu m$.

strategies for cell ensembles of hundreds or thousands of cells (cell pellets), and detect the differences between normal and diseased populations quite easily. This was the rational for the early diagnostic attempts by Wong and coworkers.[8,9] However, samples of exfoliated cells are always highly heterogeneous, often containing erythrocytes, lymphocytes and leukocytes, bacteria, etc., in addition to the desired cells. Consequently, we believe that the analysis of cells, with the aim of detecting disease, will be possible only if spectra of individual cells are collected, and analyzed statistically. However, even the spectra of very homogeneous cell samples show surprisingly large variations. We are reporting here the results of several model systems, which we believe will shed light into the problem of spectral heterogeneity, and will eventually permit the analysis of exfoliated cell samples. The first of these model systems is exfoliated human oral mucosa

(buccal) cells. These are very similar to cervical cells in that they are stratified squamous epithelial cells. Furthermore, they are easy to collect by swiping the inside of the cheek of volunteers, and they contain mostly one cell type. We also report data for an animal model of ectocervical cells, and compare their spectra with those from oral mucosa cells. This comparison was the first success in distinguishing very similar cell types.

The next cell type to be discussed is from the lining of the urethra and the bladder. These cells are found in normal human urine. This project is somewhat more complicated than the analysis of buccal cells in that a sample of cells collected from urine consists of both squamous cells from the urethra, and urothelial cells from the lining of the bladder. The occurrence of two cell types demonstrates the ability to distinguish these cells spectroscopically by methods of multivariate data analysis.

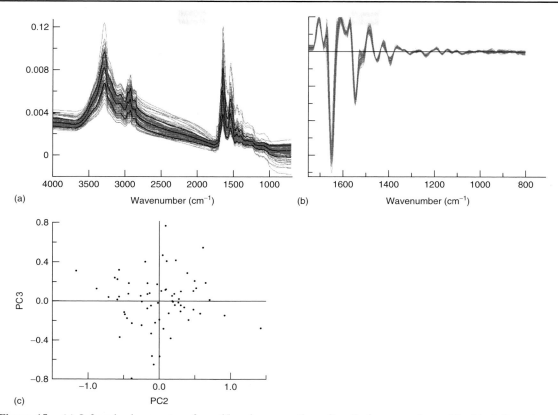

**Figure 15.** (a) Infrared microspectra of ca. 65 oral mucosa (buccal) cells from one donor. The black lines denote the mean spectrum and $\pm1$ standard deviation from the mean. (b) Second-derivative, expanded and vector-normalized spectra of data set shown in (a). (c) Scores plot (PC2 vs. PC3) of data shown in (b). Each dot represents the scores of one spectrum.

Finally, we report studies on cultured cancerous cervical (HeLa) cells, grown onto spectroscopic substrates, with the goal of detecting spectral changes of cells during the cell division cycle.

### 3.2.1 Oral mucosa cells

We have used oral mucosa (buccal) cells as one of the most easily obtainable exfoliated cell types from volunteer donors. In Sections 3.1.1 and 3.1.2, we already introduced high spatial resolution data from buccal cells. In this section, spectra collected from entire individual cells are used to introduce the possibility of using IR microspectral results to distinguish cell types and determine their state of health. In particular, the

heterogeneity of observed spectra, and methods to overcome it, are discussed.

Figure 15(a) shows individual spectra of about 65 oral mucosa (buccal) cells, along with their mean spectrum, and the standard deviation from the mean. We observed large variations in the overall amplitude of the spectra, slight Mie-scattering contributions (between 1800 and $2500\,\mathrm{cm}^{-1}$) and distinct spectral variations in the low wavenumber region. We shall discuss these spectral observations in turn. The broad, undulating features observed relatively weakly in the spectra of entire cells have been attributed to Mie scattering, discussed above. For instrument apertures approaching the size of the nucleus, the Mie scattering due to the nucleus becomes more prevalent in the overall spectra. In spectra of entire cells, such as the ones shown in Figure 15, the Mie scattering of the nuclei is averaged, or diluted, over the much larger area of the cell (nuclear area $<100\,\mu m^2$, cellular area ca. $4000\,\mu m^2$), and is observed with much lower amplitude as compared to the absorption features of the cell.

The variations of spectral patterns in the low wavenumber ($<1400\,cm^{-1}$) region is most likely due to variations in cellular spectral features observed for cells undergoing the early stages of cell division (G1, S, and G2), or apoptosis. This aspect is discussed in Section 3.2.4.

We now turn to the discussion of the overall intensity variations observed for spectra of cells. These variations can have various reasons, which are related to aspects of the areas of cells and nuclei from which spectral data are collected. If spectra of cells of different sizes were acquired with constant apertures much larger than the cells, the spectra would be averaged over the entire area of the aperture from which the data were acquired. If the cell dimensions varied greatly, large intensity differences would be created. Therefore, the cell spectra reported here were acquired by matching the spectrometer aperture to the size of the cell as closely as possible. Thus, we believe that the overall dimensions of the cells examined do not contribute significantly to the variance in observed spectral intensities.

However, the thickness of a cell and the ratio of areas covered by nucleus and cytoplasm affects the overall spectral intensities. From biochemical considerations, we may exclude the possibility that the concentration of the major cellular components differ to a large extent. Thus, it appears that the thickness (path length) of cells contributes most of the variance in spectral intensities. The dried cytoplasm of an epithelial cell is very thin (as determined by confocal Raman microspectral techniques, Section 3.1.2, typically ca. $1-2\,\mu m$). In the amide I mode, typical absorbance values observed are ca. 0.05 AU. The nucleus maintains a more spherical shape, with an average thickness of about $3-4\,\mu m$. The spectra collected from the nucleus alone are dominated

by protein features, and often exhibit absorbance values of 0.3–0.5 AU in the amide I band. This accounts for the much higher spectral quality observed for spectra collected from nuclei. Thus, one would expect that the nucleus of a cell contributes significantly to the protein spectrum of an entire cell.

The size of the nucleus then influences the overall amplitude of the protein spectral intensity of a cell in two possible ways. The area of the cytoplasm may be about $4000\,\mu m^2$ for an oral mucosa cell, and the area occupied by the nucleus may be at most $100\,\mu m^2$ for a metabolically inactive cell. A simple calculation shows that the area of the nucleus, exhibiting an absorbance of 0.5 AU, averaged over the area occupied by a cell's cytoplasm with an absorbance of 0.05 AU, contributes about 25% of the total protein signal, although the area of the nucleus is only about 2.5% of the cell's area. Thus, variations in nuclear size influence the total protein intensity to a large extent. Thus, we postulate that the nuclear size influences the overall protein spectral intensities of cells, and accounts in part for the large variance in the overall intensity of cellular spectra. This aspect is revisited during the comparison of canine cervical and human oral mucosa cells.

In spite of the apparent differences in the spectra shown in Figure 15(a), we were able to interpret them via appropriate preprocessing and multivariate statistical methods. The preprocessing applied uniformly to all spectra includes computation of second derivatives of the absorption spectra with respect to wavenumber, and subsequently vector-normalizing the derivative spectra (see Section 2.7.1). This methodology was pioneered by the Naumann group in Berlin for the analysis of spectral data of prokaryotic cell colonies.[45]

The derivative spectra are devoid of any low wavenumber baseline variations and/or baseline offset. Furthermore, the second-derivative spectra exhibit sharper peaks and better resolved bands (see Section 2.7.1), and are much more homogeneous and amenable to statistical analysis (cf. Figure 15b). A PCA scores plot of these data is shown in Figure 15(c). This plot reveals a uniform distribution of the spectra and indicates that there is no systematic variation of the spectra.

The preprocessing procedures discussed above are even more important if oral mucosa spectra from different donors are compared. We show in Figure 16(a) averaged oral mucosa spectra from six donors. Each of the traces represents the average of between 60 and 80 individual cell spectra. The variation in these average spectra is surprisingly large, seemingly indicating that the intensities of oral mucosa spectra vary significantly from donor to donor. However, these "donor-to-donor" variations can be minimized using vector normalized and second-derivative spectra (Figure 16b). PCA of this data set yields a scores plot, shown in Figure 16(c), in which the

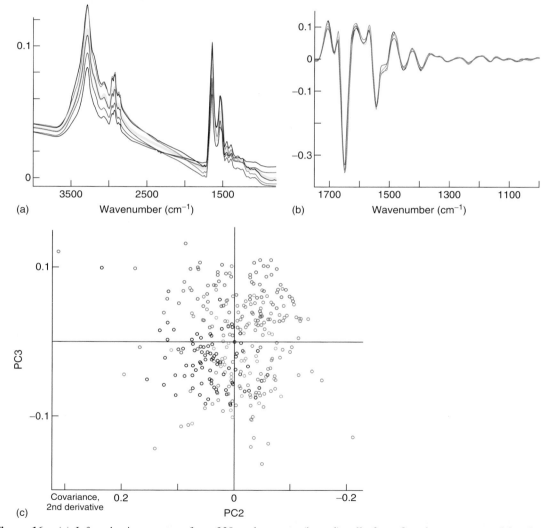

**Figure 16.** (a) Infrared microspectra of ca. 320 oral mucosa (buccal) cells from five donors, averaged by donor. The heavy yellow line denotes the mean spectrum of all spectra in the data set. (b) Second-derivative, expanded and vector-normalized spectra of data set shown in (a). (c) Scores plot (PC2 vs. PC3) of data shown in (b). Each circle represents the scores of one of the original 320 spectra.

variance along PC2 and PC3 are symmetric about the origin, although there still exist some systematic differences between the cells marked with blue and black circles. The origin of this spectral difference is not clear at this point. Furthermore, these differences are smaller than those observed when another cell type is added, for example, the canine cells that are discussed below (Section 3.2.2).

Spectral variations, similar in magnitude to the donor-to-donor variations shown in Figure 16, were found when cells of the same donor were collected at different times. PCA was found to be sufficiently sensitive to cluster spectra from the same donor by collection date. Thus, the "donor-to-donor" variations described in Figure 16 might have been more appropriately called *sampling-to-sampling* variations. These data demonstrate that it is absolutely essential to carry out any comparative studies such that all unwanted variations are minimized, or employ sufficiently large data sets to average out unwanted variations.

Although this sensitivity of IR spectroscopy to distinguish donors and methods of sample preparation may seem a bit disconcerting, we found that the systematic variations between different cell types are larger than the deviations described above between donors and/or acquisition date. However, to ensure that no spurious effects are observed, cells must be treated in exactly the same way (i.e., use of the same media, and identical washing, fixation, and sample preparation methods) to produce reliable results. These aspects are discussed in Section 3.2.2.

### 3.2.2 Canine cervical cells

Here, we report results from another model system of epithelial cells. Canine cervical cells were selected as a model for a number of reasons. First, they are a logical animal model for human cervical cells. Second, these cells are devoid of glycogen, which allows the entire spectral range to be analyzed without interference from strong glycogen signals. Thus, they also may be compared with oral mucosa cells, which are devoid of glycogen, and methods to differentiate

the very similar spectra can be devised. Canine cervical cells can be obtained by gently scraping the cervix from spayed dogs with a miniature dental brush, and vortexing the cells from the brushes in BSS. Visual inspection of these cells was used to distinguish mature, ectocervical cells from other cell types. Sample preparation and data collection proceeded as in the case of oral mucosa cells.

The variation in the spectra of cells collected from different animals is similar to that observed between human donors. After data preprocessing described above, PCA was performed, which reveals good homogeneity of the spectra, and very small systematic variations between dogs. In view of the previous discussion, this result can be expected.

Next, we present a comparison between human and canine mucosal cells.[4] The aim of this analysis was to establish that very similar cell types from different species exhibit spectral differences that are larger than those from individual donors. Morphologically, human oral mucosa cells and canine cervical cells are very similar. Furthermore, in terms of their biochemical composition, which ultimately determines the observed IR spectra, these cells are expected to be very similar. Thus, we thought it an important test of the methodology whether or not the spectra of these cells could be distinguished. We included all available data subsets to minimize factors such as different collection times from influencing the analysis. In total, seven sets of human oral mucosa cells (comprising 427 spectra), and five sets from animals (560 spectra) were combined into one data set, and analyzed via PCA. The scores plots (PC2 vs PC3) of uniformly preprocessed data are shown in Figure 17(a), and the PC3 versus PC4 plot in Figure 17(b). Thus, the separation between canine cervical and human buccal cells is excellent, both in the PC2 versus PC3 and the PC3 versus PC4 plots. However, at this point, the spectral features responsible for the differentiation are not understood.

When the spectra of cells from dogs in estrus were incorporated into the PCA, they were found to fall exactly within the range of the human

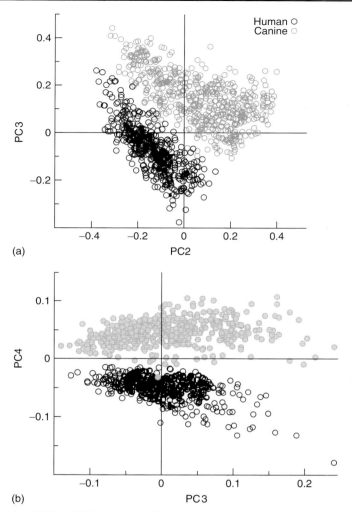

**Figure 17.** (a) Scores plot (PC2 vs. PC3) and (b) PC3 versus PC4 of human buccal and canine cervical data sets.

buccal cells, and were well separated from the spectra of cells of nonestrus dogs (Figure 18). The explanation of this spectral differentiation may be found in the level of maturity of the cells examined. The human buccal cells, exfoliated from the oral mucosa, are mature squamous epithelial cells. These cells are typically between 50 and 60 μm on edge. For nonestrus dogs, the cervical cells are not completely mature, and are generally classified as intermediate squamous cells.[46] These cells are significantly smaller, about 25–35 μm on edge. During estrous, hormonal changes cause

maturation of the canine cervical cells from intermediate to superficial cells. In this process, the cells enlarge to about the same size as mature human cervical or human buccal cells.

We believe that the nucleus-to-cytoplasm ratio (N/C) discussed in Section 3.2.1 above is responsible for the spectral differentiation shown in Figures 17 and 18. Since the nuclear proteins contribute as much as 25% to the overall spectrum of the cell, variations in the N/C ratio between the (intermediate) canine cervical cells on the one hand, and mature human buccal cells and mature

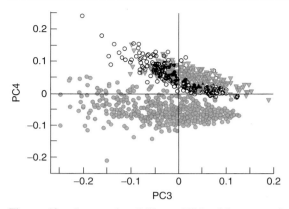

**Figure 18.** Scores plot (PC3 vs. PC4) of human oral buccal cells (black circles), canine cervical (gray circles), and canine cervical cells (gray triangles) from estrous dogs. Notice that this is the same plot as Figure 17(b), but including the estrous dogs.

canine cervical cells on the other, greatly affect the observed spectra. Thus, we believe that the patterns observed for human buccal, immature canine, and mature canine cervical cells are a direct consequence of the N/C ratio of these cells.

### 3.2.3 Human urothelial cells

We present in this section a more complex system of exfoliated cells, namely, the cells collected from human urine. The predominant cell types in urine from healthy donors are the urothelial cells from the lining of the bladder, as well as squamous cells from the distal end of the urethra. This latter type may be further subdivided into glycogen-containing and glycogen-free squamous cells. The distinction between glycogen-containing and glycogen-free cells is possible in standard cytology only after special staining for glycogen, but occurs automatically in spectral cytology due to the absence or presence of glycogen signals (see, for example, Figure 1c, trace A).

In this discussion, we establish that squamous and urothelial cells can be distinguished automatically by IR-MSP. We intend to establish this methodology as a low cost, automated test for bladder cancer.

Voided urine samples were routinely collected using a "clean catch" methodology, whereby only midstream urine is captured. This procedure helps eliminate contamination from bacterial flora that is often present in urine. In addition, specimens were only collected from male donors to alleviate possible contamination from vaginal or perineal squamous epithelial cells.[47] In this study, multiple samples were collected from four volunteers over a period of 3 months. To avoid degenerative cellular changes from proteolytic enzymes and bacterial cytolysins, specimens were immediately prepared for spectroscopic analysis. Cells were recovered from the urine using a membrane-filtration technique, which involved passing the urine sample through a nylon net filter that was held in a polypropylene holder (11-μm pore size, 47-mm diameter; Millipore, Billerica, MA, USA).

The nylon filter was then directly placed into a centrifuge tube that contained 20 mL of buffered formalin (10% formaldehyde in water) solution. Cells were shaken off the filter using a vortex mixer, and the resulting suspension was allowed to stand for 20 min, thus permitting the fixation and further preservation of the cells. The filter was then removed, and the remaining specimen centrifuged (600 g for 25 min) to concentrate the diagnostic cells. After resuspending the cells, samples were prepared on "low-e" glass slides (Section 2.4) using the CytoSpin method described in Section 2.3.2.

Slides were placed onto the sample stage of the microscope and a visual image captured from the entire sample spot. This was directly referenced against an etched mark on the slide to enable the accurate relocation of cells after spectroscopic analysis and subsequent cytological staining. Transflection spectra were collected from 100 individual cells per slide using a point-mapping method of data acquisition. IR microspectral data were collected using a fixed aperture of $25 \times 25$ μm, co-adding 128 interferograms at 4 cm$^{-1}$ spectral resolution. After spectroscopic data collection, sample slides were stained using standard cytological protocols, using "Pap" stain within our own laboratory. Cells that were examined in the analysis were then relocated on the slide and visual images captured at high magnification (40×) to allow cytological diagnosis.

Figure 19 displays photomicrographs captured from the three main types of epithelial cell identified in our urine sediments after Pap staining. Figure 19(a) is from a glycogen-containing squamous cell, Figure 19(b) for a glycogen-free squamous cell, and Figure 19(c) for an urothelial cell. (Glycogen-free, in this context, implies that spectral contributions due to glycogen could not be detected in the IR spectra). Both glycogen-free and glycogen-containing squamous cells are from the distal end of the urethra, and constitute the majority of cells found in urine, and cannot be distinguished by visual inspection, although it appears that the glycogen-free cells generally exhibit a smaller (more pyknotic) nucleus. Figure 19(d) displays spectra for the three cell types presented as vector-normalized absorbance spectra. Voided urine collected from healthy individuals contain very small numbers of transitional

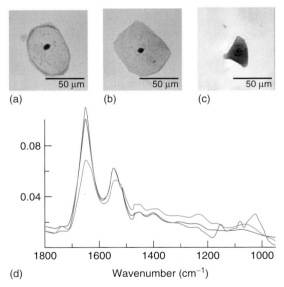

**Figure 19.** Photomicrographs of (a) glycogen-containing, (b) glycogen-free squamous cells, and (c) urothelial cells found in urine. (d) Spectra of glycogen containing (red trace), glycogen free (blue trace), and urothelial cell (green trace).

(urothelial) epithelial cells,[47] an observation that was confirmed by our spectroscopic studies. For the squamous cells, a large variation in the overall amplitude of the spectra is observed, in accordance with previously discussed results of the oral mucosa cells (see Section 3.2.1 and Figure 18). Absorbance values recorded for the amide I band vary between 0.05 and 0.8 AU for the squamous cells. We believe that these variations are largely due to deviations in both cell thickness and the nucleus-to-cytoplasm ratio within the spatial area sampled.[4,48]

Spectra recorded from urothelial cells display particularly large absorbance intensities. This type of cell can range in size from ca. $10-40\,\mu m$[47,49] and typically feature one or more large nuclei. We hypothesize that this observed intensity difference for urothelial cells is caused by both an increased cell thickness and a strong nuclear contribution to the spectra. Confocal Raman depth profiles collected from both squamous (Section 3.1.2) and urothelial epithelial cells in the authors' laboratory provide strong evidence to support such assumptions. Despite the observed large variation in the overall intensity of the recorded spectra, correct statistical treatment of the data allows the common or diagnostic features of the spectra to be emphasized.

Figure 20 shows second-derivative spectra for these three cell types. The red and blue profiles characterize the glycogen-rich and glycogen-free squamous cells respectively. These two spectra are almost identical in the amide I/amide II region $(1700-1450\,cm^{-1})$ and differ only in the $1200-1000\,cm^{-1}$ region, where the strong spectral bands characteristic for glycogen occur (marked by asterisks). The green profile describes the spectrum of urothelial cells. Within the spectral region marked (A), distinct spectral features are noticeable that are not apparent in the squamous cell spectra. This region comprises both the amide I and amide II vibrational modes of proteins, and indicate a significant change in protein composition in these cells. Within the spectral region marked (B), the profile is similar to that of the glycogen-free squamous cells

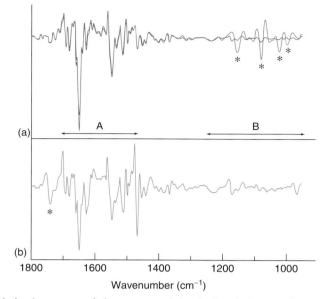

**Figure 20.** (a) Second-derivative spectra of glycogen-containing (red) and glycogen-free squamous cells isolated from urine. The arrows marked A and B designate the amide I/amide II and the glycogen/phosphate regions, respectively. Peaks marked by (*) are due to glycogen. (b) Second-derivative spectra of urothelial cells isolated from urine. The peak marked by (*) is due to the ester linkage of phospholipids.

(blue spectrum). However, the bands between 950 and 1400 cm$^{-1}$ are more pronounced and display a marked increase in intensity, in particular, the symmetric and antisymmetric phosphate stretching vibration at ca. 1085 and 1235 cm$^{-1}$, which are present in both phospholipids and nucleic acids. However, an additional spectral band, marked by an asterisk in the green trace of Figure 20, is characteristic exclusively of phospholipids. Thus, we speculate that the increased absorption of the phosphate spectral bands is not only due to an increased DNA contribution from the nucleus but also from an increased number of phospholipids contained in a thickened cell membrane. In addition, sharp peaks at 1468–1379 cm$^{-1}$ could be attributed to the aliphatic side chains of phospholipids. We are presently undertaking confocal Raman spectroscopic studies upon both squamous and urothelial epithelial cells to help verify this assumption.

Figure 21 presents results obtained from a PCA analysis of the entire spectral data set that comprised both squamous and urothelial epithelial cells. The data has been projected onto the second and third principal component dimensions, where each dot represents an individual spectrum collected from a cell. As can be seen in the diagram, the main types of normal epithelial cells identified in the urine divide into three clusters. The green cluster, representative of urothelial cells, displays a distinct separation from the squamous cells and indicates a substantial spectral difference among these cells. Furthermore, the glycogen rich (red) and glycogen absent (blue) squamous cells are also clearly differentiated. These results are among the first ever to distinguish different cell types in exfoliated samples by completely automatic spectral methods at the level of individual cells, and constitutes the groundwork for an eventual application of IR-MSP for the diagnosis of disease.

### 3.2.4 Cell cycle dependence of cellular spectra

In our quest for developing a spectroscopic method to screen cells for the presence of cervical

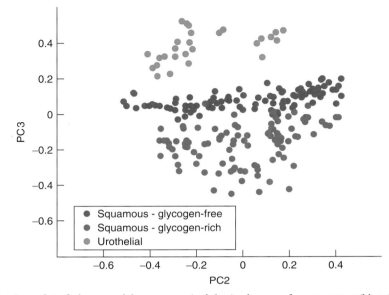

**Figure 21.** PCA scores plot of glycogen-rich squamous (red dots), glycogen-free squamous (blue dots), and urothelial cells (green dots). Each dot represents one spectrum.

cancer, cultured cervical cancer (HeLa) cells were a logical choice, since we anticipated that the spectra of these cells would be sufficiently different from those of normal cervical cells. In order to explain the enormous spectral heterogeneity of samples of cultured as well as exfoliated cells, we embarked on a study to investigate the spectral heterogeneity of cells as a function of their stage in the cell's division cycle. Some of the biological background of the cell cycle was discussed in Section 2.2, and the "mitotic shake-off" and staining procedures used to identify the stage of the cell cycle were introduced in Section 2.5.2.

Figure 22(a) shows the typical staining pattern for each of the three cell populations (G1, S, and G2) of cells within the division cycle. Positive red staining for cyclin E indicates that the cells were in the G1 phase[25] at the time of fixation. Cyclin E reaches its maximum concentration at the end of G1, and is quickly degraded during the early S phase. Since the thymidine analog BrdU is integrated into nuclear DNA only during DNA synthesis during the S phase, the green fluorescence staining for BrdU indicates that the

cell was actively synthesizing DNA at the time of BrdU incubation. These cells were assigned to the mid-late S-phase population, found typically 11–15 h after mitotic shake-off. Cells that exhibit a double-staining positive for both incorporated BrdU and cyclin E are in the early stage of the S phase when they were fixed—typically 6–8 h after mitotic shake-off. The final group of cells exhibited only positive yellow fluorescence staining for cyclin B1, and were assigned to the G2 phase. Cyclin B1 moves into or near the nucleus toward the end of the G2 phase and reaches its maximum concentration after about 18–20 h. The blue hues in Figure 22(a) result from counterstaining with DAPI to visualize the location of the nucleus.

In IR spectra of these cells, the differences at the various time points of the cell cycle manifest themselves mostly in changes of the amide I and II bands: there is a 10-cm$^{-1}$ shift in the amide I peak position toward higher wavenumber, accompanied by a shift in an amide I shoulder, as the cells progress through the G1, S, and G2 phases, indicative of changes in protein structures.[23] Similar frequency shifts are observed in the

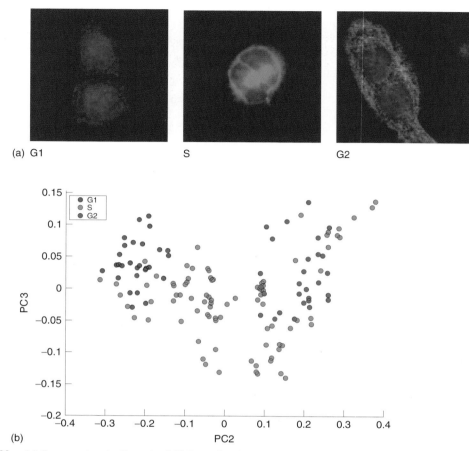

**Figure 22.** (a) Immunochemically stained HeLa cell pairs at different stages of the cell cycle. See text for detail. (b) PCA scores plot of cells at different stages of the cell cycle: G1(red), S (green), and G2 (blue). For cell sizes, see Figure 9.

amide II vibrations at 1538, 1540, and 1543 cm$^{-1}$ for the G1, S, and G2 phases respectively. Furthermore, the amide II peak shows a low frequency shoulder at about 1500 cm$^{-1}$ that is barely evident in the absorption spectra but becomes more prominent between G1, S, and G2 phase.

In Figure 22(b), scores plots of cellular populations are presented. The separation between G1 and G2 stages is very good, indicating that there are systematic variations in the protein spectra between G1 and G2 stages. This result is not too surprising, considering that in G1 the proteins for DNA replication are synthesized, whereas in G2 proteins for mitosis are synthesized.

However, the S-phase spectra do not form a distinct cluster, but span the entire PC3/PC2 space, with much larger variance than the G1/G2 cells. Thus, we conclude that the heterogeneity of the S-phase spectra is large, and S-phase spectra cannot easily be separated from G1 and G2 phase spectra. The reasons for the spectral heterogeneity in the S phase are not clear at this point. It is well known that the DNA replication during the S phase is not a continuous process, but occurs at different locations along the DNA, and at various instances. Spectrally, some S-phase cells appear very similar to G1 cells, while others appear similar to G2-phase cells. Still other S-phase cells have quite different spectra from both G1 and G2-phase cells. A possible hypothesis for this might be that there is a gradual change from G1 to G2 spectral properties during the S phase, which is interrupted by time periods where spectra are very different, presumably when actual DNA replication occurs.

## 3.3  Infrared spectral studies of live cells

Here, we wish to discuss the first IR spectra of individual human cervical cancer (HeLa) cells suspended in buffer or cell culture medium reported in 2004.[50] Although we did not establish at first whether or not these cells were viable at the time of spectral data acquisition, later results, and a comparison with the results from other groups, suggest that the cells were living, and confirm that this method is applicable for studying live cells. As pointed out in Sections 3.1.2 and 3.1.3, similar studies can be carried out much more readily using RA-MSP, since the water signal is much weaker as compared to the cellular features, and since water immersion optics can be utilized.

The IR data reported here were collected for entire cells, using $25-40\,\mu$m apertures. Measurements were carried out both in transflection and transmission modes. IR spectroscopy of cells in an aqueous environment is difficult because of the very strong water vibrations that interfere with the protein amide I, II, and A vibrations. However, the sensitivity of modern IR microspectrometers is sufficiently high to permit observation of the cellular spectra even in the presence of the high water background signals. The results reported here might have far-reaching implication for the use of IR microspectroscopy to monitor cell proliferation, drug response, and other cell biological parameters in live cells.

Figure 23(a) shows a visual image of this cell taken through the objective used for IR data acquisition, and Figure 23(b), trace A, shows the transflection spectrum of a cell suspended in BSS. Although this cell may have been compressed in the sample preparation process, we believe that the spectrum is representative of live cells.

The negative water band at $3400\,\text{cm}^{-1}$, and the distorted 1 amide I/II band profile shown in trace A, are artifacts due to overcompensation for the water background. This artifact is due to the fact that the cell contains less water than the surrounding buffer. Consequently, the calculated absorption spectra are overcompensated for the liquid water background.[50] This can be avoided by scaling the amplitude of the buffer spectrum when the sample absorbance spectrum is computed, or by adding a scaled buffer spectrum to the raw absorption spectrum of the sample. Since the exact difference in water content of the cell and the medium is unknown, both procedures are based on visual fitting. The corrected spectrum for the cell was obtained by adding a scaled water spectrum to the raw cell spectrum until the corrected spectrum displays positive water and amide A band envelopes. This is shown in Figure 23(b), trace B. In the corrected spectrum, the amide I and II profiles also appear with more normal band profiles, as shown in Figure 23(c).

The spectra of the cell in growth medium show a sharp band in the symmetric phosphate stretching region ($1085\,\text{cm}^{-1}$) with relatively large intensities throughout the nucleus and the cytoplasm. Although this band was observed for all cells in growth medium (and not for cells in BSS), it is not due to the growth medium itself, which has no spectral features in this region, but appears to be a signature of cell metabolism. (The spectral features of growth medium are nearly identical to those of water, since the most abundant component, glucose, is present in such low concentrations that its spectrum is barely discernible with the naked eye. Since these experiments were carried by ratioing against the background spectra of the growth medium, we can eliminate the medium as a source of this band.) The origin of the band at $1085\,\text{cm}^{-1}$ is presently unknown, but its presence in both cytoplasm and nucleus suggests that it is due to a short-lived component containing phosphate groups, such as ATP, phosphorylated carbohydrates, or proteins that are particularly prevalent in metabolically active cells.

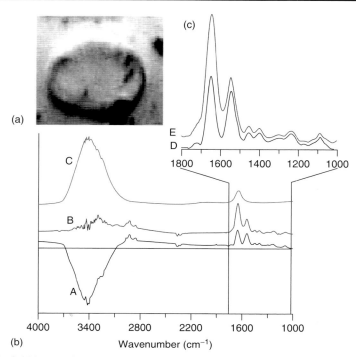

**Figure 23.** (a) Bright-field image of HeLa cell in growth medium. The cell measures about 40 μm in diameter. (b) IR spectrum of cell (trace A) ratioed against spectrum of medium; spectrum of pure medium (trace C), and corrected spectrum (trace B). See text for detail. (c) Expanded region of cellular spectrum before (trace D) and after (trace E) background correction.

# 4   CONCLUSIONS

In this chapter, we have reviewed some of the most recent findings on "spectral cytology" from our laboratories. Over the course of the past 4 years, we have collected spectra from in excess of 10 000 individual cells—both for high spatial resolution studies, in an effort to understand the spectral consequences of cell biological events, and for establishing databases for eventual spectral diagnostic algorithms.

The spectra of cultured cells present a fairly consistent picture, and the variance in spectral data from cultured cells can be reasonably well explained in terms of the different cell biological events. However, the variance of spectral results for exfoliated cells still presents some problems. The variance in spectral patterns for exfoliated cells is particularly surprising given the fact that the cells that are easily collected—be it by collecting voided urine or by swiping the surface of the oral cavity—are mature squamous cells that should be very similar. In these cells, protein synthesis and general metabolic activity has slowed down enormously, as evidenced by the pyknotic nucleus. Nevertheless, large differences exist in overall spectral intensity, and abundance of glycogen. These differences make analysis of the spectral data quite difficult.

One consequence of the large spectral variance is that the spectrum obtained from one single measurement cannot be trusted, and that any conclusions must be based on analysis of thousands of measurements from individual cells. However, the results from such large data sets reveal very good distinction between cell types, states of disease, and cellular activity.

The authors of this contribution hope that this review will aid in future studies to establish spectral cytology as a viable method of medical diagnosis, and will prevent repetition of careless studies, which are abundant in the literature.

## ACKNOWLEDGMENT

Partial support of this research from grants CA 81675 and CA 090346 from the National Institutes of Health is gratefully acknowledged.

## END NOTES

[a.] Work performed prior to Jan 1, 2006 was carried out in the Department of Chemistry, City University of New York, Hunter College, New York, NY 10021.
[b.] DNA packing ratio is defined as the length of a free, double-helical DNA molecule to the length of the DNA–protein complex in chromatin.
[c.] Ten milliwatts of laser power focused into a spot size of about $1 \, \mu m^2$ is the limiting power density before the sample is destroyed. The use of a water or oil immersion objective will alleviate this problem somewhat.

## ABBREVIATIONS AND ACRONYMS

| | |
|---|---|
| ANNs | Artificial Neural Networks |
| ATCC | American Tissue and Cell Culture Corporation |
| AU | Absorbance Units |
| BrdU | 5-Bromo-2′-deoxyuridine |
| BSS | Balanced Saline Solution |
| CCD | Charge-Coupled Device |
| DAPI | 4′,6-diamidino-2-phenylindole |
| DMEM | Dulbecco's Modified Eagle's Medium |
| EDTA | Ethylene Diamine Tetraacetate |
| FBS | Fetal Bovine Serum |
| FITC | Fluorescein Isothiocyanate |
| FSD | Fourier Self-Deconvolution |
| FT-IR | Fourier-Transform Infrared |
| FWHH | Full Width at Half Height |
| HCA | Hierarchical Cluster Analysis |
| H&E | Hematoxylin/Eosin |

| | |
|---|---|
| IR | Infrared |
| IRB | Institutional Review Board |
| IR-MSP | Infrared Microspectroscopy |
| MCT | Mercury-Cadmium-Telluride |
| N/C | Nucleus-to-Cytoplasm Ratio |
| NSLS | National Synchrotron Light Source |
| Pap | Papanicolaou |
| PBS | Phosphate Buffered Saline |
| PCA | Principal Component Analysis |
| RA-MSP | Raman Microspectroscopy |
| SNR | Signal-to-Noise Ratio |

## REFERENCES

1. J.W. Chan, D.S. Taylor, T. Zwerdling, S.M. Lane and K. Ihara, *Biophys. J.*, **90**, 648–656 (2006).

2. C.M. Krishna, G. Kegelaer, I. Adt, S. Rubin, V.B. Kartha, M. Manfait and G.D. Sockalingum, *Biopolymers*, **85** (5), 462–470 (2006).

3. M. Diem, M. Romeo, S. Boydston-White, M. Miljkovic and C. Matthaeus, *Analyst*, **129** (10), 880–885 (2004).

4. M. Romeo, B. Mohlenhoff, M. Jennings and M. Diem, *Biochim. Biophys. Acta*, **1758** (7), 915–922 (2006).

5. S. Boydston-White, M.J. Romeo, T. Chernenko, A. Regina, M. Miljkovic and M. Diem, *Biochim. Biophys. Acta*, **1758** (7), 908–914 (2006).

6. C. Krafft, T. Knetschke, R.H.W. Funk and R. Salzer, *Anal. Chem.*, **78**, 4424–4429 (2006).

7. D. Helm, H. Labischinski, G. Schallen and D. Naumann, *J. Gen. Microbiol.*, **137**, 69–79 (1991).

8. P. Wong, R. Wong, T. Caputo, T. Godwin and B. Rigas, *Proc. Natl. Acad. Sci.*, **88**, 10988–10992 (1991).

9. P. Wong, S. Lacelle, M. Fung, K. Fung and M. Senterman, *Biospectroscopy*, **1**, 357–364 (1995).

10. G.N. Papanicolaou and H.F. Traunt, *Am. J. Obstet. Gynecol.*, **42** (2), 193–206 (1941).

11. B. Wood, M. Quinn, F. Burden and D. McNaughton, *Biospectroscopy*, **2**, 01–11 (1996).

12. B. Wood, M. Quinn, B. Tait, M. Ashdown, T. Hislop, M. Romeo and D. McNaughton, *Biospectroscopy*, **4**, 75–91 (1998).

13. M.J. Romeo, F. Burden, M. Quinn, B. Wood and D. McNaughton, *Cell. Mol. Biol.*, **44** (1), 179–187 (1998).

14. L. Chiriboga, P. Xie, H. Yee, V. Vigorita, D. Zarou, D. Zakim and M. Diem, *Biospectroscopy*, **4**, 47–53 (1998).

15. R. Sindhuphak, S.I. Issaravanich, V. Udomprasertgul, P. Srisookho, S. Warakamin, S. Sindhuphak, R. Boonbundarlchai and N. Dusitsin, *Gynecol. Oncol.*, **90**, 10–14 (2003).

16. L. Chiriboga, P. Xie, H. Yee, D. Zarou, D. Zakim and M. Diem, *Cell. Mol. Biol.*, **44** (1), 219–229 (1998).

17. M. Diem, S. Boydston-White and L. Chiriboga, *Appl. Spectrosc.*, **53** (4), 148A–161A (1999).

18. D.M. Haaland, H.D.T. Jones and E.V. Thomas, *Appl. Spectrosc.*, **51** (3), 340–345 (1997).

19. L.E. Hood, J.H. Wilson and W.B. Wood, 'Molecular Biology of Eukaryotic Cells', W.A. Benjamin, Menlo Park (1974).

20. M. Jackson and H.H. Mantsch, *Crit. Rev. Biochem. Mol. Biol.*, **30** (2), 95–120 (1995).

21. P. Lasch, A. Pacifico and M. Diem, *Biopolymers*, **67** (4–5), 335–338 (2002).

22. L. Chiriboga, H. Yee and M. Diem, *Appl. Spectrosc.*, **54** (4), 480–485 (2000).

23. S. Boydston-White, T. Chernenko, A. Regina, M. Miljkovic, C. Matthaeus and M. Diem, *Vib. Spectrosc.*, **38** (1–2), 169–177 (2005).

24. J.A. Aten, P.J.M. Bakker, J. Stap, G.A. Boschman and C.H.N. Veenhof, *Histochem. J.*, **24** (5), 251–259 (1992).

25. A. Koff, F. Cross, A. Fisher, J. Schumacher, K. Leguellec, M. Philippe and J.M. Roberts, *Cell*, **66** (6), 1217–1228 (1991).

26. V. Deckert and W. Kiefer, *Appl. Spectrosc.*, **46**, 322–328 (1992).

27. C. Otto and J. Greve, *Internet J. Vib. Spectrosc.*, **2** (3), (1998).

28. N. Everall, J. Lapham, F. Adar, A. Whitley, E. Lee and S. Mamedov, *Appl. Spectrosc.*, **61** (3), 251–259 (2007).

29. A. Savitzky and M.J.E. Golay, *Anal. Chem.*, **36** (8), 1627–1639 (1964).

30. M.J. Adams, 'Chemometrics in Analytical Spectroscopy', 2nd edition, Royal Society of Chemistry, Cambridge (2004).

31. B.R. Wood, L. Chiriboga, H. Yee, M.A. Quinn, D. McNaughton and M. Diem, *Gynecol. Oncol.*, **93** (1), 59–68 (2004).

32. J.H. Ward, *J. Am. Stat. Assoc.*, **58** (301), 236–244 (1963).

33. N. Jamin, P. Dumas, J. Moncuit, W.H. Fridman, J.L. Teillaud, G.L. Carr and G.P. Williams, *Cell. Mol. Biol.*, **44** (1), 9–14 (1998).

34. M. Diem, L. Chiriboga, P. Lasch and A. Pacifico, *Biopolymers*, **67** (4–5), 349–353 (2002).

35. B. Mohlenhoff, M.J. Romeo, M. Diem and B.R. Wood, *Biophys. J.*, **88** (5), 3635–3640 (2005).

36. N. Jamin, L. Miller, J. Moncuit, W.H. Fridman, P. Dumas and J.L. Teillaud, *Biopolymers (Biospectroscopy)*, **72**, 366–373 (2003).

37. C. Matthäus, T. Chernenko, J.A. Newmark, C.M. Warner and M. Diem, *Biophys. J.*, **93**, 668–673 (2007).

38. J.K. Kauppinen, D.J. Moffatt, H.H. Mantsch and D.G. Cameron, *Appl. Spectrosc.*, **35** (3), 271–276 (1981).

39. G. Mie, *Ann. Phys. (Leipzig)*, **25**, 377–452 (1908).

40. M.J. Romeo, B. Mohlenhoff and M. Diem, *Vib. Spectrosc.*, **42**, 9–14 (2006).

41. P. Laven, http://www.philiplaven.com/ 2004.

42. P. Walstra, *Brit. J. Appl. Phys.*, **15**, 1545–1552 (1964).

43. H.N. Holman, M.C. Martin, E.A. Blakely, K. Bjornstad and W.R. McKinney, *Biopolymers (Biospectroscopy)*, **57**, 329–335 (2000).

44. M. Diem, M.J. Romeo, C. Matthaus, M. Miljkovic, L. Miller and P. Lasch, *Infrared Phys. Technol.*, **45**, 331–338 (2004).

45. D. Naumann, V. Fijala, H. Labischinski and P. Giesbrecht, *J. Mol. Struct.*, **174**, 165–170 (1988).

46. P.A. Holst, 'Canine Reproduction: A Breeder's Guide', Alpine Publications, Loveland (1985).

47. N.A. Brunzel, 'Fundamentals of Urine and Body Fluid Analysis', W. B. Saunders, Philidelphia (1994).

48. M.J. Romeo, B. Mohlenhoff and M. Diem, *Vib. Spectrosc.*, **42** (1), 9–14 (2006).

49. P. Rathert, S. Roth and M.S. Soloway, 'Urinary Cytology, Manual and Atlas', 2nd edition, Springer-Verlag, New York (1991).

50. M. Miljković, M.J. Romeo, C. Matthaeus and M. Diem, *Biopolymers*, **74** (1–2), 172–175 (2004).

# Infrared Spectroscopy in the Identification of Microorganisms

## Mareike Wenning[1], Siegfried Scherer[1] and Dieter Naumann[2]

[1] Technische Universität München, Freising, Germany
[2] Robert Koch-Institut, Berlin, Germany

## 1 INTRODUCTION

Identification of microorganisms is a major task in medical diagnosis. The unambiguous identification of a microorganism determines its pathogenic potential and is a crucial prerequisite for efficient therapy. In most cases, the identification procedure in the clinical laboratory relies on phenotypic characterization of microbes. This implies cultivation on selective or nonselective media, followed by microscopic examination and Gram stain. These data, in combination with a few additional tests, such as oxidase or catalase reactions, enable the assignment of the organism to larger taxa such as fermenting or nonfermenting gram-negative bacilli, staphylococci, or yeasts. This information is needed to select the appropriate identification system. In clinical microbiology, commercially available test kits exist, which are specifically designed to identify pathogenic yeast or bacteria. These are widely distributed and fully automated systems like the VITEK system (BioMérieux) or BD Phoenix (Becton Dickinson) have been developed to increase sample throughput and reduce work effort. Molecular biology tools allow identification of microbes by analyzing their DNA sequence or specific restriction patterns of the genome or DNA fragments, or by recognizing specific DNA targets using specific primers or probes. The major advantage is an increase in speed caused by the omission or reduction of an enrichment procedure because of the reduced detection limit. In principle, such methods also allow for the detection of noncultivable pathogens. Fully automated devices to identify microbes at a molecular basis, such as the RiboPrinter system (DuPont) or the MicroSeq system (Applied Biosystems), are available.

Since Fourier transform infrared (FT-IR) spectroscopy was introduced as a technique to identify microbes by the group of Dieter Naumann,[1–4] it has gained growing interest for identifying these species. The absorption of IR radiation by cell components results in fingerprint-like spectra that reflect the overall chemical make-up of the cells under investigation. Consequently, in contrast to many other methods, identification is not based on a restricted set of characteristics

*Vibrational Spectroscopy for Medical Diagnosis*. Edited by Max Diem, Peter R. Griffiths and John M. Chalmers.
© 2008 John Wiley & Sons, Ltd. ISBN 978-0-470-01214-7.

such as enzyme activities or DNA sequence, but is achieved by selecting discriminating features from a large amount of spectral information. As a physicochemical technique, FT-IR spectroscopy benefits from the fact that operating costs are extremely low as practically no consumables are required, while at the same time spectra contain a huge amount of information, which can be exploited to help in solving different kinds of identification problems. By comparison with large reference data sets, spectra of microbial cells can be analyzed for identification purposes, or to reveal certain characteristics or even strain identity. Over the last 10 years, application of IR spectroscopy to identify microorganisms has undergone a remarkable evolution. Highly sophisticated spectrometers have been designed for high sample throughput, generating high quality data in a short time, and requiring only limited sample handling due to their simple protocols and low running costs; FT-IR spectroscopy seems to be an attractive alternative to conventional techniques.

Many articles that deal with the application of vibrational spectroscopy in biology and microbiology give an overview on technical items and different applications.[4–7] This chapter focuses on the establishment and use of reference databases in the identification of microorganisms, which represents the most crucial step in the application of this technique in microbiology. The reader is introduced to the methodology of FT-IR spectroscopy and to problems arising when reliable reference data sets are to be generated.

## 2 METHODOLOGY

### 2.1 Nature of spectral information

FT-IR spectroscopy is a whole-organism fingerprinting technique. All components of cells contribute with their individual characteristic absorption of IR light to the spectral pattern generated. The microbial cell is composed of a huge variety of different components that can be assigned to proteins, lipids, polysaccharides or derivatives and combinations thereof. More than 95% of dry cell

**Table 1.** Quantitative composition of growing *E. coli* cells (only macromolecules).

| Cell component | Percentage of dry cell mass |
| --- | --- |
| Proteins | 55 |
| Polysaccharides | 5 |
| Lipids | 9 |
| Lipopolysaccharides | 3 |
| DNA | 3 |
| RNA | 21 |

Reproduced from Madigan *et al.*[8] © Prentice Hall, 2001.

mass is organized in macromolecules; only 3% are monomers or low molecular compounds.[8] Table 1 lists the quantitative composition of actively growing *Escherichia coli* cells. Clearly, proteins make up the main part of dry cell mass and are found in all parts of the cell. They have structural functions in the cell wall and membranes as well as catalytic functions in the cytoplasm. RNA is the second most frequently found component in growing or dividing microbial cells; however, its contribution to cell mass can vary enormously. As RNA is involved in protein synthesis, the amount of RNA in the cell is strongly dependent on the growth phase. Bacterial growth can be divided into at least three phases: lag phase (adaptation), logarithmic phase (exponential growth), and stationary phase (constant cell number) and as a response to the different physiological state the composition of cells varies. In rapidly growing cells, the content of RNA is high due to increased protein synthesis. Cultures entering stationary phase cut down their active metabolism and, as a consequence, the content of RNA is also dramatically reduced. Lipids and polysaccharides are mainly components of the cell membrane and cell wall and can vary in composition according to the physiological state of the cells.[a] Furthermore, they may serve as storage materials like glycogen or polyhydroxybutyric acid and can therefore accumulate in the cells.

Many studies have been conducted to reveal absorption patterns of different substances; representative spectra are shown in the Appendix to this book. Band assignments to specific

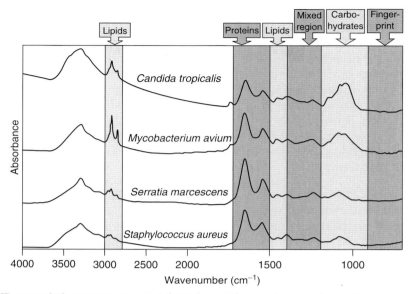

**Figure 1.** FT-IR transmission spectra recorded from four different microorganisms. The shaded spectral regions contain exploitable information. The biochemical components that are dominating the absorptions in the respective range are indicated above the spectra. For clarity, spectra are displayed vertically shifted and the region between 1800 and 700 cm$^{-1}$ is expanded.

vibrations of functional groups may be found in the literature.[9-12] The absorption spectrum of specific classes of substances can be interpreted from these vibrational patterns, and it becomes obvious that characteristic absorption bands of the most important classes of biomacromolecules appear dominantly in relatively narrow ranges of the spectrum. In the region of 3100–2800 cm$^{-1}$ and 1500–1400 cm$^{-1}$, the main absorptions of modes associated with the C−H vibrations of lipids are found (Figure 1), with the C=O stretching mode appearing between 1750 and 1730 cm$^{-1}$. Between 1700 and 1500 cm$^{-1}$, two large peaks, the so-called amide I and II bands, are caused by the protein absorptions. The most characteristic features of carbohydrates can be found between 1200 and 900 cm$^{-1}$. Absorptions below 900 cm$^{-1}$ are only weak and highly variable, and assignments to particular molecules have not yet been achieved.

As IR spectroscopy is a quantitative technique, the relative amount of each cell component is represented in the spectra. Figure 1 displays typical transmission spectra of four different organisms. As can easily be seen from the intensities of the amide I and II bands, proteins make up a prominent part of bacterial spectra. Yeast spectra also show strong absorptions of carbohydrates that may be as dominant as the protein features, since their cell wall is mainly composed of mannan and glucan. The cell envelope is also the main distinguishing feature to differentiate gram-positive from gram-negative bacteria. Whereas gram-positive organisms may have up to 25 layers of peptidoglycan that can make up 90% of the cell wall, gram-negative organisms only have a single or a few peptidoglycan layers, accounting for ∼10% of the cell wall. The main components in the gram-negative cell wall are lipopolysaccharides (LPSs). The cell wall of some organisms such as mycobacteria contains mycolic acids, as indicated by increased intensities of the C−H stretching bands of lipids between 2800 and 3000 cm$^{-1}$ (Figure 1). Differences in fatty acid composition (such as the amount of saturated or unsaturated fatty acids), the composition of polysaccharides or structures on the cell surface, to mention a few, represent a large pool of potential discriminating features that can be exploited by FT-IR spectroscopy to differentiate and classify microorganisms according to their phenotype. However, as the expression of phenotypic characters is sensitive to environmental changes, the success of FT-IR spectroscopy as a technique to identify microorganisms strongly depends on a strict and completely standardized protocol.

## 2.2 Measurement techniques and sampling of microorganisms

The three most commonly used techniques for recording mid-IR spectra of biological materials are (i) transmission, (ii) transflection, and (iii) attenuated total reflection (ATR). For the first two, samples have to be transferred to a support material prior to the measurement. This implies some kind of sample preparation in order to bring the microorganism into a suitable condition for measurement. The choice of the support material depends on the actual technique used: for transmission measurements, IR transparent materials like ZnSe, Si, or $CaF_2$ are used; for transflection measurements, light-reflecting materials like polished metals or low-e glass microscope slides (see **Introduction to Spectral Imaging, and Applications to Diagnosis of Lymph Nodes**) are required. In all cases, it is essential to use water-resistant materials, since water is the main component of live microbial cells. Some manufacturers offer multisample systems allowing for high sample throughput, such as the 96-well plate for the HTS-XT spectrometer from Bruker Optics. In microbial diagnostics, microorganisms generally have to be cultivated to obtain a sufficient amount of cell material to be deposited on the sample support. Measurements from both liquid cultures or from solid media are possible; however, cultivation on solid media is less labor intensive because cells can be harvested directly from the agar surface, whereas liquid cultures need to be washed to get rid of the medium.

In order to avoid strong water absorptions, the samples should be dried as effectively as possible prior to the measurement.

ATR measurements can be performed directly on samples, enabling in situ measurements of microbes. Here, the internal reflection element (IRE) of the ATR accessory is brought into direct contact with the cells under investigation, eliminating sample preparation steps. Samples for ATR spectroscopy that have been grown in a liquid medium may be analyzed without the need for centrifugation; cells harvested from solid media may be simply smeared onto the IRE. Because cells to be investigated do not necessarily have to be removed from their environment and do not need to be dried, the ATR technique is particularly useful for kinetic studies in continuous microbial systems like proliferating cultures or biofilms.[13-15]

With the development of FT-IR microspectroscopy, a new field of IR spectroscopy emerged. Here, the spectrometer is coupled to a light microscope equipped with Cassegrain objectives for IR measurements. The sensitive liquid-nitrogen-cooled mercury cadmium telluride (MCT) detector requires much less sample material in order to record high quality spectra, which leads to a remarkable reduction of the incubation times from ~24 to 6–10 h. This makes the identification of unknown microorganisms possible within one working day.[16-18] The microcolonies grown on solid media are transferred to the sample carrier by replica stamping. Since this procedure results in a spatially accurate imprint on the sample holder and each transferred colony can be measured separately, it permits analysis of mixed cultures with respect to qualitative and quantitative composition without the need of producing pure cultures.[19]

# 3 ESTABLISHING REFERENCE DATABASES

Besides the identification of microbes, FT-IR spectroscopy can be applied without having reference databases available by using the whole cell

IR signatures to quickly sort high numbers of isolates[20] or characterize certain properties of microorganisms.[21-23] However, the main application of mid-IR spectroscopy in microbiology is identification. After more than 15 years of identifying microbes by FT-IR spectroscopy, the high potential of the technique for identification purposes is evident and has been demonstrated many times. Numerous approaches have been made to set up systems for the identification and differentiation of yeasts and bacteria[17,24-44] and even a protocol for filamentous fungi has been published.[45] However, because the identification success always depends on the quality of the corresponding reference databases, generation of these has to be made with caution concerning reproducibility of measurements, choice of reference strains, identification of reference strains, and data analysis.

## 3.1   Reproducibility

For all identification methods, high reproducibility is a key factor for reliability and success, and the degree of standardization needs to be particularly high when the method relies on variable characteristics such as whole cell fingerprints or physiological parameters. The biochemical composition of microbial cells is dependent on the growth phase and differs between younger and older cells. Furthermore, as the medium supplies nutrients for growth, it directly influences cell composition and hence a cell's IR spectrum. Consequently, to achieve high reproducibility in FT-IR spectroscopy, the culture of microorganisms with respect to medium, temperature, time, and fitness of the culture used for inoculation must be carefully controlled. The influence of most of these parameters on identification and discrimination has been investigated in detail for some microorganisms.[2,26,46-48]

Before starting to compile a reference database, reproducibility of strain measurements has to be tested under the conditions chosen. Some organisms have characteristics that may make recording of highly reproducible spectra a difficult task. This is the case, for example, for endospore-forming bacilli, where highly sporulated stock cultures delay growth of the standardized culture. In such cases, a defined preculture helps in standardizing the starting point of the incubation process. In general, when conditions for incubation are kept constant, very high reproducibility can be achieved. Figure 2 depicts a cluster analysis of three independent repetitive measurements

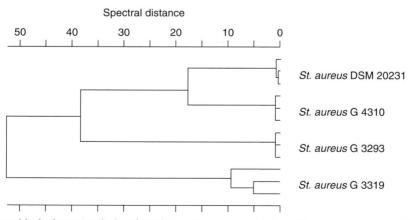

**Figure 2.** Hierarchical cluster analysis of twelve spectra measured from three independent cultivations of four *Staphylococcus aureus* strains, respectively. Spectral ranges used for calculation of spectral distances: 3000–2800, 1800–1500, 1500–1200, 1200–900, and 900–700 cm$^{-1}$. Distances were calculated from first derivative spectra according to equation (2). Average linkage algorithm was used for cluster analysis.

of four different strains of *Staphylococcus aureus* incubated on tryptic soy agar (TSA) for 24 h at 30 °C. The high discriminative power of the technique enables differentiation at the strain level, and spectra of all four strains are well separated with three strains exhibiting very good reproducibility. A fourth strain shown in the bottom cluster exhibits adequate reproducibility but is clearly worse than the others. This dependence of reproducibility on individual strains is frequently observed among microorganisms. Furthermore, different strains react differently to changes in incubation conditions. Oust *et al.*[47] have investigated the effects of small variations in cultivation time and temperature on *Lactobacillus* spectra, as these may occur on a daily basis. Different strains showed different levels of reproducibility, but spectra showed only minor changes that did not affect the overall differentiation performance.

The growth medium, however, may induce significant variation to spectra. Figure 3 displays a comparison of spectra recorded from strains grown on three different media. Four *St. aureus* strains were incubated on brain heart infusion agar (BHIA), Columbia blood agar (CBA) or TSA, at 30 °C for 24 h. As can be clearly seen, the growth medium significantly influences cell composition, thus leading to separate groupings in the cluster analysis. Spectra recorded from cells grown on TSA cluster into different groups than the other spectra, with differences twice as high as between BHIA and CBA. Obviously, organisms express more discriminating features on TSA than on BHIA and CBA, which increases the differentiation capacity. Furthermore, the spectra from strain G 3293 inoculated on CBA show higher similarity to the spectra from strain G 4310 inoculated on CBA than to the spectra of strain G 3293 inoculated on BHIA, indicating that differences induced by different media may exceed spectral differences observed between strains. Similar effects have been demonstrated by Helm *et al.*[2] for staphylococci, van der Mei *et al.*[48] for streptococci, and Oust *et al.* for lactobacilli.[47] The influence of such changes on identification depends on the composition of the database and on the

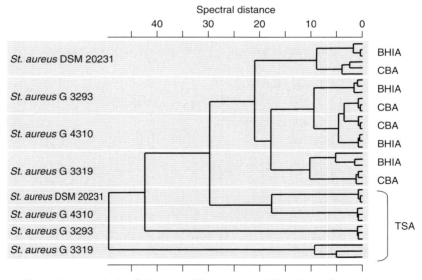

**Figure 3.** Hierarchical cluster analysis of three repetitive spectra of four *Staphylococcus aureus* strains grown on three different media: BHIA, brain heart infusion agar; CBA, columbia blood agar; and TSA, tryptic soy agar. Spectra belonging to the same strain are indicated by shading. Spectral ranges used for calculation of spectral distances: $3000-2800 \, \mathrm{cm}^{-1}$, $1800-1500 \, \mathrm{cm}^{-1}$, $1500-1200 \, \mathrm{cm}^{-1}$, $1200-900 \, \mathrm{cm}^{-1}$, and $900-700 \, \mathrm{cm}^{-1}$. Distances were calculated from first derivatives according to equation 2, average linkage algorithm was used for cluster analysis.

organisms under investigation. The more closely related are the organisms included in the database, the higher is the possibility of false or conflicting identification results, since it becomes more likely that differences induced by the medium may influence species identification.

## 3.2 Choice and identification of reference strains

The main prerequisite for a reliable identification system is a well-composed database that fulfills a number of requirements: (i) it should cover a large number of different species with (ii) a high number of reference strains per species, but (iii) the composition should be well balanced with no species dominating the others, and (iv) all reference strains should be exactly identified.

Since FT-IR spectroscopy distinguishes phenotypes, and since phenotypic variance between strains of the same species—but from different habitats—may be extremely large, differences between strains need to be included in the database in order to ensure reliable identification. In genotypic methods, reference strains from official culture collections like ATCC (American Type Culture Collection), DSMZ (Deutsche Sammlung von Mikroorganismen und Zellkulturen), LMG (Belgian Coordinated Collections of Microorganisms), etc., are often sufficient to set up an identification procedure. FT-IR databases, however, need to include many more strains,[36,38,41] which should be isolated from different habitats. The number of strains per species is called the depth of a spectral reference database. Figure 4 illustrates how the number of strains included in a reference database influences the identification success for the five species of *Listeria*.[38] The same set of 277 strains was identified against an artificial neural network (ANN) trained with 100, 171, and 243 reference strains, respectively. The more were the strains

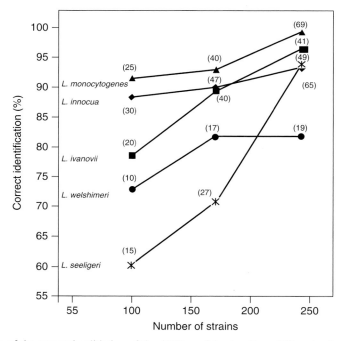

**Figure 4.** Comparison of the external validation of the ANN model using three different reference data sets including 100, 171, and 243 strains. The number of strains per species included in each data set is indicated in parentheses. [Reproduced from Rebuffo *et al.*[38] with permission from the American Society for Microbiology, © 2006.]

included, the higher were the identification accuracies, reaching 99.2% correct identification in the case of *Listeria monocytogenes*. However, when only few strains are available, as it was the case for *Listeria welshimeri*, identification rates leveled off at a relatively low value (around 80%). For microbiologically important species, the availability of strains is usually not a problem, whereas strains from less common species are rare. When compiling a database, the number of strains per species should be comparable for all species included in order to avoid statistical dominance of certain species, since species represented with many more references tend to cause false-positive identifications.

Besides including sufficient numbers of strains per species, the number of species per genus, which is called the width of a spectral reference database, is another important parameter. Although in most applications only a fraction of the existing species is of importance, their close relatives also need to be included in the reference data set. As identification models are trained with given reference data, only very limited or no outlier detection is performed. In order to avoid misidentifications and to ensure identification of even uncommon species, it is pivotal that databases contain rare species as well.

The third, and by far most important, aspect to be considered when generating reference databases is the correct identification of reference strains. Commercial phenotypic identification kits are less suitable for this purpose since they are subject to considerable errors and have been designed for a limited number of species only. Instead, identification of isolates to be used as reference strains requires molecular tools like DNA sequencing, restriction analysis, or specific polymerase chain reaction (PCR) systems. In fact, identification of strains may be the most challenging part in setting up a reference database of IR spectra since it is in many cases complicated by a fuzzy taxonomy. Sooner or later, when dealing with large taxa, such as enterobacteria, microbiologists need to deal with taxonomy, which is in many cases an art of its own. To make things even worse, taxonomy is ever changing. In this respect, one big advantage of FT-IR databases over commercially available systems is the fact that they can be easily adapted to recent changes in taxonomy, such as altered names or the description of novel species. However, if *E. coli* with a special toxin gene is no longer considered *E. coli* but *Shigella*, the limit of FT-IR spectroscopy has probably been reached. However, all microbiological identification systems are limited by taxonomic knowledge and definitions.

## 3.3 Techniques for data analysis

The last step in generating an identification system based on FT-IR spectra is the analysis of the recorded spectra, which is a key factor for successful identification. Unlike in chemistry, where substances may be identified by assignment of their IR absorption band frequencies and their relative intensities, microbial spectra are far too complex to perform visual spectra analysis. Instead, spectra are used as fingerprints and are analyzed by a variety of statistical methods. The difficulty in the data analysis is to reach sufficient generalization in order to attain reliable identification on the species level. This is much more difficult than differentiating one strain from another because species-specific features present in all strains of a species have to be extracted from the total amount of variance. Describing data analysis in detail would go far beyond the scope of this review, but some techniques are introduced as examples. For an overview of techniques frequently used in FT-IR spectroscopy on microorganisms, the reader is referred to recent publications and review articles.[6,49–51]

### 3.3.1 Principal component analysis

The choice of a method for data analysis is dependent on the classification problem. If only a few classes need to be distinguished, principal component analysis (PCA) is a popular and useful technique.[24,47,52] PCA is an unsupervised method requiring no a priori knowledge of class assignment, since data do not have to be fit to a model but are analyzed without any assumption. This makes the method also applicable to data sets with unknown composition; however, for identification purposes known spectra have to be included to assign species names to the groups unraveled. The main goal of PCA is the compression of information into few variance-weighted variables. From the whole data set, only those components are extracted that explain most of the variance. Results are presented for pairs or triplets of components in a two- or three-dimensional PCA map with each point representing one spectrum. If more components are needed to describe a data set, a second map is drawn representing additional dimensions. Figure 5 shows a typical PCA scatter plot for the differentiation of two classes. One hundred and forty-three spectra acquired from 13 different species and strains belonging to both gram-positive and gram-negative microorganisms were analyzed in the spectral ranges between 3000–2800 and 1500–1400 cm$^{-1}$, which are known to comprise information on membrane lipids. In the scatter plot, two clusters can be differentiated with each one comprising exclusively spectra from either gram-positive or gram-negative organisms.

PCA maps are a comfortable way of getting an overview of the structure of data sets; however, as data are presented in a two-dimensional way, PCA is difficult to apply to complex identification problems that would require multidimensional maps. Alternatively, PCA can be used for data reduction prior to application of other techniques.[17,53]

### 3.3.2 Cluster analysis

Hierarchical clustering methods (see **Infrared and Raman Microspectroscopic Studies of Individual Human Cells**) also analyze intrinsic group structures within large and mixed data sets. These unsupervised techniques do not need any a priori knowledge of class assignment or portioning of the data into a training and test data set. The history of hierarchical cluster analysis is

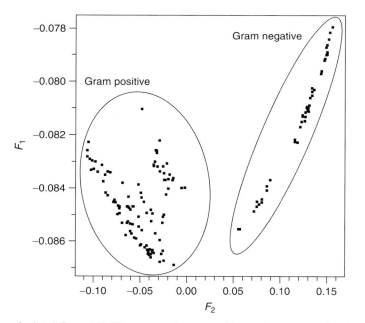

**Figure 5.** PCA map calculated from 143 IR spectra of gram-positive and gram-negative bacteria from repetitive measurements on 13 different species and strains. For projection of data, the factorial coordinates (factor loadings) $F_1$ and $F_2$ were used. The normalized first derivatives in the spectral ranges of 2800–3000 cm$^{-1}$ and 1400–1500 cm$^{-1}$ were used as input data.

generally represented by a minimal spanning tree, also called *dendrogram*, by which the merging process of classes can be visually followed. Adequate distance measures for calculating the similarity/dissimilarity between the spectra and to obtain the distance matrix as input for hierarchical cluster analysis are, for example, the Euclidian or Mahalanobis distance measures, or the Pearson's product momentum correlation coefficient (see below). The decision on the optimal distance measure, as well as the optimal clustering algorithm, has to be made experimentally. Ward's and average linkage algorithms have frequently been used in the literature and may provide satisfactory results.[19,31,32,39,54] The classification schemes obtained cannot provide an objective criterion of best class portioning. Thus, the operator has to determine the number of classes on the basis of some a priori knowledge about the inherent class structure in the data set.

### 3.3.3 Identification of unknown samples by distance measures

One simple possibility of identifying unknown strains by comparison with spectral reference databases of any size is to use a distance measure. For this purpose, Helm and coworkers[1] have defined a measure of dissimilarity, which they called D-value. For each pair of unknown and reference spectra of the library, the Pearson's product moment correlation coefficient, defined as,

$$r_{y1y2} = \frac{\sum\limits_{i=1}^{n} y_{1i} \cdot y_{2i} - n \cdot \overline{y_1} \cdot \overline{y_2}}{\sqrt{\sum\limits_{i=1}^{n} y_{1i}^2 - n \cdot \overline{y_1}^2} \cdot \sqrt{\sum\limits_{i=1}^{n} y_{2i}^2 - n \cdot \overline{y_2}^2}} \tag{1}$$

is calculated. In case of IR spectra, $y_{1i}$ and $y_{2i}$ are the individual absorbance values at discrete wavenumbers of the two spectra to be compared; $n$ is the number of data points in the given wavenumber range; and $\overline{y_1}$ and $\overline{y_2}$ are the arithmetic mean values of $y_1$ and $y_2$.

From the correlation coefficient $r_{y1y2}$, the so-called differentiation index, or spectral distance, $D$ is calculated where

$$D = (1 - r_{y1y2})1000 \tag{2}$$

$D$ may adopt values between 0 and 2000, with 0 for identical spectra (or spectral ranges), 1000 for completely noncorrelated, and 2000 for completely negatively noncorrelated spectra. As a result of the identification procedure, a list comprising the most similar spectra found in the database is reported with calculated D-values serving as distance measures. Table 2 displays such an identification result for an unknown *Staphylococcus* strain. The 12 most similar spectra from the database all belong to different *St. aureus* strains and are listed by increasing D-values. The very low D-values indicate high similarity between the query spectrum and the reference spectra.

For this kind of analysis, a database simply consists of a collection of reference spectra and no calibration is performed in advance of the identification. Hence, a very important step in the procedure is to define the spectral regions used for calculation of the D-values. Often the whole spectral range does not necessarily produce the

**Table 2.** Hit list of an identification report based on calculation of D-values (OPUS software, Bruker, Ettlingen, Germany). Spectral ranges used for calculation were 3000–2800, 1800–1500, 1500–1200, 1200–900, and 900–700 cm$^{-1}$. Distances were calculated from first derivatives with scaling to first range (3000–2800 cm$^{-1}$).

| Hit number | D-value | Sample name |
|---|---|---|
| 1 | 0.684 | *Staphylococcus aureus* |
| 2 | 0.765 | *Staphylococcus aureus* |
| 3 | 0.783 | *Staphylococcus aureus* |
| 4 | 0.817 | *Staphylococcus aureus* |
| 5 | 0.876 | *Staphylococcus aureus* |
| 6 | 0.889 | *Staphylococcus aureus* |
| 7 | 0.896 | *Staphylococcus aureus* |
| 8 | 0.946 | *Staphylococcus aureus* |
| 9 | 0.972 | *Staphylococcus aureus* |
| 10 | 0.974 | *Staphylococcus aureus* |
| 11 | 0.987 | *Staphylococcus aureus* |
| 12 | 0.993 | *Staphylococcus aureus* |

best identification results because parts of the spectra contain nondiscriminating information, which reduces the identification performance. It is, therefore, advisable to only analyze reference spectra for those spectral regions that are most discriminative. The multiclass problems are almost impossible to do by eye. Helm *et al.*[1] have described a stepwise approach to determine the contribution of the different spectral windows (see Figure 1) to the identification performance and to find the best combination of these.

For users without a detailed knowledge of statistics, application of this method may be the first encounter with databases. This algorithm is embedded in the OPUS software package (Bruker Optics, Ettlingen, Germany) provided with a spectrometer that is dedicated to the identification of bacteria; no additional software is needed for application in this instrument. Furthermore, as all spectra serving as references are merely compiled with no calibration process involved, databases can be extended by simply adding spectra to the reference data set. Even for very large databases comprising high numbers of different species, the calculation of D-values usually gives good results.[32,36,38] Kümmerle *et al.*[32] constructed a spectral data library for yeasts comprising 322 strains from 73 species and achieved 97.5% identification accuracy in a study with 722 unknown yeast strains isolated from food samples. Oberreuter *et al.*[36] have compiled a database for different species of actinomycetous genera with 544 reference strains from 54 species and obtained 87% identification accuracy using the leave-one-out method. However, although identification rates are high with this method, the potential of FT-IR spectra still has not been fully exploited, because a large amount of information may be lost during the process of averaging all differences into only one distance measure. This procedure may lead to difficulties in discriminating closely related organisms and, since the same spectral regions are used for identification of all species included, there is no flexibility in the system with respect to complex identification problems.

### 3.3.4 Identification of unknown samples by artificial neural networks

Much more flexibility is provided by ANNs, a technique belonging to the systems of artificial intelligence. ANNs undergo a learning process where training data are used to calibrate the model. During this iterative and supervised learning process, weights of input variables are changed so that input data (absorption values at specific wavenumbers) are related to predefined output data (classes, e.g., species). In advance, a reduction of input data by covariance analysis, PCA, or related techniques is required. If the number of changeable weights exceeds the number of samples, supervised methods tend to overfit the model.[55,56] This means that systematic (class specific) differences between classes are not determined because too much information is put into the calibration process. Instead, the model memorizes differences between the spectra used for calibration, which may lead to difficulties when unknown spectra are to be classified. Thus, feature selection and data reduction are important steps in the development of optimized ANNs. In FT-IR spectroscopy, this implies searching for, and selecting absorbance values at, those wavenumbers that contribute maximally to the discrimination between the objects in predefined classes. This has to be done prior to the training step by independent data analysis using dedicated feature selection software algorithms.[55,57]

Through the application of ANNs, complex identification problems can be solved by dividing the identification procedure into several steps, resulting in an architecture of several ANNs used as modules to form a hierarchical identification system. Each ANN is trained with its own selected spectral features as input data and therefore addresses specifically the differentiation of the given classes. Figure 6 illustrates an ANN divided into four levels.[17] The model was set up to identify medically relevant organisms from blood cultures of typical clinical settings and included 89 strains of gram-negative bacteria from the genera *Enterobacter*, *Escherichia*, and *Pseudomonas* as well as gram-positive cocci from the

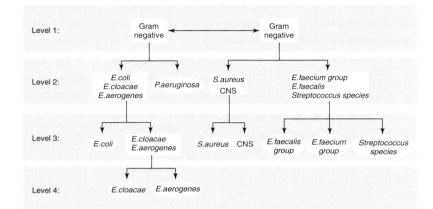

**Figure 6.** Artificial neural network classification scheme for the discrimination of gram-positive and gram-negative bacteria. The identification procedure was split into four hierarchically organized levels. Spectral ranges used as input data were 780–1200 and 2800–3000 cm$^{-1}$ for the first level and 780–1500 cm$^{-1}$ for levels two to four.

three genera *Staphylococcus, Enterococcus,* and *Streptococcus.* In a prospective case study, the authors were able to correctly identify 112 from 114 (98.2%) isolates from clinical origin using their ANN model.

The power of ANNs has been recognized by many groups and ANNs are increasingly used for differentiation problems in microbiology.[17,28,38,55,56,58–60] Udelhoven *et al.*[55] developed an ANN to classify bacilli, staphylococci, pseudomonads, and *Candida* at the species level and to identify fluconazole-resistant or sensitive strains of *Candida albicans.* In total, their ANN consisted of three layers with the top-level differentiating between the different genera, the second layer determining species identity within the genera, and the third level fluconazole-resistant or sensitive strains of *C. albicans.*

The superiority of ANNs over other methods was recently shown by Rebuffo *et al.*[38] They developed an ANN for the identification of *Listeria* including 243 reference strains from five species. Owing to the close relationship between the species an identification of 277 unknown strains applying the "D-value approach" resulted in only 85.2% correct classifications. An optimized two-level ANN, however, assigned 96.0% of the unknown spectra to the correct classes. For

the human pathogen *L. monocytogenes,* an accuracy as high as 99.2% was achieved.

## 3.4 Validation of reference databases

After having established an identification system, detailed assessment of its accuracy is an important prerequisite for routine application, as high reliability is essential for medical applications. Furthermore, the correctness of the resulting models needs to be checked in order to avoid artifacts such as overfitting produced by the statistical techniques used.

For this purpose, two validation types must be distinguished: (i) internal validation and (ii) external validation. For internal validation, new spectra of strains contained in the reference data set are used. Here, the only difference between spectra used for calibration and for validation is the variation introduced by cultivation and sample preparation and, thus, identification rates should be high. Otherwise, if identification accuracy in the internal validation is low, the robustness of the model is insufficient and it needs modification.

More significant than the internal validation is an external validation, since spectra of unknown strains not included in the reference data set are used as independent test cases. In addition to the variance of sample cultivation, preparation, and measurement, spectra of strains unknown to the model represent intraspecies biological variance of phenotypic characters. Hence, identification accuracy is usually lower than is obtained by internal validation. Thus, the identification accuracy expected under practical conditions can be predicted. When optimizing the model, the biggest problem in executing an independent external validation is in most cases the lack of a sufficient number of strains. Alternatively, the leave-one-out method can be performed, where one strain is left out from the calibration and is used for validation. This is done for all strains individually and the identification accuracy can be calculated as percentage of correct assignments.

## 4   APPLICATION OF REFERENCE DATABASES

One valuable advantage of FT-IR spectroscopy in microbiology is the fact that microbial IR patterns may be analyzed for the solution of several different problems. Unlike conventional phenotypic characterization or molecular differentiation techniques where each characteristic needs to be determined by a separate test or analysis, FT-IR spectra express spectral features of all biochemical cellular constituents present in the cell, and the same spectrum can be used to determine several properties of the microorganism.

Needless to say, the objective of clinical microbiology in the first place is the identification of unknown microorganisms at the species level. But, if extensive spectral libraries are available that comprise organisms from many different taxa, this goal can be achieved by a step-wise identification procedure,[17,55] which can also replace microscopic examination of the samples and basic phenotypical tests. Besides determining the Gram behavior,[2,6,61] further differentiation into families like *Enterobacteriaceae* or *Pseudomonadaceae*[2] can substitute the Gram stain and test for the presence of cytochrome oxidase. Similarly, differentiation of gram-positive cocci according to G + C (guanine + cytosine) content in their DNA or catalase reaction is comparably simple. Figure 7 depicts a dendrogram calculated from spectra of 22 strains of the

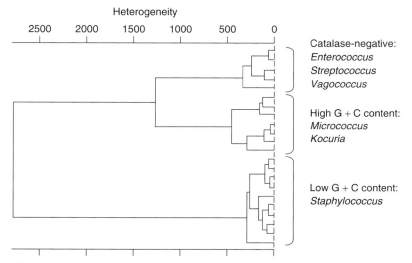

**Figure 7.**   Hierarchical cluster analysis calculated from 22 gram-positive strains. Three spectra per strain were taken from three independent cultivations. For calculating the dendrogram, first derivatives and the spectral ranges from 3000–2800, 1800–1500, 1500–1200, 1200–900, and 900–700 cm$^{-1}$ were used. Ward's algorithm was used for cluster analysis using the D-value matrix as input data for cluster analysis. Heterogeneity is a similarity factor as defined by Ward. [J.H. Ward, *J. Am. Stat. Ass.*, **58**, 236–244.] Heterogeneity is defined in the Glossary.

catalase-negative (facultative anaerobic) genera *Enterococcus, Streptococcus,* and *Vagococcus* as well as catalase-positive (aerobic) genera *Micrococcus, Kocuria,* and *Staphylococcus,* which can be further differentiated phylogenetically into high G + C (*Micrococcus, Kocuria*) and low G + C (*Staphylococcus*) gram-positives. The discrimination in Figure 7 was achieved by hierarchical cluster analysis without selecting any specific spectral ranges and is therefore quite an easy task to be solved by reference databases.

Another possible application of reference databases is the recognition of spore-forming bacilli by the detection of specific spectral markers

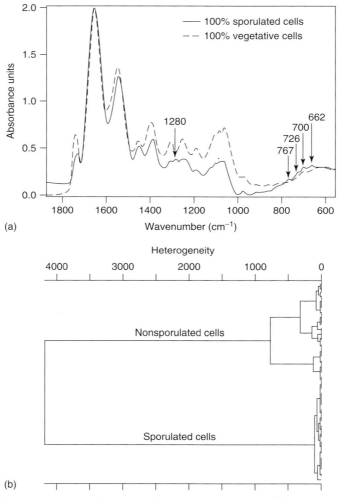

**Figure 8.** Detection of spore-forming bacilli by specific spectral markers and differentiation of bacilli with high and low degree of sporulation (M. Fricker, Zentralinstitut für Ernährungs- und Lebensmittelforschung, unpublished data). (a) Transmission spectra of vegetative cells (dotted line) and highly sporulated cells (solid line) of *B. cereus* normalized to a maximum absorption of 2 AU. Spectral features specific for endospore formation are indicated by arrows with corresponding wavenumbers. (b) Differentiation of *B. cereus* strains with high and low degree of sporulation by cluster analysis. For calculating the dendrogram, first derivatives and the spectral ranges from 1500–1200 cm$^{-1}$ and 900–650 cm$^{-1}$ were used. Ward's algorithm was used for cluster analysis.

for endospores.[62,63] Figure 8(a) depicts FT-IR transmission spectra of sporulated and vegetative *Bacillus cereus* cells. Several bands that are characteristic of endospores can easily be seen by eye and may be used for (i) the recognition of spore formers and (ii) the differentiation between sporulated and nonsporulated cells as demonstrated in Figure 8(b). The cluster analysis in Figure 8(b) was performed with four independent measurements of 12 *B. cereus* strains, 6 of which show a high degree of sporulation after 24-h incubation and 6 a low degree of sporulation. For calculation of the distances, only the spectral regions from $1500-1200\,\mathrm{cm}^{-1}$ and $900-650\,\mathrm{cm}^{-1}$ exhibiting the most discriminating features for endospores have been taken into consideration, leading to a very clear separation of sporulated and nonsporulated cells.

Several databases for the identification of microorganisms are already available. Bruker Optics offers data sets for Gram testing and for identification of water-borne microorganisms. Furthermore, a number of libraries for identification of *Listeria*, bacilli, staphylococci, actinomycetous genera, *Enterobacteriaceae, Pseudomonadaceae*, lactic acid bacteria, and yeasts comprising more than 700 different species have been compiled by the Department of Microbiology, Technical University of Munich (http://www.wzw.tum.de/micbio) and can be purchased from Bruker Optics.

Even though it is a valuable technique for identification of microorganisms with pronounced differentiation capacity, FT-IR spectroscopy should be used as a taxonomic tool only with caution. Oberreuter *et al.*[64] have investigated intraspecies diversity among numerous strains of three species of *Corynebacterium, Rhodococcus*, and *Brevibacterium* using FT-IR spectroscopy and 16S rDNA sequence data. They found that species with high similarity in 16S sequences could differ markedly in their FT-IR spectra and vice versa. Hence, application of FT-IR spectroscopy for taxonomic studies should always be accompanied by molecular analyses.

# 5 TYPING OF MICROORGANISMS

Identification of microorganisms at the species level is an important task in microbiology. However, for epidemiological studies and analysis of contamination routes, subtyping of microbes below the species level with respect to serovar,[2,65] genotype,[59,66] or resistance to certain antibiotics or bacteriocins[22] is required.

## 5.1 Serotyping

For the ubiquitous human food pathogen *L. monocytogenes*, 13 serovars are known, of which only a few are of clinical relevance. To facilitate serotyping, Rebuffo-Scheer *et al.*[65] have developed a serotyping scheme based on microbial FT-IR spectra and ANNs. Spectra from 106 *L. monocytogenes* strains covering all known serovars were used to construct the model illustrated in Figure 9. Differences in the structure of teichoic acids result in specific spectral patterns in the carbohydrate region that can be used for differentiation of the four serogroups 1/2, 3, 4, and 7 (Figure 10). In the internal validation at the serogroup level, 100% of the spectra were assigned to the correct serogroup. At the serovar level, 94.3% correctly classified spectra could be obtained. An external validation using 166 strains from 12 serovars achieved 98.8% correct classification at the serogroup, and 91.6% at the serovar level. Combining the ANN model for serotyping with the previously established ANN model for species identification[38] enables the identification of *Listeria* species and subtyping of *L. monocytogenes* strains using only one FT-IR spectrum. No other technique available to date is able to provide similarly detailed results from a single measurement within such a short time.

Previous studies[2,67] had already indicated that different antigenic structures could be revealed by FT-IR spectroscopy. Helm *et al.*[2] analyzed 23 strains of *E. coli* belonging to three different

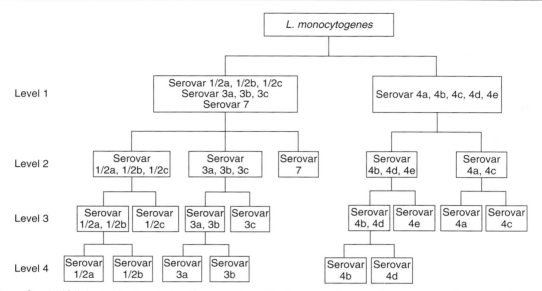

**Figure 9.** Artificial neural network classification scheme for the discrimination of serogroups and serovars of *Listeria monocytogenes*. The identification procedure was divided into four steps. The first two steps classify spectra into four serogroups 1/2, 3, 4, and 7, and by the third and fourth step differentiation to serovar level is obtained.

O-antigenic groups (O18, O25, O114) by hierarchical cluster analysis considering the extended carbohydrate region from 1200 to 700 cm$^{-1}$. The spectra were grouped into three distinct clusters each containing exclusively spectra of one particular serogroup (Figure 11). Similar results have been recently obtained from an FT-IR study on the relationship between O-antigen subtypes in *E. coli* O123 and *E. coli* O4 strains carrying genes for Shiga toxins and intimin.[68]

Kim *et al.*[67] undertook serotyping of six *Salmonella enterica* strains of different serovars using both whole cells and LPS extracts. Applying canonical variate analysis (CVA) on the whole spectrum as well as solely on the polysaccharide region of the spectrum, they were able to distinguish the spectra of all six LPS extracts, although spectra of whole cells did not express enough differences to obtain discrimination by CVA. However, contrary to these findings, Baldauf *et al.*[69] were able to separate different *S. enterica* serovars based on whole cell FT-IR spectra.

Recently, Kirkwood *et al.*[70] investigated *Clostridium botulinum* strains belonging to group I and

II, producing neurotoxin of type A, B, E, and F. Hierarchical cluster analysis clearly distinguished between group I and II strains, and within each group subclusters were formed corresponding to the different toxin types.

## 5.2 Differentiation of genotypes

For organisms where no serological typing schemes exist or serotyping is not sufficient, differentiation by molecular techniques like randomly amplified polymorphic DNA (RAPD) analysis, enterobacterial repetitive intergenic consensus (ERIC) PCR or pulsed field gel electrophoresis (PFGE) is common. However, as molecular methods are less suitable for routine application, there is a demand for simple and cost-efficient alternatives. Several reports describe the use of FT-IR spectroscopy as a typing method.[59,66,71] Mouwen *et al.*[59] applied FT-IR spectroscopy in combination with ANNs to perform subtyping of *Campylobacter jejuni* and *C. coli* strains according to four different ERIC-PCR patterns. As in most other studies, these

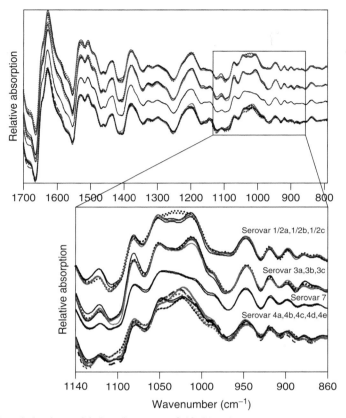

**Figure 10.** Typical first derivatives of infrared spectra of 12 *Listeria monocytogenes* serovars. Black dotted line, 1/2a; black solid line, 1/2b; gray solid line, 1/2c; black dotted line, 3a; black solid line, 3b; gray solid line, 3c; black dotted line, 4a; black solid line, 4b; gray solid line, 4c; gray dotted line, 4d; black dashed line, 4e; black solid line, 7. Each strain is represented by two independent spectra to show the reproducibility of the measurements. Spectra are stacked to clearly show the spectral differences among the four serogroups (1/2, 3, 7, and 4). The polysaccharide region ($1,200$–$900 \, cm^{-1}$) has been expanded to show the subtle differences between and within the serogroups. [Reproduced from Rebuffo-Scheer *et al.*[39] with permission from the American Society for Microbiology, © 2007.]

authors used the so-called carbohydrate region between 1200 and $900 \, cm^{-1}$ of the microbial IR spectra for discrimination, confirming that this region has a particular potential for subtyping studies. The resulting ANN was internally validated, achieving 99.16% correct classification.

Sandt *et al.*[66] analyzed 79 *C. albicans* strains isolated from nine intensive care unit patients over a 4-month period by FT-IR spectroscopy and RAPD. RAPD profiles revealed that each patient was colonized by one individual strain and that strains from different patients had

different patterns. Cluster analysis over the whole spectral range except for the lipid ester region ($1760$–$1720 \, cm^{-1}$) performed with several FT-IR spectra for each strain could clearly differentiate all nine strains with only one spectrum grouped to another strain.

Seltmann *et al.*[71] employed FT-IR spectroscopy for the epidemiological analysis of 135 *S. enterica* strains of the same sero- and phage type. Hierarchical cluster analysis on the basis of the spectral range between 1185 and $1120 \, cm^{-1}$ revealed two clearly separated groups.

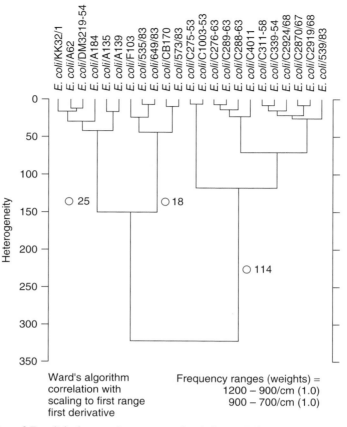

**Figure 11.** FT-IR typing of *E. coli* isolates at the serogroup level. Spectral distances (D-values) were calculated from first derivatives in the spectral ranges 900–1200 cm$^{-1}$ and 700–900 cm$^{-1}$. For calculating the dendrogram, the Ward's algorithm was used.

## 5.3 Determination of resistance or susceptibility to antibiotics

In clinical microbiology, antibiotic susceptibility testing is in many cases as important as the identification of the microorganism itself, since efficient and early antimicrobial therapy may be life saving. The conventional way of testing resistance to antibiotics is the disk diffusion assay, where inhibition zones of microbial growth on solid media around little tablets coated with an antibiotic are determined. Furthermore, automated systems for phenotypic identification of microorganisms also offer the possibility of simultaneous antibiotic susceptibility testing.

Although not very much has been published on the application of FT-IR spectroscopy to determine resistance or susceptibility of microorganisms to certain antibiotics, some early results have documented the potential of this technique.[4,6,18,72,73] Bouhedja *et al.*,[72] Sockalingum *et al.*,[73] and Naumann[6] have demonstrated that, even in untreated cells (grown in absence of the antibiotic) of *E. coli, Pseudomonas aeruginosa,* and *C. albicans*, differences between resistant and susceptible strains can be observed and exploited for differentiation. Naumann[6] studied 14 *C. albicans* strains sensitive or susceptible to fluconazole. Including the information of only two narrow ranges of the fingerprint region into the analysis, they were able to distinguish sensitive from resistant strains. Alternatively, Naumann *et al.*[4] and Ngo Thi *et al.*[18] compared control cells without treatment with cells grown in the presence of the antibiotic in order to gain knowledge on the molecular changes triggered by antibiotic treatment. Naumann *et al.*[4] investigated changes in *St. aureus* cells induced by chloramphenicol. They found several peaks with reduced intensities due to inhibition of protein synthesis and increased intensity because of accumulating sugar compounds, most probably cell wall components. Ngo Thi *et al.*[18] applied FT-IR microspectroscopy in combination with the disk diffusion assay to analyze effects of oxacillin on one sensitive and one resistant strain of *St. aureus*. The advantage of microspectroscopy in this experimental set up was that from the same agar plate microcolonies could be sampled that were exposed to different oxacillin concentrations due to the gradient of antibiotic from the center toward the edge of the Petri dish. As a result, differences between spectra from cells grown under influence or absence of the antibiotic could be extracted from spectra obtained from cells grown on the same agar plate, which facilitated the detection of slight spectral changes. In the polysaccharide region, the authors were able to identify two bands that could serve as marker bands to monitor the effect of oxacillin.

# 6 FT-IR MICROSPECTROSCOPY

In clinical microbiology, time matters. The faster an organism can be identified, the sooner appropriate antimicrobial therapy can be given to the patient to accelerate recovery. This leads to decreasing mortality rates and decreasing variable costs for the hospital,[74] an aspect worthy of consideration in times of increasingly limited health budgets.

Therefore, FT-IR microspectroscopy, i.e., the combination of microscopy and FT-IR spectroscopy, is a promising tool for identifying microbes in the clinical microbiological laboratory. Improved spectroscopic technology with

higher spatial resolution and sensitive liquid-nitrogen-cooled MCT detectors made it possible to record high quality spectra from very small amounts of samples.[75,76] Unlike with conventional spectroscopy, microorganisms analyzed by microspectroscopic techniques need an incubation step of only a few hours, since microcolonies between 60 and 100 μm in diameter provide a sufficient amount of biomass for the FT-IR measurement.[16,27,40,77] Prior to the measurement, microcolonies are transferred from the solid growth medium to an IR transparent sample carrier by means of a stamping device.[18] In Figure 12, several of such imprints of microcolonies from bacteria and yeasts are displayed.

During the last few years, several reports have been published reporting the characterization of microcolonies[16,18,54] and identification and typing of yeast and bacteria[17,19,27,31,39–41,78,79] via FT-IR microspectroscopy. The technique seems to be particularly advantageous where a significant reduction in cultivation time is possible. Rebuffo-Scheer *et al.*[39] have analyzed typical slow-growing microorganisms like mycobacteria using FT-IR microspectroscopy and were able to reduce culture time from 10–15 days to 40–50 h for faster growing species and from 4–6 weeks to 10–14 days for slow-growing species. After this considerably reduced cultivation time, microcolonies had reached diameters of at least 80 μm allowing the measurement of high quality spectra. Ten *Mycobacterium* species represented by 28 strains were differentiated on the basis of hierarchical cluster analysis and even species hard to separate by other methods, e.g., *Mycobacterium avium* and *M. intracellulare*, could be discriminated.[39]

Another advantage of FT-IR microspectroscopy is the possibility of analyzing mixed populations without the need to isolate and purify the cultures as nutrient agar plates from dilution plating can be used for replica stamping. A first application was published by Wenning *et al.*,[19] who investigated the composition of microbial cheese consortia. After having established a reference database comprising 18 species with relevance to surface-ripened cheeses, population analyses

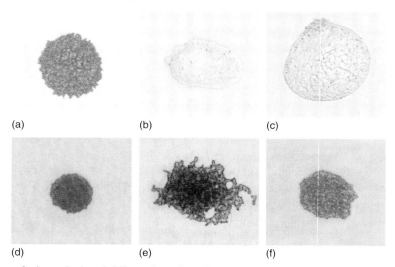

**Figure 12.** Shapes of microcolonies of different bacteria and yeast species after replica stamping. (a) *Corynebacterium casei*; (b) *Microbacterium gubbeenense*; (c) *Corynebacterium ammoniagenes*; (d) *Torulaspora delbrueckii*; (e) *Candida intermedia*; and (f) *Rhodotorula mucilaginosa*.

were performed on two different consortia. Since around 200 spectra could be recorded per day, a quantitative analysis of a mixed population otherwise requiring at least 1 week could be accomplished within 1 day.

# 7 CONVENTIONAL FT-IR OR FT-IR MICROSPECTROSCOPY?

The decision on which technique to use strongly depends on the problem to be solved. FT-IR microspectroscopy certainly has several advantages over conventional FT-IR spectroscopy, especially the decreased incubation time and the opportunity to rapidly analyze mixed populations. But the fundamental question to be clarified is the accuracy of microspectroscopic characterizations based on microcolonies in comparison to conventional spectroscopy.

Undoubtedly, the sensitivity of FT-IR microspectroscopy as applied to microbial microcolonies is comparable to conventional spectroscopy using much larger amounts of sample, as both methods are able to discriminate microbes at the strain level.[2,39,41,60,77,79] However, as far as

identification accuracy is concerned, the picture is not so clear, since conflicting results have been reported.[27,40,41] Wenning *et al.*[41] and Essendoubi *et al.*[27] applied FT-IR microspectroscopy for the identification of yeasts and found both methods to be suitable approaches for these microorganisms. Sandt *et al.*[40] set up an identification scheme for clinically relevant gram-positive and gram-negative bacteria using microspectroscopy. They achieved good identification results for the gram-positive species; however, identification of gram-negative microorganisms proved to be difficult. By comparing spectra of 6–10 h old microcolonies to spectra recorded with conventional spectroscopy after 24 h incubation, they revealed that spectra from 24-h cultures exhibited more complex patterns especially in the carbohydrate region, whereas microcolony spectra were less complex leading to less spectral interspecies diversity. For the generation of a reference database, one would thus expect that identification accuracy with FT-IR microspectroscopy would be lower than with conventional spectroscopy, which is exactly what Sandt *et al.*[40] claim to have found, although they did not test this hypothesis systematically.

A more systematic comparison was performed by Rebuffo-Scheer *et al.*[80] who evaluated the performance of conventional and microspectroscopy with a set of 25 *Listeria* strains. For the five species, *L. innocua, L. ivanovii, L. monocytogenes, L. seeligeri,* and *L. welshimeri,* a database was compiled applying both conventional and microspectroscopy. For measurement with conventional FT-IR spectroscopy, cells were cultivated 24 h at 30 °C, for FT-IR microspectroscopy 18 h at 25 °C proved to be optimal. Five strains per species were included in the reference data set, and an internal validation was performed using the method of calculating D-values. The overall identification accuracy for conventional spectroscopy was 92.8%, whereas FT-IR microspectroscopy identified only 79.2% of the spectra. As one reason for this large discrepancy Rebuffo-Scheer *et al.*[80] found that spectra of older cells grown 24 h for conventional spectroscopy showed more discriminative features in the polysaccharide region than young cells from microcolonies, confirming the findings of Sandt *et al.*[40] Interestingly, clear differences in identification accuracy observed when identification performances were compared for each species separately. Conventional spectroscopy proved to be clearly superior for *L. innocua, L. ivanovii,* and *L. monocytogenes,* while for *L. seeligeri* and *L. welshimeri* FT-IR microspectroscopy yielded slightly better results. Rebuffo-Scheer

*et al.* postulate that microcolony heterogeneity is a potential cause for this difference. They performed mapping measurements across microcolonies of all five species according to a procedure described elsewhere[39] and determined the microheterogeneity within microcolonies by hierarchical cluster analysis. While the heterogeneity values for *L. innocua, L. ivanovii,* and *L. monocytogenes* were relatively high, indicating larger differences in the biochemical composition between various zones of the microcolonies, *L. seeligeri* and *L. welshimeri* generally exhibited considerably lower heterogeneity values. This is exemplified in Figure 13, which depicts cluster analyses of mapping measurements across colonies of 120 µm in diameter of *L. welshimeri* (a) and *L. ivanovii* (b). The spectra of *L. welshimeri* show a considerably lower heterogeneity than spectra from *L. ivanovii.*

As all five *Listeria* species are genetically closely related, it seems that the higher heterogeneity of microcolonies of *L. innocua, L. ivanovii,* and *L. monocytogenes* contributes to a decreased identification accuracy of microcolony spectra. Possibly, results could be improved by applying multivariate statistical tools for data analysis; however, taking into consideration that young cells exhibit fewer spectral characteristics than older cells, it seems to be unlikely that the discrepancy of more than 13% can be compensated.

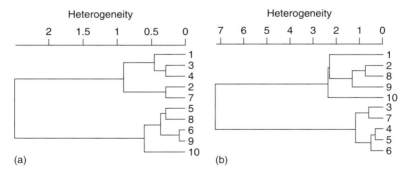

**Figure 13.** Cluster analyses of mapping measurements across microcolonies of 120 µm diameter of (a) *L. welshimeri* and (b) *L. ivanovii* performed by FT-IR microspectroscopy. Spectra are numbered from position 1 (first edge of the colony) to position 10 (opposite edge of the colony). For calculating the dendrogram, first derivatives and the spectral range from 1800 to 900 cm$^{-1}$ were used. The algorithm applied was Ward's algorithm with scaling to first range.

From the data published so far, it can be concluded that FT-IR microspectroscopy is a valuable tool in the identification of microorganisms and has speeded up identification procedures. However, it appears that for some organisms its application may be less suitable. In particular, very closely related species like *Listeria* or enterobacteria have proven to be difficult to distinguish. Although excellent results for the identification of clinically relevant bacteria applying microspectroscopy have been reported,[17] there were merely four gram-negative species included in the study of which only two belonged to the same genus. Thus, more studies are needed to finally evaluate the potential of FT-IR microspectroscopy in comparison to conventional FT-IR spectroscopy concerning the identification of closely related species. Nevertheless, many studies have proven that FT-IR microspectroscopy is undoubtedly a powerful technique in the identification of many organisms. Very good results have been obtained for yeasts and a variety of gram-positive cocci or mycobacteria, and the possibility to analyze mixed populations is unique among all phenotypic methods.

# 8   PERSPECTIVES OF THE IR TECHNIQUE: NEW DEVELOPMENTS

The use of mid-IR spectroscopy for microbial characterization, differentiation, and identification is presently the best developed and most frequent application of biomedical FT-IR spectroscopy. It is especially remarkable that FT-IR technologies are applied in microbiological laboratories not only for research purposes but also for routine analysis at this time. In the food industry, it is already used for microbiological quality control to guide adequate production measures. This situation has been greatly promoted by dedicated high-throughput FT-IR instrumentation and data-processing software available in the market nowadays.

One particularly attractive aspect of vibrational spectroscopy in microbiology is the possibility of achieving subspecies differentiation and the ability to analyze all kind of cells that can be grown in culture. No other technique is currently available that is able to achieve subspecies differentiations and trace microbiological contaminations, e.g., in food microbiology, or to perform epidemiological investigation in clinical microbiology similarly quickly and easily. It is interesting to note that this potential is currently being evaluated in several laboratories and dedicated instrumentations are being designed for microbial subspecies differentiations in collaboration with industrial partners.

New avenues of microbiological applications can be expected from the use of IR microscopes, whether it is for the analysis of mixed populations or to speed up the identification of microorganisms. This technology may not only help in scaling down the number of cells needed for analysis and to investigate mixed cultures but also to detect light-microscopic and spectroscopic features simultaneously. However, a fully automated IR microscopic system that combines detection, enumeration, and identification of microorganisms has still to be developed.

A novel application for FT-IR microspectroscopy in microbiology was published in 2004 by Kirkwood *et al.*[81] who combined FT-IR microspectroscopy with focal-plane-array (FPA) detectors and microarray printing of samples. FPA detectors are usually employed in histology for mapping measurements of tissue sections and are assemblies of multiple single detectors covering pixel areas of $16 \times 16$, $32 \times 32$, $64 \times 64$, or higher, generating many hundred spectra simultaneously. Microarray printing carried out by highly automated robots originates from molecular biology and permits the deposition of nanoliter-volume droplets of microbial suspensions onto IR transparent microscopic slides. Owing to the little space required by the small sample amount, several hundred spots can be placed onto one single slide and measured automatically.

Probably the whole arsenal of multivariate bioinformatic techniques has been used by the

FT-IR spectroscopic community, and multivariate statistical analysis of spectroscopic data constitutes a new discipline of its own within the scientific area of biomedical spectroscopy. As for any other scientific discipline, these techniques can not only be used to better evaluate existing data sets but also to address completely new problem solutions. The prospect of completely new applications arose when it was realized that determining structural relations between different large data matrices obtained from the same sample object, but with fundamentally different techniques, is not only a challenge per se but will also provide completely new insights into the interlink between biological structures that are not easily achieved by any other means. One of these new applications recently published[82] was the use of genetic algorithms in combination with partial least squares regression analysis to correlate genes selected from gene expression profiles of the genome to given metabolic markers from spectral data sets obtained by FT-IR spectroscopy. The analysis of covariance patterns in these very complex mixed data sets helped in rapidly recognizing and visualizing the interrelationships and trends in a developing and changing biological system.

# ACKNOWLEDGMENTS

This work was partly supported by the FEI (Forschungskreis der Ernährungsindustrie e.V., Bonn), the AiF (Arbeitskreis für industrielle Forschung), and the Ministry of Economics and Technology, project No. 14126N. Furthermore, the generous support of this work by the Robert Koch-Institut and the former Ministry of Research and Technology is gratefully acknowledged. The technical assistance of Angelika Brauer in preparing some figures is acknowledged. Many people who have contributed to the work in the laboratories of the authors are acknowledged in the original literature. The authors wish to thank M. Fricker for kindly contributing unpublished data.

# END NOTE

a. The physiological state is strongly, but not exclusively, connected to the growth phase. It is also influenced by the kind of nutrients supplied, atmosphere, etc.

# ABBREVIATIONS AND ACRONYMS

| | |
|---|---|
| ANN | Artificial Neural Network |
| ATCC | American Type Culture Collection |
| ATR | Attenuated Total Reflection |
| BHIA | Brain Heart Infusion Agar |
| CBA | Columbia Blood Agar |
| CVA | Canonical Variate Analysis |
| DSMZ | Deutsche Sammlung von Mikroorganismen und Zellkulturen |
| ERIC | Enterobacterial Repetitive Intergenic Consensus |
| FPA | Focal-Plane-Array |
| FT-IR | Fourier Transform Infrared |
| IRE | Internal Reflection Element |
| LMG | Belgian Coordinated Collections of Microorganisms |
| LPSs | Lipopolysaccharides |
| MCT | Mercury Cadmium Telluride |
| PCA | Principal Component Analysis |
| PCR | Polymerase Chain Reaction |
| PFGE | Pulsed Field Gel Electrophoresis |
| RAPD | Randomly Amplified Polymorphic DNA |
| TSA | Tryptic Soy Agar |

# REFERENCES

1. D. Helm, H. Labischinski and D. Naumann, *J. Microbiol. Methods*, **14**, 127–142 (1991).

2. D. Helm, H. Labischinski, G. Schallehn and D. Naumann, *J. Gen. Microbiol.*, **137**, 69–79 (1991).

3. D. Naumann, D. Helm and H. Labischinski, *Nature*, **351**, 81–82 (1991).

4. D. Naumann, D. Helm, H. Labischinski and P. Giesbrecht, 'The Characterization of Microorganisms by Fourier-Transform Infrared Spectroscopy (FT-IR)', in "Modern Techniques for

Rapid Microbiological Analysis", ed W.H. Nelson, VCH, New York (1991).

5. K. Maquelin, L.P. Choo-Smith, C. Kirschner, N.A. Ngo Thi, D. Naumann and G. Puppels, 'Vibrational Spectroscopic Studies of Microorganisms', in "Handbook of Vibrational Spectroscopy", eds J.M. Chalmers and P.R. Griffiths, John Wiley & Sons, Chichester, Vol. 5 (2002).

6. D. Naumann, 'Infrared Spectroscopy in Microbiology', in "Encyclopedia of Analytical Chemistry", ed R.A. Meyers, John Wiley & Sons, Chichester (2000).

7. D. Naumann, 'FT-Infrared and FT-Raman Spectroscopy in Biomedical Research', in "Infrared and Raman Spectroscopy of Biological Materials", eds H.U. Gremlich and B. Yan, Marcel Dekker, New York (2001).

8. M.T. Madigan, J.M. Martinko and J. Parker, 'Brock Mikrobiologie', Spektrum Akademischer Verlag, Heidelberg (2001).

9. N.B. Colthup, L.H. Daly and S.E. Wiberley, 'Introduction to Infrared and Raman Spectroscopy', 3rd edition, Academic Press, Boston (1990).

10. D. Li-Vien, N.B. Colthup, W.G. Fateley and J.G. Grasselli, 'The Handbook of Infrared and Raman Characteristic Frequencies of Organic Molecules', Academic Press, Boston (1991).

11. F.S. Parker, 'Applications of Infrared, Raman, and Resonance Spectroscopy in Biochemistry', Plenum Press, New York (1983).

12. B. Schrader, 'Infrared and Raman Spectroscopy', VCH, Weinheim (1995).

13. H.Y. Cheung, J. Cui and S. Sun, *Microbiology*, **145**, 1043–1048 (1999).

14. M. Gue, V. Dupont, A. Dufour and O. Sire, *Biochemistry*, **40**, 11938–11945 (2001).

15. R.M. Donlan, J.A. Piede, C.D. Heyes, L. Sanii, R. Murga, P. Edmonds, I. El-Sayed and M.A. El-Sayed, *Appl. Environ. Microbiol.*, **70**, 4980–4988 (2004).

16. L.P. Choo-Smith, K. Maquelin, T. van Vreeswijk, H.A. Bruining, G.J. Puppels, N.A. Ngo Thi, C. Kirschner, D. Naumann, D. Ami, A.M. Villa, F. Orsini, S.M. Doglia, H. Lamfarraj, G.D. Sockalingum, M. Manfait, P. Allouch and H.P. Endtz, *Appl. Environ. Microbiol.*, **67**, 1461–1469 (2001).

17. K. Maquelin, C. Kirschner, L.P. Choo-Smith, N.A. Ngo Thi, T. van Vreeswijk, M. Stammler, H.P.

Endtz, H.A. Bruining, D. Naumann and G.J. Puppels, *J. Clin. Microbiol.*, **41**, 324–329 (2003).

18. N.A. Ngo Thi, C. Kirschner and D. Naumann, 'FT-IR Microspectrometry: A New Tool for Characterizing Micro-organisms', in "Biomedical Spectroscopy: Vibrational Spectroscopy and Other Novel Techniques", eds A. Mahadevan-Jansen and G.J. Puppels, Proceedings of SPIE, Bellingham, Vol. 3918 (2000).

19. M. Wenning, V. Theilmann and S. Scherer, *Environ. Microbiol.*, **8**, 848–857 (2006).

20. B.J. Tindall, E. Brambilla, M. Steffen, R. Neumann, R. Pukall, R.M. Kroppenstedt and E. Stackebrandt, *Environ. Microbiol.*, **2**, 310–318 (2000).

21. I. Adt, D. Toubas, J.M. Pinon, M. Manfait and G.D. Sockalingum, *Arch. Microbiol.*, **185**, 277–285 (2006).

22. A. Oust, T. Moretro, K. Naterstad, G.D. Sockalingum, I. Adt, M. Manfait and A. Kohler, *Appl. Environ. Microbiol.*, **72**, 228–232 (2006).

23. C. Santivarangkna, M. Wenning, P. Foerst and U. Kulozik, *J. Appl. Microbiol.*, **102**, 748–756 (2007).

24. N.M. Amiali, M.R. Mulvey, J. Sedman, M. Louie, A.E. Simor and A.A. Ismail, *J. Microbiol. Methods*, **68**, 236–242 (2007).

25. S.H. Beattie, C. Holt, D. Hirst and A.G. Williams, *FEMS Microbiol. Lett.*, **164**, 201–206 (1998).

26. M.C. Curk, F. Peladan and J.C. Hubert, *FEMS Microbiol. Lett.*, **123**, 241–248 (1994).

27. M. Essendoubi, D. Toubas, M. Bouzaggou, J.M. Pinon, M. Manfait and G.D. Sockalingum, *Biochim. Biophys. Acta*, **1724**, 239–247 (2005).

28. R. Goodacre, E.M. Timmins, R. Burton, N. Kaderbhai, A.M. Woodward, D.B. Kell and P.J. Rooney, *Microbiology*, **144**, 1157–1170 (1998).

29. F. Guibet, C. Amiel, P. Cadot, C. Cordevant, M.H. Desmonts, M. Lange, A. Marecat, J. Travert, C. Denis and L. Mariey, *Vib. Spectrosc.*, **33**, 133–142 (2003).

30. H. Haag, H.U. Gremlich, R. Bergmann and J.J. Sanglier, *J. Microbiol. Methods*, **27**, 157–163 (1996).

31. C. Kirschner, K. Maquelin, P. Pina, N.A. Ngo Thi, L.P. Choo-Smith, G.D. Sockalingum, C. Sandt, D. Ami, F. Orsini, S.M. Doglia, P. Allouch, M. Mainfait, G.J. Puppels and D. Naumann, *J. Clin. Microbiol.*, **39**, 1763–1770 (2001).

32. M. Kümmerle, S. Scherer and H. Seiler, *Appl. Environ. Microbiol.*, **64**, 2207–2214 (1998).

33. H. Lamprell, G. Mazerolles, A. Kodjo, J.F. Chamba, Y. Noel and E. Beuvier, *Int. J. Food Microbiol.*, **108**, 125–129 (2006).

34. A. Mayer, H. Seiler and S. Scherer, *Ann. Microbiol.*, **53**, 299–313 (2003).

35. M.A. Miguel Gomez, M.A. Bratos Perez, F.J. Martin Gil, A. Duenas Diez, J.F. Martin Rodriguez, P. Gutierrez Rodriguez, A. Orduna Domingo and A. Rodriguez Torres, *J. Microbiol. Methods*, **55**, 121–131 (2003).

36. H. Oberreuter, H. Seiler and S. Scherer, *Int. J. Syst. Evol. Microbiol.*, **52**, 91–100 (2002).

37. A. Oust, T. Moretro, C. Kirschner, J.A. Narvhus and A. Kohler, *J. Microbiol. Methods*, **59**, 149–162 (2004).

38. C.A. Rebuffo, J. Schmitt, M. Wenning, F. von Stetten and S. Scherer, *Appl. Environ. Microbiol.*, **72**, 994–1000 (2006).

39. C.A. Rebuffo-Scheer, C. Kirschner, M. Staemmler and D. Naumann, *J. Microbiol. Methods*, **68**, 282–290 (2007).

40. C. Sandt, C. Madoulet, A. Kohler, P. Allouch, C. De Champs, M. Manfait and G.D. Sockalingum, *J. Appl. Microbiol.*, **101**, 785–797 (2006).

41. M. Wenning, H. Seiler and S. Scherer, *Appl. Environ. Microbiol.*, **68**, 4717–4721 (2002).

42. S. Lai, R. Goodacre and L.N. Manchester, *Syst. Appl. Microbiol.*, **27**, 186–191 (2004).

43. A. Maoz, R. Mayr and S. Scherer, *Appl. Environ. Microbiol.*, **69**, 4012–4018 (2003).

44. H. Oberreuter, A. Brodbeck, S. von Stetten, S. Goerges and S. Scherer, *Eur. Food Res. Technol.*, **216**, 434–439 (2003).

45. G. Fischer, S. Braun, R. Thissen and W. Dott, *J. Microbiol. Methods*, **64**, 63–77 (2006).

46. D. Lefier, D. Hirst, C. Holt and A.G. Williams, *FEMS Microbiol. Lett.*, **147**, 45–50 (1997).

47. A. Oust, T. Moretro, C. Kirschner, J.A. Narvhus and A. Kohler, *FEMS Microbiol. Lett.*, **239**, 111–116 (2004).

48. H.C. van der Mei, D. Naumann and H.J. Busscher, *Infrared Phys. Technol.*, **37**, 561–564 (1996).

49. L. Mariey, J.P. Signolle, C. Amiel and J. Travert, *Vib. Spectrosc.*, **26**, 151–159 (2001).

50. R.M. Jarvis and R. Goodacre, *Bioinformatics (Oxford, England)*, **21**, 860–868 (2005).

51. O. Preisner, J.A. Lopes, R. Guiomar, J. Machado and J.C. Menezes, *Anal. Bioanal. Chem.*, **387**, 1739–1748 (2007).

52. M. Lin, M. Al-Holy, H. Al-Qadiri, D.H. Kang, A.G. Cavinato, Y. Huang and B.A. Rasco, *J. Agric. Food Chem.*, **52**, 5769–5772 (2004).

53. P. Geladi, *Spectrochim. Acta, Part B*, **58**, 767–782 (2003).

54. N.A. Ngo Thi and D. Naumann, *Anal. Bioanal. Chem.*, **387**, 1769–1777 (2007).

55. T. Udelhoven, D. Naumann and J. Schmitt, *Appl. Spectrosc.*, **54**, 1471–1479 (2000).

56. R. Goodacre, E.M. Timmins, P.J. Rooney, J.J. Rowland and D.B. Kell, *FEMS Microbiol. Lett.*, **140**, 233–239 (1996).

57. J. Schmitt and T. Udelhoven, 'Use of Artificial Neural Networks in Biomedical Diagnosis', in "Infrared and Raman Spectroscopy of Biological Materials", eds H.U. Gremlich and B. Yan, Marcel Dekker, New York (2001).

58. T. Udelhoven, M. Novozhilov and J. Schmitt, *Chemom. Intell. Lab. Syst.*, **66**, 219–226 (2003).

59. D.J. Mouwen, R. Capita, C. Alonso-Calleja, J. Prieto-Gomez and M. Prieto, *J. Microbiol. Methods*, **67**, 131–140 (2006).

60. K. Tintelnot, G. Haase, M. Seibold, F. Bergmann, M. Staemmler, T. Franz and D. Naumann, *J. Clin. Microbiol.*, **38**, 1599–1608 (2000).

61. P.L. Lang and S.C. Sang, *Cell. Mol. Biol.*, **44**, 231–238 (1998).

62. R. Goodacre, B. Shann, R.J. Gilbert, E.M. Timmins, A.C. McGovern, B.K. Alsberg, D.B. Kell and N.A. Logan, *Anal. Chem.*, **72**, 119–127 (2000).

63. D. Helm and D. Naumann, *FEMS Microbiol. Lett.*, **126**, 75–80 (1995).

64. H. Oberreuter, J. Charzinski and S. Scherer, *Microbiology*, **148**, 1523–1532 (2002).

65. C.A. Rebuffo-Scheer, J. Schmitt and S. Scherer, *Appl. Environ. Microbiol.*, **73**, 1036–1040 (2007).

66. C. Sandt, G.D. Sockalingum, D. Aubert, H. Lepan, C. Lepouse, M. Jaussaud, A. Leon, J.M. Pinon, M. Manfait and D. Toubas, *J. Clin. Microbiol.*, **41**, 954–959 (2003).

67. S. Kim, B.L. Reuhs and L.J. Mauer, *J. Appl. Microbiol.*, **99**, 411–417 (2005).

68. L. Beutin, Q. Wang, D. Naumann, W. Han, G. Krause, L. Leomil, L. Wang and L. Feng, *J. Med. Microbiol.*, **56**, 177–184 (2007).

69. N.A. Baldauf, L.A. Rodriguez-Romo, A. Mannig, A.E. Yousef and L.E. Rodriguez-Saona, *J. Microbiol. Methods*, **68**, 106–114 (2007).

70. J. Kirkwood, A. Ghetler, J. Sedman, D. Leclair, F. Pagotto, J.W. Austin and A.A. Ismail, *J. Food Prot.*, **69**, 2377–2383 (2006).

71. G. Seltmann, W. Voigt and W. Beer, *Epidemiol. Infect.*, **113**, 411–424 (1994).

72. W. Bouhedja, G.D. Sockalingum, P. Pina, P. Allouch, C. Bloy, R. Labia, J.M. Millot and M. Manfait, *FEBS Lett.*, **412**, 39–42 (1997).

73. G.D. Sockalingum, W. Bouhedja, P. Pina, P. Allouch, C. Mandray, R. Labia, J.M. Millot and M. Manfait, *Biochem. Biophys. Res. Commun.*, **232**, 240–246 (1997).

74. J. Barenfanger, C. Drake and G. Kacich, *J. Clin. Microbiol.*, **37**, 1415–1418 (1999).

75. A.A. Christy, Y. Ozaki and V.G. Gregoriou, 'Modern Fourier Transform Infrared Spectroscopy', Elsevier Science, Amsterdam (2001).

76. D.L. Wetzel and S.M. LeVine, 'Biological Applications of Infrared Microspectroscopy', in "Infrared and Raman Spectroscopy of Biological Materials", eds H.U. Gremlich and B. Yan, Marcel Dekker, New York (2001).

77. N.A. Ngo Thi, C. Kirschner and D. Naumann, *J. Mol. Struct.*, **661–662**, 371–380 (2003).

78. M. Kansiz, P. Heraud, B. Wood, F. Burden, J. Beardall and D. McNaughton, *Phytochemistry*, **52**, 407–417 (1999).

79. D. Toubas, M. Essendoubi, I. Adt, J.M. Pinon, M. Mainfait and G.D. Sockalingum, *Anal. Bioanal. Chem.*, **387**, 1729–1737 (2007).

80. C.A. Rebuffo-Scheer, J. Dietrich, M. Wenning and S. Scherer, Identification of five *Listeria* species by FTIR macrospectroscopy is superior over FTIR microspectroscopy, submitted.

81. J. Kirkwood, S.F. Al-Khaldi, M.M. Mossoba, J. Sedman and A.A. Ismail, *Appl. Spectrosc.*, **58**, 1364–1368 (2004).

82. A. Oust, B. Moen, H. Martens, K. Rudi, T. Naes, C. Kirschner and A. Kohler, *J. Microbiol. Methods*, **65**, 573–584 (2006).

# Antemortem Identification of Transmissible Spongiform Encephalopathy (TSE) from Serum by Mid-infrared Spectroscopy

**Peter Lasch, Michael Beekes, Heinz Fabian and Dieter Naumann**
*Robert Koch-Institut, Berlin, Germany*

## 1 INTRODUCTION

The past decade has witnessed substantial progress toward the application of infrared (IR) spectroscopy as a useful analytical tool in biomedical research. Biomedical applications of IR spectroscopy include research studies for the characterization of prokaryotic and eukaryotic cells,[1–5] body fluids,[6–11] and, in combination with microscopy, the characterization of tissues[12–15] to mention a few. Prerequisites for this progress have been new technical developments and the implementation of modern multivariate concepts of data analysis.

Biomedical IR spectroscopy typically relies on the absorption properties of the samples under investigation. Biomedical IR spectra constitute highly complex superpositions of spectral contributions from all of a sample's vibrationally active biochemical molecular species. Hence, broad and superimposed band contours are observed, which impede a comprehensive understanding of the IR spectra in an analytical sense. For this reason, the concepts of spectral analysis fall into two distinct categories. The first approach aims at extracting quantitative sample information, typically the concentration of analytes such as cholesterol, glucose, urea, or albumin in human serum,[8,16] or of glucose, lactate, and lipids in amniotic fluid.[17] In this quantitative type of spectral analysis, a minimal understanding of the vibrational spectra is beneficial. The other concept relies on more qualitative data analysis methods. Here, IR spectra are referred to as *fingerprints*, which can be identified or classified by pattern-matching techniques. The fingerprint approach assumes that characteristic disease-related structural and/or compositional changes in cells, tissues, or serum cause a multiplicity of reproducible spectral alterations. It is implied that the sum of these changes constitutes another spectral fingerprint, which is characteristic of the disease. The classification of disease-related spectral patterns is also called *disease pattern recognition* (DPR). This DPR approach differs from quantitative concepts of spectral analysis in that no understanding of the underlying sample biochemistry is required. This is certainly an advantage, but implies at the same time an important conceptual drawback. First and foremost, the DPR approach requires an experimental proof of the specificity of the IR fingerprint for a particular disease. DPR studies should, therefore, cover samples not only from specific "diseased" and "healthy" individuals but should also ideally contain the complete differential diagnoses of the disease under study. Certainly, this is the largest challenge the DPR approach is faced with and the discussion of this important problem will be a major issue of this chapter.

Despite these problems, both data analysis concepts have been successfully applied over the last decade to establish mid-IR-based blood or serum assays.[9,16,18–23] The prospects of carrying out analyses on blood or blood fractions, such as serum, are attractive from several perspectives. First of all, blood can be easily isolated from the body and its biochemical changes are known to be associated with numerous kinds of diseases. Furthermore, IR analyses of serum are reagent free and the quantitative or qualitative sample information is available simultaneously from a single IR spectrum. Finally, an IR serum assay is typically based on a very small sample volume of only a few microliters.

Serum is the liquid component of blood from which clotting factors have been removed. It contains proteins at high concentration $(60-80 \, \text{mg mL}^{-1})$ in addition to various small molecules, including salts, lipids, amino acids, and sugars.[24] The protein composition of serum is complex; more than 9000 different proteins varying in concentration over at least nine orders of magnitude are known to occur in serum.[24–26] High abundance proteins include albumin, immunoglobulins, complement factors, transferrin, and thyroxin-binding protein. In addition to these major constituents, serum contains many other low abundance proteins such as receptor ligands, tissue leakage products, or aberrant secretions.[25] The concentrations of most of the blood constituents as well as other important parameters (pH, oncotic pressure, partial

pressures, etc.) are physiologically maintained within tight boundaries. The stable composition of blood, and also of serum, isolated from the healthy human body under homeostatic conditions makes serum an ideal candidate for the diagnosis of disease.

With these aspects in mind, our research group started a program in the late 1990s for the detection of various forms of transmissible spongiform encephalopathy (TSE) from serum. TSE such as scrapie in sheep, bovine spongiform encephalopathy (BSE) in cattle, chronic wasting disease (CWD) in deer and elk, or the new variant of Creutzfeldt–Jakob disease (vCJD) in humans belong to a group of fatal neurodegenerative disorders that are caused by misfolded and apparently aggregated isoforms of the prion protein (PrP$^{sc}$, sc: scrapie) and their accumulation in the central nervous system (CNS).[27] The BSE crisis in the 1990s and the emergence of vCJD in 1996 demonstrated that transmission of TSE, e.g., by food products, pose a serious threat to human health, requiring effective surveillance measures of animal livestock. Many countries have implemented test systems in which the pathological prion protein PrP$^{sc}$ is directly detected postmortem. These tests provide a high diagnostic sensitivity and specificity but require the analysis of CNS tissue.

In the last few years, a number of blood-based diagnostic screening tests, either specifically targeting the causative agent of TSE infections (PrP$^{sc}$) or surrogate markers, have been suggested.[28,29] However, so far none of the proposed antemortem tests has been put into practice under "real life conditions" (March, 2007). Therefore, the development of rapid and reliable diagnostic screening methods that permit antemortem identification of TSE remains highly desirable for the surveillance and control of TSE.

The research program launched at the Robert Koch-Institut (RKI) for the development of a serum-based technique for the antemortem detection of TSE is aimed at addressing several aspects. One of the key objectives was to test whether IR spectroscopy of blood or serum can be generally used to distinguish animals suffering from a TSE infection and individuals of a control group. Having shown this in an experimental scrapie study in the hamster model, we have addressed the question for the earliest time point at which a scrapie infection is detectable, and applied the technique to diagnose other TSE infections such as BSE in cattle. Last but not least, the research program also covered technical developments and triggered improvements in spectral data analysis. The program that is the main subject of this chapter was carried out in close collaboration with other researchers namely Jürgen Schmitt and Thomas Udelhoven from Synthon GmbH and Wolfgang Petrich from Roche Diagnostics GmbH.

## 2 TECHNICAL ASPECTS OF TSE SERUM ANALYSIS BY FT-IR SPECTROSCOPY

Body fluids such as blood, or blood components, cerebrospinal, synovial, or amniotic fluid are ideal candidates for diagnostic analyses because biochemical changes associated with pathophysiological processes influence their composition. Many of these changes are, in principle, available to discrimination by vibrational spectroscopy since the IR spectrum of a biofluid represents the superposition of spectral contributions from all components of sufficiently high concentration.

One of the major technical problems of biomedical IR spectroscopy of living cells, in vivo spectroscopy of tissues or of biofluids is the presence of water. Water exhibits very strong absorption features in the diagnostically relevant mid-IR region ($1000-4000 \, cm^{-1}$) and may mask weak signals of proteins, lipids, amino acids, or carbohydrates in this region. One of the simplest solutions to overcome the water absorption problem is to dry the sample. For mid-IR spectroscopy of tissues, for example, it is common practice to examine cryo-sections, that is, dry thin sections of thaw-mounted tissue

slices. After cryo-sectioning and drying, these sections form thin films that can be mounted on an IR transparent or reflective supporting substrate. A similar approach is used for serum analyses. Here, a small volume of serum is transferred onto a sample carrier and allowed to dry. In this way, the serum forms a thin film suitable for transmission- or transflection-type measurements. The disadvantage of this procedure is the need to precisely control the conditions of the drying process because most of the IR absorption features are strongly affected by changes in the relative humidity of the film.

Another approach for characterizing biofluids in the mid-IR range utilizes IR transmission cells with optical path lengths in the range $6-8 \, \mu m$. This liquid-film technique requires measurements of both the liquid sample and the water, and the final sample spectrum can be produced by digital subtraction of the water background absorbance spectrum from the absorbance spectrum of the sample or division of the single-beam spectra and subsequent logarithmic conversion to absorbance units. Short path lengths of only $6-8 \, \mu m$ limit the intensities of the IR bands of interest and require relatively high sample concentrations (which is fortunately the case in the example of serum). It must be noted, however, that these kinds of measurements are technically challenging and require a very high spectral signal-to-noise ratio (SNR) because of the relatively weak sample signals.

Biofluids can also be studied by attenuated total reflection (ATR) spectroscopy. In ATR spectroscopy, the IR beam is typically guided through an IR transparent crystal in a way that one or several total internal reflections take place at a surface of the crystal. This creates an evanescent, or near-field, standing wave, at the ATR crystal–sample boundary that interrogates the sample. Samples are prepared directly on the crystal surface and the absorption properties of the samples can be observed as a result of attenuation of the evanescent IR wave. The penetration depth of the IR radiation depends on the refractive indices of the sample and the internal reflection element (IRE), the wavelength, $\lambda$, and the angle of incidence, and usually varies between 0.07 and $0.25\lambda$. An ATR spectrum thus contains only information on a thin sample layer in the immediate vicinity of the IRE's surface. The ATR approach allows measurements of liquid samples, such as serum, without too much interference from IR absorption of bulk water. A major drawback of the ATR technique is the deposition of serum proteins on the surface of the IRE. Furthermore, ATR crystals are relatively expensive and not easy to clean. These problems and the advantages of IR transmission spectroscopy in terms of SNR have prompted us to omit systematic tests of the ATR technique.

The following discussion, therefore, covers a description of the technical concepts of the dry and liquid film techniques, and provides a systematic comparison of both data acquisition techniques. For this purpose, spectra from BSE positive and BSE negative control cattle (see Section 4 for sample description) have been obtained and compared in terms of "hard" spectral parameters, such as spectral reproducibility or SNR, and also by comparing "soft" diagnostic parameters describing the accuracy of IR-based reclassification (sensitivity, specificity).

## 2.1 Dry film technique

IR measurements of dry serum films were carried out by means of a multisample cuvette (sample wheel) with 15 measurement and one background position, which was originally developed for the analysis of microbial samples (see Figure 1). Serum aliquots of $2.6\,\mu L$ were pipetted onto the sample spots and allowed to air-dry at $37\,^\circ C$ for 15 min. For IR spectroscopy, the cuvette can be sealed by a KBr window, which ensures stable environmental conditions, particularly of the relative humidity, during the measurements. IR spectra of the dry films were recorded in transmission mode with a Bruker (Bruker Optics, Ettlingen, Germany) IFS 28/B Fourier-transform infrared (FT-IR) spectrometer. The instrument was equipped with a pyroelectric deuterated triglycine sulfate (DTGS) detector operating at room temperature. This and the extended linear range at high absorbance are advantageous for routine use under practical conditions. Typically, three replicate samples were measured for each serum specimen. The sample chambers and optics were purged with dry air in order to keep the water vapor level constantly low.

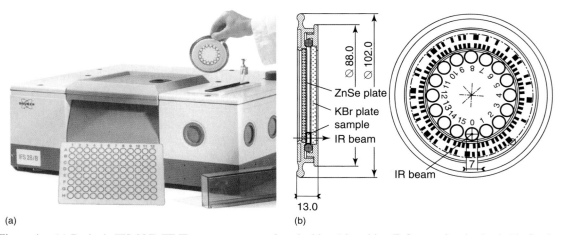

(a)                                                                 (b)

**Figure 1.** (a) Bruker's IFS 28/B FT-IR spectrometer equipped with a 16 position ZnSe sample wheel suitable for dry serum film measurements (shown schematically in (b)). Bruker now offers a microplate compatible module (HTS-XT) that facilitates automatic performance of measurements employing ZnSe, or Si standard microplates of a 96-, 384-, or 1596-well format.

## 2.2   Liquid film technique

IR measurements of sera in the liquid state have the advantage of a simple and rapid sample preparation procedure. Since drying is not required, the sample preparation involves only filtering to eliminate solid contaminants from the serum fraction to be measured. We have observed that the use of disposable high-performance liquid chromatography (HPLC) microfilters of a pore size of $0.5\,\mu m$ significantly reduce the risk of occlusions of the IR flow cell and the HPLC tubing system. The central part of our apparatus for measurements of liquid samples consists of a flow-through cell that is specifically designed for transmission-type IR measurements of aqueous solutions (AquaSpec cell, microbiolytics, Freiburg, Germany). The AquaSpec cell is microfabricated and consists of two $CaF_2$ disks and a spacer made of a sealant material of defined elasticity, which ensures high-pressure stability and fast relaxation for a constant effective sample thickness.[30] Both the shape and the thickness of the spacer can be customized to fit the experimental requirements. Typical optical path lengths in our experiments were $6–8\,\mu m$. Liquid samples were measured using an IFS 28/B spectrometer from Bruker that was equipped with a sensitive liquid nitrogen cooled mercury cadmium telluride (MCT) detector.

Single-beam spectra of the AquaSpec flow-through cell filled with water or serum are shown in Figure 2(a). Both spectra demonstrate major spectral contributions from bulk water with the H−O−H bending vibration at $1646\,cm^{-1}$. Furthermore, the water absorption centered near $3400\,cm^{-1}$ (O−H stretching vibrations) and an optical low-pass filter with a cutoff at approximately $3500\,cm^{-1}$ result in an optically opaque spectral coverage above $3100\,cm^{-1}$. Below $1000\,cm^{-1}$, the optical material of the AquaSpec cell ($CaF_2$) gives rise to total absorption. Spectra from serum are obtained subsequently by the division of single-beam spectra and logarithmic conversion to absorbance. Typical absorption features of serum constituents are found in the amide I and amide II region at 1548 and

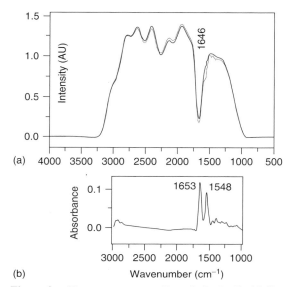

**Figure 2.**   IR serum spectra collected via the liquid film technique (optical path length: $6–8\,\mu m$). (a) Single-beam background spectrum (solid line) and single-beam sample spectrum of bovine serum (dashed line). (b) Absorbance spectrum of bovine serum obtained by division of single-beam spectra. Window material of the cell: $CaF_2$. Note that an optical low-pass filter with a cutoff at ca. $3500\,cm^{-1}$ was used.

$1653\,cm^{-1}$ respectively (see Figure 2b). These two bands clearly dominate the IR serum spectra, indicating that proteins with $\alpha$-helical structure (e.g., albumin) constitute the major class of serum components. Spectral contributions from substances with lower abundance such as lipids, amino acids, or sugars are also detectable but typically exhibit less intense IR absorptions.

The use of the commercially available AquaSpec cell allows collection of high-quality IR spectra of biofluids. However, the processing of hundreds of samples would be laborious and time consuming due to the manual mode of operation. This, and the need to collect spectra from hundreds of TSE serum samples, prompted us to design an apparatus that enables high sample throughput for serial IR measurements of liquid samples. The schematic diagram of this apparatus is given in Figure 3. The system is based on the AquaSpec cell, an HPLC pump, and a manual

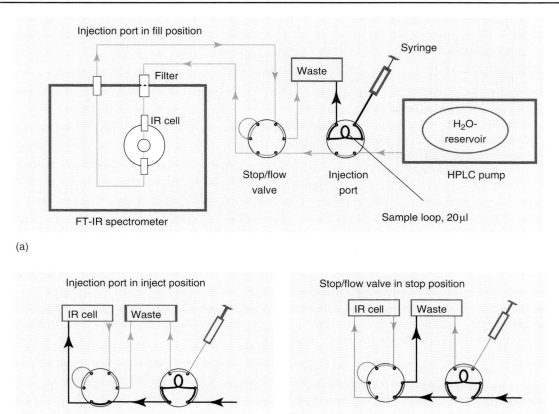

(a)

(b)                                                    (c)

**Figure 3.** Schematic diagram of the apparatus developed for series measurements of biofluids. (a) Injection port in fill position and stop/flow valve in flow position. (b) Injection port in inject position and sample is transported into the flow cell. (c) Stop/flow valve in stop position and sample is measured.

injector, both from Agilent Technologies (see Reference 31 for details). In the upstream of the AquaSpec cell, a replaceable particle filter with a pore size of $2\,\mu m$ is mounted, which traps remaining solid contaminants. Switching of the stop/flow valve, or the injection port, as well as control of the HPLC pump are fully computer controlled (see Figure 3 and Reference 31 for details).

## 2.3  IR spectroscopy of serum samples

Typical mid-IR spectra of two serum samples measured in the dry and the liquid state are shown in Figure 4. This figure illustrates that $\sim 2.6\,\mu L$

serum dried onto a ZnSe window provides approximately $20\times$ higher absorbance values than an equivalent liquid sample measured in a flow-through cuvette with an approximately 8-$\mu m$ optical path length. The intense absorptions, particularly in the amide I and II regions, are potential sources of error due to nonlinear IR signal response. One could avoid these problems by dispensing smaller sample volumes, but the reproducible deposition of only $2\,\mu L$ of highly viscous liquids such as serum is technically challenging.

The general level of spectral reproducibility can also be estimated from Figure 4, where varying absorbance values indicate variations of

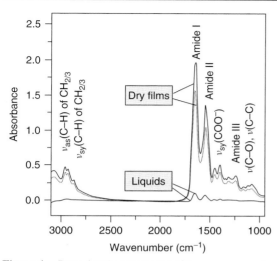

**Figure 4.** Raw absorbance spectra of two serum samples measured as a film dried onto ZnSe windows (upper traces) and as a liquid (lower traces).

**Table 1.** Comparison of measurement reproducibility for the liquid and dry film technique in the fingerprint ($1050-1500\,\text{cm}^{-1}$), amide I ($1600-1700\,\text{cm}^{-1}$), and the CH-stretching ($2800-3100\,\text{cm}^{-1}$) regions. Reproducibility was determined on the basis of Pearson's product-momentum correlation coefficient calculated for parts of IR spectra of three independent sample preparations.

| Spectral window ($\text{cm}^{-1}$) | Liquid film technique (D-values $\pm$ standard deviation)[a] | Dry film technique (D-values $\pm$ standard deviation)[a] |
|---|---|---|
| $1050-1500$ | $1.31 \pm 0.16$ | $1.62 \pm 0.81$ |
| $1600-1700$ | $0.33 \pm 0.09$ | $8.87 \pm 8.8$ |
| $2800-3100$ | $10.00 \pm 2.14$ | $0.69 \pm 0.46$ |

[a] $D = (1 - \alpha) \times 1000$; $\alpha$: Pearson's product-momentum correlation coefficient with $-1 \leq \alpha \leq 1$.

film thickness between the two dried samples (compare solid and dotted curves of dry serum films in Figure 4). In case of the liquid technique, these variations are much less pronounced, as indicated by nearly identical traces of the liquid films in Figure 4. However, many spectral effects caused by film thickness variations can be compensated computationally by spectral preprocessing (normalization) applied to the raw IR spectra. In order to compare and quantify the reproducibility of IR measurements of dry and liquid films, we have introduced the so-called differentiation indices that are based on Pearson's correlation coefficients (see **Infrared Spectroscopy in the Identification of Microorganisms** and Reference 32 for details) obtained between pairs of spectra from three independent sample preparations. The term *reproducibility* refers to successive measurements of different preparations of one individual serum sample and differs from repeatability, which compares repeated measurements of only one preparation.

Differentiation indices, or D-values, can vary between 0 and 2000 (0–1000 for positively correlated data).[33] The smaller the D-values, the higher the reproducibility. D-values have been

routinely obtained after spectral preprocessing in spectral regions of interest. A systematic comparison of spectral reproducibility for the liquid and dry film technique in the fingerprint ($1050-1500\,\text{cm}^{-1}$), amide I ($1600-1700\,\text{cm}^{-1}$), and the CH-stretching ($2800-3100\,\text{cm}^{-1}$) regions is given in Table 1. From these numbers, it can be seen that the reproducibility varies largely as a function of measurement technique and spectral region. In the amide I region, e.g., the reproducibility of the liquid film technique is superior to the dry film technique, most likely due to the occasional nonlinear response in the spectra of dried serum samples. Just the opposite situation can be observed in the CH-stretching region, where spectra of liquid samples display only weak spectral features, which are additionally superimposed by strong $v(\text{O}-\text{H})$ water absorption bands. The reproducibility in the fingerprint region between 1050 and $1500\,\text{cm}^{-1}$ is comparable for both sampling techniques.

Another "hard" parameter with potential impact on classification accuracy is the SNR. We have determined this key parameter for serum spectra of liquid and dry films in several spectral regions (see Table 2). Noise was obtained as the standard deviation in the respective wavelength range from absorbance spectra obtained by processing two single-beam spectra from the same serum sample. Our results indicated a comparable SNR

**Table 2.** Mean signal, mean noise, and mean signal-to-noise ratios (SNR) of IR serum spectra measured as liquid and as dry films, respectively. Spectral noise was determined from processing two single-beam spectra from one and the same sample. The amide I peak intensities and the peak intensities at $1400 \, cm^{-1}$ ($\nu_{sy}COO^-$) were taken from the absorbance spectra. Mean values $\pm$ standard deviations are given.

|  | Liquid film technique | Dry film technique |
|---|---|---|
| Noise, region $1600-1700 \, cm^{-1}$ | $(1.140 \pm 0.227) \times 10^{-4}$ | $(1.811 \pm 1.420) \times 10^{-3}$ |
| Amide I peak intensity | $0.110 \pm 0.013$ | $1.500 \pm 0.353$ |
| SNR, amide I region | $965 \pm 223$ | $828 \pm 677$ |
| Noise, region $1350-1450 \, cm^{-1}$ | $(1.987 \pm 0.321) \times 10^{-5}$ | $(9.345 \pm 2.160) \times 10^{-5}$ |
| Peak intensity at $1400 \, cm^{-1}$ | $0.0236 \pm 0.0028$ | $0.378 \pm 0.092$ |
| SNR at $1400 \, cm^{-1}$ ($\nu_{sy}COO^-$) | $1192 \pm 238$ | $4047 \pm 1355$ |

for both techniques in the amide I region and an approximately four times better SNR between 1350 and $1450 \, cm^{-1}$ for spectra acquired by the dry film technique. A comparison of the absolute noise levels, however, demonstrated the advantages of the liquid film technique (see rows one and four in Table 2), which are most likely due to the more specialized equipment. For example, IR spectra of liquid samples were acquired by using a sensitive MCT detector, which, together with the dedicated optical low-pass filter, enabled data collection at a much lower absolute noise level compared to that of a DTGS detector at comparable measurement times.

## 2.4 Comparative classification analysis of dry and liquid film spectra

We now turn to a comparative classification analysis of spectra acquired by the dry and liquid film techniques. These analyses were carried out on spectral data sets containing triple replicates of IR spectra from identical serum samples of BSE negative controls, or BSE positive cattle. In this comparison, the basic idea was to establish binary classification models (classifiers), which are able to determine class memberships, using both the dry and liquid film techniques. This class membership information, and the BSE status as revealed by a "gold standard" method (histopathology, immunocytochemistry), was subsequently employed to determine statistical measures of binary classification such as

the diagnostic sensitivity or specificity. Thus, the performance of a binary diagnostic test system was used to evaluate the capability of the data acquisition technique.

Owing to the highly complex classification tasks, binary classification analyses were performed by supervised (concept-driven) classification techniques, namely multilayer perceptron artificial neural networks (MLP-ANN).[34] These types of artificial neural networks (ANNs) need to be trained and internally validated before they can accomplish class assignments. While training and internal validation of neural network models is usually performed with data of a training and internal validation subset, respectively, the actual effectiveness of classification can be determined only on the basis of independent data. It is important to keep these independent spectra of the external validation subset totally separate from training and internal validation subsets.

Pattern recognition by ANNs requires weight adjustments in the training phase, wherein the output of the net is constrained to converge to the right target value.[35] Because these weights are initialized randomly, ANNs may find different solutions for identical problems. The results given in the three rows labeled ANN I–III in Table 3 illustrate this dilemma, since they show varying values of sensitivity and specificity for one and the same classification task (identical spectral database, no modifications of network topology, learning rules, etc.). Nevertheless, the mean values of sensitivity and specificity indicated that more than 90% of the BSE positive

**Table 3.** Diagnostic sensitivity and specificity for BSE determined by external validation of ANNs with serum spectra from 26 BSE positive and 21 BSE negative animals. The first three rows (ANN I–III) display specificity and sensitivity values obtained by challenging each time three specific ANNs. The last row shows the average of these values. For ANN model development, the information content of the following spectral regions was considered: without amide I region: 950–1600, 1700–1750, and 2800–3100 cm$^{-1}$; amide I region included: 950–1750 and 2800–3100 cm$^{-1}$; fingerprint region included: 650–1600, 1700–1750, and 2800–3100 cm$^{-1}$; SENS: sensitivity (%); SPEC: specificity (%).

| | Liquid film technique | | | | Dry film technique | | | | | |
| | Without amide I | | Amide I incl. | | Without amide I | | Amide I incl. | | Fingerprint incl. | |
| | SENS | SPEC | SENS | SPEC | SENS | SPEC | SENS | SPEC | SENS | SPEC |
|---|---|---|---|---|---|---|---|---|---|---|
| ANN I | 96.2 | 85.7 | 92.3 | 95.8 | 92.3 | 90.5 | 88.5 | 95.8 | 95.8 | 92.3 |
| ANN II | 92.3 | 81.0 | 92.3 | 85.7 | 92.3 | 95.8 | 88.5 | 90.5 | 96.2 | 90.5 |
| ANN III | 88.5 | 85.7 | 84.6 | 95.8 | 92.3 | 90.5 | 96.2 | 90.5 | 96.2 | 90.5 |
| ANN average | 92.3 | 84.1 | 89.7 | 92.4 | 92.3 | 92.3 | 91.1 | 92.3 | 96.1 | 91.1 |

animals and approximately 90% of the BSE negative controls were identified in accordance with the true disease status of the donors. Furthermore, the similar classification results suggest that the specific methodology of data collection has no major impact on the differentiation between samples originating from BSE positive and BSE negative animals. It is also interesting to note that the rate of misclassifications of some individual samples clearly surpassed the average misclassification rate, regardless of the data acquisition technique used (and vice versa). These interesting details further suggest that indeed the sample's intrinsic compositional and structural properties, and not the data acquisition technique, determine the class assignment.

Under the experimental conditions used in our laboratory, both the dried and the liquid film technique are equally useful for mid-IR spectroscopy of serum. The dry film technique has some practical advantages, as it requires a less complicated sample preparation, minute amounts of sample volume, and only moderate technical expertise. Moreover, FT-IR spectrometers with multisample cuvette accessories for high sample throughput measurements of dried samples are commercially available (see, for example, the 96-well plate of Figure 1), while the liquid sample technique requires expensive hardware such as a flow-through cell and HPLC

equipment. A principal disadvantage of the dry film technique is, however, the fact that the spectral features of many IR bands are sensitive to changes in the humidity of the environment. This requires very careful control of environmental conditions during the measurements. In addition, the amide I region is often not available for quantitative analysis due to considerable sample heterogeneity and nonlinearity effects when the peak absorbance of the amide I band is significantly larger than 1.5 AU. Here lies the major strength of the liquid film technique, which allows IR spectra to be obtained in the conformation-sensitive amide I region with exceptionally high reproducibility.

# 3 IDENTIFICATION OF SCRAPIE INFECTION FROM SERUM BY INFRARED SPECTROSCOPY AND CHEMOMETRICS

## 3.1 Detection of terminal stages of a scrapie infection

The first study dealing with the application of IR spectroscopy to diagnose TSE from serum was published in 2002 by our laboratory at the Robert Koch-Institut.[22] In this study, IR spectra were acquired from sera of 312 hamsters using the dry film technique. All individual serum samples were routinely measured in triplicate, independently taken from three different sample preparations. The database established in this scrapie study comprised IR serum spectra from control hamsters and animals in the terminal stage of a scrapie infection. Animals were inoculated via oral, intracerebral, or intraperitoneal infection routes with brain material containing PrP$^{sc}$. In order to mimic potential defense mechanisms of the host against inoculated material, most of the control hamsters were similarly mock-infected using brain tissue from noninfected hamsters. In the terminal stages of the disease, all scrapie-infected animals showed neurological and behavioral symptoms of an advanced 263 K scrapie infection such as head bobbing, generalized tremor or ataxia of gait.

Classification analysis of the database of IR serum spectra was carried out by means of MLP-ANNs. This type of ANNs belongs to the group of supervised, or concept-driven, pattern recognition techniques in the sense that labeled subsets of patterns are required from which ANNs can "learn" how to perform classification.[34,35] Thus, supervised classification requires a teaching, or training phase in which given sets of patterns (spectra) and the class assignments are analyzed by the classifier. Thereafter, the classification model can be challenged by a set of independent samples, which should not be included during training and optimization of the model. It is important to stress that the accuracy of prediction should be determined only on the basis of these independent data.

Within the context of the scrapie study on terminally diseased animals, an ANN model was trained and optimized on the basis of two subsets of IR spectra, the training and internal validation data sets. The accuracy of classification was then determined by challenging the ANN model with a third spectral data set of independent external validation spectra.

The importance of preprocessing and feature extraction prior to classification is often overlooked. One of the results of the terminal scrapie study was that spectral preprocessing (derivatives,

**Table 4.** ANN classification results of the experimental scrapie study (terminal stage of disease). IR spectra acquired from serum samples of 184 hamsters (external validation data) were classified by an ANN that was trained and validated using spectra of another 128 animals. The following values of diagnostic test parameters have been calculated from the confusion matrix: accuracy: 98.9%, sensitivity: 97.5%, and specificity: 100%.

| | | Prediction by FT-IR spectroscopy | |
| --- | --- | --- | --- |
| | | − | + |
| Actual scrapie | − | 105 | 0 |
| status | + | 2 | 77 |

normalization) as well as adequate feature selection are essential prerequisites for classification analysis.[22] The classification results for animals of the external validation subset are given in the confusion matrix of Table 4. Among 79 hamsters with a positive scrapie status, 77 could be identified as terminally affected with scrapie. This corresponds to a true positive rate (sensitivity) of 97.5%. The true negative rate (specificity) was 100%, since all of the 105 control animals were classified correctly. These findings demonstrated for the first time the potential of FT-IR spectroscopy as a rapid and reliable method for the antemortem diagnosis of scrapie from serum, and possibly other TSE from blood. Although a number of important issues such as the question for the earliest time point at which spectral signs of a scrapie infection become detectable, or for the specificity of the spectral differences for a scrapie infection remained open, this initial feasibility study demonstrated the presence of pathological alterations in the sera of scrapie-infected hamsters.

What is the nature of the spectral differences on which the classification was based? At the time of publication of our results (2002), we could only speculate about this. From different attempts of data analysis by ANNs, we could learn that the classification accuracy was reduced when the C−H stretching region (2800–3100 cm$^{-1}$) was omitted from spectral analysis. We have interpreted this finding as evidence for differences in lipid composition in the sera of terminally infected and control animals.[22] Apart from these findings, we realized that the ANN approach is, on the one hand, particularly suitable for classification purposes. On the other hand, the approach is of only limited value to identify discriminative spectral markers. Since the spectral differences between serum spectra of scrapie positive and scrapie negative animals are rather small and distributed over many spectral regions, the identification of discriminative spectral features required further systematic investigations in which alternative concepts of spectral analysis needed to be employed.

The complexity of the feature identification problem is illustrated in Figure 5. This figure shows two representative spectra (normalized second derivatives) of the classes "scrapie positive" and "scrapie negative". The spectra exhibit only minute spectral differences and indicate that a visual identification of discriminative spectral features is virtually impossible. The "bar code"

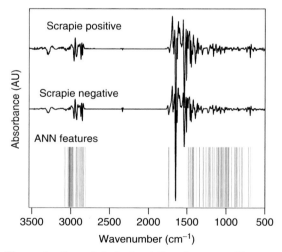

**Figure 5.** Second derivative infrared spectra of hamster serum from a scrapie-infected animal in the terminal stage of infection (scrapie positive) and a noninfected control animal (scrapie negative). Discriminative spectral features allowing disease identification by ANN are shown at the bottom. Spectra were acquired by using the dry film technique.

pattern in the lower part of Figure 5 shows the discriminant spectral features as revealed by the COVAR (vide infra) feature selection routine prior to ANN classification (see Reference 22 for details).

In the years following the first report, we have expanded our research efforts in several directions. One of these was the molecular identification of scrapie-associated spectral features. Another goal was to answer the question of whether and at which stage the IR-based approach allows preclinical detection of scrapie following uptake of the scrapie agent via the gastrointestinal tract. In the next section, we present the results of these studies.

## 3.2  Detection of preclinical stages of a scrapie infection in hamsters

Our efforts within the scope of a preclinical or time-course scrapie study focused on the detection of early spectral markers of an infection in sera from orally infected hamsters. In the hamster model, the initial target sites of a scrapie infection, the brain and spinal cord, displayed signs of PrP$^{sc}$ accumulation 70 days after peroral administration of the 263 K agent.[36] Our studies therefore involved analyses of sera from animals at incubation times of 70, 100, and 130 days post infection (dpi), of terminally ill animals (approximately 165 dpi) and of age-matched mock-infected control hamsters. Furthermore, in order to address molecular changes in serum possibly resulting from a presumed catabolic type of metabolism, a potential covariate of the late disease stages, we also analyzed spectra of control hamsters that had been deprived of food for 48 h.[37]

Similar to the approach employed in the terminal scrapie study (Section 3.1), the data analysis strategy for the time-course study was aimed at two goals. First, we attempted to establish an ANN model for classification. ANNs are powerful tools in this respect, but have the drawback that decision rules are not available in practice. Thus, it is basically impossible to get access to the specific spectral information on which the ANN classification relies. Thus, the second goal was to utilize univariate statistical methods for the detection of discriminative spectral features.

### 3.2.1 ANN classification

A preliminary inspection of sera from 70 dpi animals did not reveal significant spectral changes that would allow scrapie diagnosis at this disease stage. This and the specific requirements of ANN model developments (high sample numbers) prompted us to define the following classes in the time-course study:

(i) The "control" class included spectra from mock-infected and age-matched control hamsters. Additionally, this class contained spectra from noninfected individuals and hamsters deprived of food.
(ii) The "preclinical scrapie" class contained spectra from orally infected animals at 100 and 130 dpi.
(iii) The "terminal scrapie" class contained spectra originating from scrapie-infected hamsters in the terminal stages of the disease.

For the preclinical, or time-course study, a hierarchically organized modular ANN classifier was developed. Modular types of ANNs allow one to break down complex multiclass classification problems into smaller, preferably two-class (binary) classification tasks. These modular network classifiers offer a number of advantages, which have been discussed in detail elsewhere.[38,39] The most significant advantage of modular over monolithic ANNs is that tailored combinations of spectral features specifically optimized for each binary classification task can be employed. The hierarchy of ANNs utilized is schematically illustrated in Figure 6. The upper-level, or top-level, ANN was specifically designed to discriminate IR spectra of terminally infected individuals (class (iii)) from serum spectra of control or preclinical hamsters (classes (i) and (ii)). Differentiation of animals in the preclinical disease stages from control hamsters was then achieved by a second classifier, the

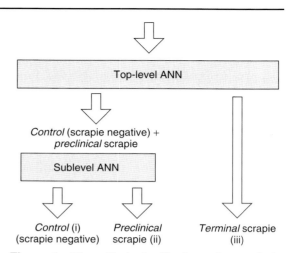

**Figure 6.** Hierarchical classification scheme of the modular ANN classifier developed for the identification of control hamsters, animals in the preclinical, and in the terminal stages of a scrapie infection. The top-level ANN was specifically designed to identify IR serum spectra from terminally ill and preclinical, or control hamsters. The discrimination between scrapie negative controls and hamsters in the preclinical stage of scrapie is subsequently achieved by a second classifier, the sublevel ANN.

sublevel network (see Figure 6). The strategies of network training and validation followed the principles already outlined in the Sections 2 and 3.1. According to these principles, the spectral data sets were divided into subsets for training, internal, and external validation with the classification results given only for the external validation subsets.

### 3.2.2 Top-level classification

The results obtained with the top-level classifier are summarized in Table 5. According to this confusion matrix, the external validation of the top-level ANN yielded a sensitivity of 97.7% and a specificity of 97.3%. These results confirmed our earlier findings in the terminal scrapie study (see Table 4 and Reference 22); in addition, they demonstrated the distinctness of serum spectra from animals at terminal and preclinical stages of scrapie.

**Table 5.** Time-course scrapie study: the confusion matrix shows the results of external validation of the top-level classifier used to identify terminally diseased hamsters. Note that the class "−" contains age-matched controls and also scrapie-infected animals in the preclinical stage of the disease (100 + 130 dpi). Accuracy: 97.7%, sensitivity: 97.3%, and specificity: 97.7%.

|  |  | Prediction by FT-IR spectroscopy | |
|  |  | − | + |
| --- | --- | --- | --- |
| Actual status | − | 380 | 9 |
| terminal scrapie | + | 1 | 36 |

**Table 6.** Time-course scrapie study: identification of hamsters in the preclinical stage of a scrapie infection (100 and 130 dpi, class "+"), age-matched controls (class "−"). The confusion matrix was obtained with the sublevel classifier on the basis of spectra from the external validation subset (see text for details). Accuracy: 93.1%, sensitivity: 87.1%, and specificity: 93.7%.

|  |  | Prediction by FT-IR spectroscopy | |
|  |  | − | + |
| --- | --- | --- | --- |
| Actual status | − | 269 | 18 |
| preclinical scrapie | + | 4 | 27 |

### 3.2.3 Sublevel classification

Classification results of the external validation data subset by the sublevel ANN are shown in Table 6. The rate of true positive classifications for preclinical animals was 27/31, which is equivalent to a sensitivity of 87.1%. The specificity obtained in the same data set equaled 93.7% (269 true negatives, 18 false positives) and the weighted accuracy of prediction was 93.1%. These results demonstrated that scrapie serum markers are detectable not only in the terminal stage of scrapie but are also present as early as 100 days post infection.

### 3.2.4 Serum markers

As pointed out above, the aim of the time-course study was not only the straightforward classification analysis but also the identification of serum constituents that form the molecular basis of the systematic differences between sera of scrapie-infected and control animals. Analyses of this type are nontrivial since mid-IR serum spectra are complex superpositions of spectral contributions of a large variety of different biomolecules. Furthermore, the differences are rather small and are detectable only if high standards of data acquisition (reproducibility, SNR) are met. For example, the class mean spectra for top- and sublevel classification exhibit a remarkably high degree of similarity (cf. Figure 7), making a visual identification of discriminative spectral features virtually impossible.

A popular strategy to overcome these problems is the use of feature selection methodologies, such as univariate t-tests or the multivariate COVAR (covariance) function of the NeuroDeveloper software, which is based on multiple covariance analysis and a partial F-value.[22] Both of these feature selection routines turned out to be particularly helpful in identifying discriminative spectral features. For example, univariate t-tests of the top-level classification problem demonstrated discriminative spectral features in the spectral regions of 1730–1765 cm$^{-1}$ ($\nu$(C=O)$_{\text{ester}}$) and 2800–3050 cm$^{-1}$ ($\nu$(C−H)), which are largely attributable to vibrational modes of lipids such as triacylglycerols and phospholipids. On the other hand, the same type of tests of sublevel classes exhibited only insignificant alterations in these regions (see also Figure 7). It turned out that the four most significant "preclinical" spectral features are found either in the amide I (∼1630 and ∼1649 cm$^{-1}$) or in the amide A or B (∼3358 and ∼3155 cm$^{-1}$, respectively) spectral regions.[37] Since these bands arise from vibrations of proteins, one could hypothesize that these findings are somehow related to the causative agent of TSE, PrP$^{sc}$. In particular, we noticed that the pathological isoform of PrP$^{sc}$ is rich in $\beta$-sheet structure elements, which are known to exhibit strong absorptions in the amide I region near 1630 cm$^{-1}$. Nevertheless, we do not believe that the discriminatory spectral features observed at 1630 cm$^{-1}$ are directly related to the presence of PrP$^{sc}$ in the sera of preclinical hamsters.

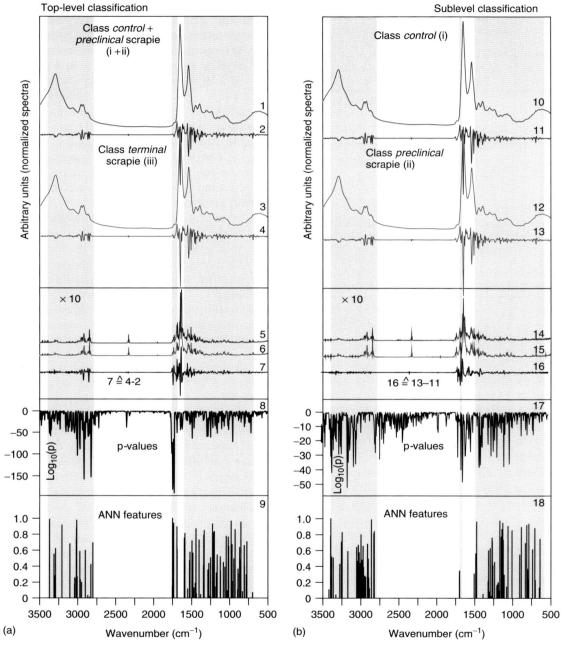

(a)

(b)

First and foremost, the analytical sensitivity of IR spectroscopy is undoubtedly too low for the specific detection of subfemtomolar concentrations of PrP$^{sc}$ in a background of highly abundant proteins such as albumin, immunoglobulins, and others. Furthermore, the 1630 cm$^{-1}$ feature was not identified in sera of terminally infected hamsters in which the concentration of PrP$^{sc}$ is expected to be much higher. Therefore, it seems more plausible that other—not yet identified—β-sheet rich serum proteins are responsible for these changes.

Another important aspect to be discussed in the context of the time-course study is the problem of the specificity of the serum alterations detected by FT-IR spectroscopy. If the causative agent of the disease cannot be directly measured, one has to verify the specificity of the method by analyzing sample spectra from individuals suffering from alternative diseases (differential diagnoses). However, the availability of these samples is often limited.

Covariates such as a systematic instrument drift, age, or nutrition status of the sample donors may additionally mislead the investigator who is trying to understand the spectral differences. Such covariates, or confounding factors, are indeed closely related to the specificity of the classification technique; therefore, it is important to investigate these factors. For example, we have observed that neurological and behavioral symptoms of an advanced 263 K-scrapie infection affect the hamsters' ability to drink and to ingest food. As a consequence, one may assume that the altered composition of serum lipids in the very late stages of a scrapie infection is in fact a result of a catabolic type of metabolism. Since the lipid metabolism is known to be associated with the actual nutrition status, it is conceivable that the chemistry of serum lipids is disturbed nonspecifically at least in the terminal stages of a scrapie infection. These considerations could be backed by serum lipid analyses using the semiquantitative technique of high-performance thin layer chromatography (hpTLC). Indeed, comparative lipid serum analyses of hamsters deprived of food for 48 h, mock-infected controls, and animals in the terminal stages of scrapie revealed a systematic decrease of triglycerides and cholesterol serum concentrations in the groups of terminally ill and nutrition-deprived animals.[37] Thus, it cannot be ruled out that some of the spectral features utilized by the ANN in the terminal scrapie study were at least in part nonspecific for TSE.

A popular strategy to effectively eliminate covariates when developing supervised classifiers is to modify the composition of the training subset in a way that the (hidden) covariates are, in terms of numbers, equally represented in the disease and comparison group. For this reason, we have added serum spectra from noninfected hamsters exhibiting catabolic serum patterns to

---

**Figure 7.** IR serum spectra of scrapie-infected hamsters. Spectra were collected from serum films dried on ZnSe crystals. (a) top-level ANN classification—discrimination between control or preclinical hamsters (classes (i) + (ii)) and animals in the terminal stage of a scrapie infection (class (iii)). (b) sublevel ANN classification—differentiation between the scrapie stage *preclinical* (class (ii)) and *controls* (scrapie negative, class (i)). Traces 1, 3, 10, 12: mean absorbance spectra of the classes (i) + (ii), (iii), (i), and (ii), respectively. Traces 2, 4, 11, 13: mean second derivative spectra of the classes (i) + (ii), (iii), (i), and (ii), respectively. Traces 5, 6, 14, 15: standard deviations of traces 2, 4, 11, and 13, respectively. Intensities are expanded by a factor of 10. Traces 7 and 16: differences between traces 4 and 2, and 13 and 11, respectively (expanded by a factor of 10). These spectra denote the main spectral differences on which the top- or the sublevel classifiers are based. Traces 8 and 17: p-values obtained by independent t-tests using the intensities of vector-normalized second derivative spectra as inputs. Small p-values cast a doubt on the null-hypothesis of equal class means. Note the logarithmic scaling. Traces 9 and 18: spectral features found by the COVAR function of Synthon's NeuroDeveloper ANN software package.[38] The height of the bars denotes the relative significance of the individual feature as obtained by the COVAR function. Identification of spectral features was carried out using information from highlighted spectral regions.

the comparison group of the top-level problem. By doing so, the covariate "catabolic IR fingerprint" could be represented in both classes of the top-level ANN, i.e., in the classes terminal scrapie (iii) and preclinical or control animals (ii+i). Interestingly, this modification of the sample composition did not reduce the classification accuracy for terminal stages of scrapie (see Tables 4 and 5). The validity of this approach was also confirmed by successful elimination of spectral markers of catabolism as discriminative spectral features (see also C—H region of trace 9 in Figure 7 and the discussion in Reference 37).

In order to advance the IR spectroscopy–based methodology for scrapie detection, and to transfer the technique to other TSE, we believe it is important to further address the problem of specificity. However, a comprehensive approach targeting this important issue would require the analysis of serum samples from individuals with a broad spectrum of alternative infectious or noninfectious diseases such as classical infections of bacterial, viral, fungal, or parasitic origins, or neurodegenerative diseases such as Alzheimer's disease. From a practical point of view, such an array of samples could not be collected within the frame of our hamster animal model. Having shown a proof-of-concept, it would be more appropriate to focus the experimental efforts in the later project stages to more relevant diseases and species: BSE in cattle, scrapie in sheep, CWD in deer and elk, and finally the Creutzfeldt–Jakob disease (CJD) in humans. Only in these species would a large-scale field study be both feasible and diagnostically relevant.

# 4 IDENTIFICATION OF BSE INFECTION IN CATTLE: RESULTS OF BLINDED VALIDATION STUDIES

Over the past few years, several research groups have initiated programs to specifically detect BSE infections in serum by a combination of FT-IR spectroscopy and chemometrics.[40–42]. Since the first and—in terms of sample number and variety—most comprehensive study was published by the principal author of this article, this section focuses mainly on the results of the RKI research group.

Contrary to the experimental scrapie studies described in the last section, the BSE study was carried out as a truly blinded field study. Blinded studies differ from openly labeled studies in that the disease status of the animals to be validated is not known to the examiner. Another important difference between our BSE field and experimental scrapie studies lies in the fact that the subsets of samples from BSE-infected and control cattle comprised unmatched samples in terms of disease stage, age, breed, and gender of the animals, and potentially other diseases. The latter aspect is important as the BSE subsets contained not only spectra from TSE-diseased (BSE positive) and healthy control animals but also included samples originating from *true* controls, i.e., cattle with a clinical suspicion for BSE, but a negative BSE test result. Furthermore, the control groups were supplemented with serum samples of animals suffering from a variety of classical viral or bacterial infectious diseases (see Reference 40 for details). The spectra of these cattle samples exhibited a number of interesting spectral changes, particularly in the amide I region, which are discussed in detail in the next section. Serum samples from animals with a confirmed BSE status (histopathology, immunocytochemistry) were supplied by the Veterinary Laboratories Agency (VLA, Weybridge, UK). Sera of BSE negative controls were either acquired from the same institution or from various sources in Germany.[40]

The experimental methods employed in the BSE studies were the same as discussed in the previous section: mid-IR spectra were acquired as triplicates from serum samples using the dry film technique. Spectra were then tested for spectral quality and preprocessed for further analysis. Splitting of the spectral data sets for training and validating supervised ANN classifiers was carried out according to the principles described in Sections 2 and 3.

In this section, we wish to summarize the results of two different BSE studies. These studies differed in the number of cattle examined as well as in the methods of data analysis. The so-called VLA study was carried out in cooperation with Synthon GmbH and the VLA, and included blinded validation with sera from 92 cattle. The second study (Roche study) involved blinded validation of a much larger data set of serum samples from 260 animals supplied by Roche Diagnostics GmbH.

## 4.1 VLA study

For this first BSE validation study, the training process was based solely on ANNs. In the VLA study, IR spectra used for training and internal validation originated from a total of 701 animals. Classification of the blinded validation subset (spectra of 92 animals) by an optimized ANN resulted in an overall classification accuracy of 84.8%. The rate of true positive classifications (sensitivity) was 85.4% (41/48) and the corresponding true negative rate (specificity) equaled 84.1% (37/44).[40] These findings confirmed results and conclusions of our previous scrapie study and demonstrated the capability of FT-IR spectroscopy as a sensitive technique for the detection of BSE infections from serum. With this proof of feasibility, we could show for the first time that the BSE status of cattle and the mid-IR fingerprint of serum are closely related. However, when aiming at implementing this method into a practical application, further improvements of the accuracy of prediction had to be achieved. Thus, we initiated a second BSE study, the Roche study, two months later, which was carried out on the basis of an expanded sample set and improved classification models.

## 4.2 Roche study

Spectral data in the Roche study were analyzed either by a single ANN or by means of a decision tree.[40] In the following discussion, we focus on the results attained by the decision tree metaclassifier. The decision tree was established and optimized on the basis of individual neural network models that had been independently developed in our laboratory at the RKI, and by our collaborating partners at Synthon. Additional classification techniques included linear and quadratic discriminant analysis (LDA/QDA) using optimally discriminating spectral subregions specifically selected by a genetic algorithm. Development and optimization of this decision tree model (see Figure 8) was carried out on the basis of the subclassifier's performance such as sensitivity, specificity, or robustness. The latter properties were obtained by challenging the ANN or LDA/QDA classifier by spectra of the internal validation subset. Spectral subsets for training and internal validation comprised triplicates of IR spectra from a total of 843 animals. Furthermore, the blinded validation subset contained IR spectra from another 260 animals. In terms of sample, or animal, numbers, the Roche study thus represents, to the best of our knowledge, the largest study in the field of biomedical vibrational spectroscopy.

The expansion of the training and internal validation subsets by additional samples and, more importantly, improvements of the classification model resulted in a significantly improved accuracy of prediction. The decision tree classifier correctly identified 85 of the 89 BSE positive animals and 158 of the 171 from the BSE negative control group, which corresponds to a numerical sensitivity of 95.5% (89.8–98.5%) and a specificity of 92.4% (89.4–93.9%) respectively with a 95% confidence interval (see Table 7). Thus, a sensitivity and specificity of larger than 89% was interpreted by us as a successful proof-of-principle of the IR-based technique. Furthermore, these findings suggest that the combination of IR spectroscopy and chemometrics can serve as a technique for fully automatic, objective, and rapid diagnosis of BSE, or other TSE, from serum.

The successful application of IR spectroscopy for antemortem detection of BSE and other TSE was confirmed a few months later by other scientists. Coworkers of microbiolytics GmbH presented an antemortem test system for the

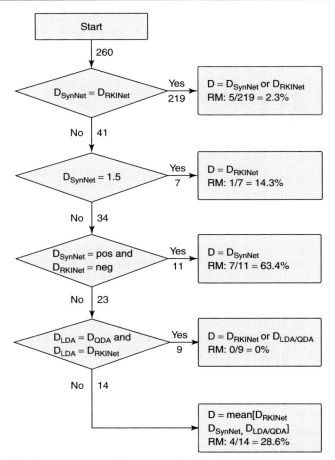

**Figure 8.** Flowchart of the decision tree used as a metaclassifier in the BSE field (Roche) study. In this study, BSE infections in cattle could be successfully identified on the basis of dried film IR serum spectra and supervised classification techniques (decision tree, see text for details). $D_{RKINet}$: class assignment by an ANN trained and optimized at RKI, SENS ~ SPEC, accuracy: high, robustness: moderate. $D_{LDA/QDA}$: class assignment by linear/quadratic discriminant analysis, SENS ~ SPEC, accuracy: moderate, robustness: high. $D_{SynNet}$: class assignment by an ANN trained and optimized by Synthon, SENS < SPEC, accuracy: high, robustness: moderate. RM: rate of misclassifications, or percentage of false classifications.

diagnosis of BSE by direct FT-IR measurements on liquid serum samples.[41] For IR serum measurements, a fully automated bench-top analyzer was used that was constructed around a proprietary flow-through cell. Using the liquid film technique, the authors acquired IR spectra from 390 animals to teach a support vector machine (SVM) classifier. The external validation subset contained spectra of another 80 cattle and a classification accuracy of 100% was reported. This "AquaSpec

BSE" rapid antemortem test was later evaluated in 2006 by the scientific expert working group of the European food safety authority (EFSA). Under more strict experimental conditions, mainly due to the presence of true control samples, the AquaSpec technology could not reproduce the perfect test results of 100% and achieved a sensitivity of 88.2% (74.4–95.7) and a specificity of 78.4% (74.3–82.3, confidence level of 0.95).[43]

**Table 7.** Confusion matrix showing the results of blinded validation obtained within a field study of diagnosing BSE infections from serum (Roche study). The IR-based classification model was established by training and validating a classification algorithm (decision tree) with 843 serum samples (see text for details). Accuracy: 93.5%, sensitivity: 95.5% (89.8–98.5)%*, and specificity: 92.4% (89.4–93.9)%*. Asterisk indicates 95% confidence interval.

|  |  | Prediction by FT-IR spectroscopy | |
|---|---|---|---|
|  |  | − | + |
| Actual BSE | − | 158 | 13 |
| status | + | 4 | 85 |

The authors of a third BSE study utilized the dry film technique to classify signatures of BSE in serum using IR spectroscopy.[42] A high throughput screening extension (HTS/XT) FT-IR spectrometer, operated in transmission mode and equipped with a room-temperature pyroelectric detector was employed for the acquisition of serum spectra. Serum samples were spotted by an automatic spotting system onto silicon wafers that served as optically transparent substrates. Again, a supervised classification approach was used to distinguish IR spectra from BSE positive and control animals. Altogether, spectra were collected from 641 animals, among them 481 were to teach the classifier and 160 were for blinded validation. For independent validation under blinded conditions, a classification model was developed that was based on four separate classification methods: LDA, robust linear discriminant analysis (r-LDA), ANNs, and SVM. The results of these individual classifiers were then combined by a voting scheme, which revealed a sensitivity of 96.1% and a specificity of 92.9% for the blinded validation data set. The reported classification accuracy was 94.4%, which did not significantly differ from the accuracy achieved in our Roche study. Furthermore, at a confidence level of 0.95 ($\alpha = 0.05$), a sensitivity of >85% and specificity of >90% was established, which confirmed our findings (>89 % for both sensitivity and specificity, see Table 7).

# 5 DETERMINATION OF THE ALBUMIN/GLOBULIN RATIO OF SERUM BY IR SPECTROSCOPY

Our IR spectroscopic studies on the detection of scrapie infections in hamsters, or BSE in cattle, differed with respect to the donor animals, strains of TSE agents, sample numbers, and the classification techniques employed. While the experimental scrapie studies were carried out on sera isolated from scrapie-infected hamsters and mock-infected healthy control animals, the BSE studies additionally involved serum analyses from cattle suffering from distinct "classical" viral or bacterial infectious diseases. This was of particular importance since it allowed us to identify and compare both BSE specific spectral changes and serum alterations associated with classical infectious diseases. Encouraged by the success of the IR-based serum technique for TSE diagnosis, and inspired by unexpected findings in the spectra of cattle with bacterial/viral infections, we initiated a separate program to explore spectral alterations associated with viral or bacterial infections.

Within the context of these studies, serum samples from cattle with known infections of bacterial (*Escherichia coli, Staphylococci, Streptococci* or *Proteus*) or viral origin (bovine leukaemia virus, bovine respiratory syncytial virus, bovine herpes virus-1, or bovine diarrhea virus) were analyzed by the liquid film technique.

A first survey of the second-derivative spectra revealed remarkable spectral changes, predominantly in the amide I/II region (see Figure 9a). Particularly, it was found that most of the second derivative IR spectra exhibit relatively high absorptions of the amide I band components at $1657\,cm^{-1}$ and less intense contributions of the band components at 1637 and $1690\,cm^{-1}$ ("class A" spectra in Figure 9a). On the other hand, we found in some spectra exactly the opposite situation with the intensities at 1637 or $1690\,cm^{-1}$ being higher than normal compared to the $1657\,cm^{-1}$ band component ("class C" spectra). Spectra of "class B" in Figure 9(a)

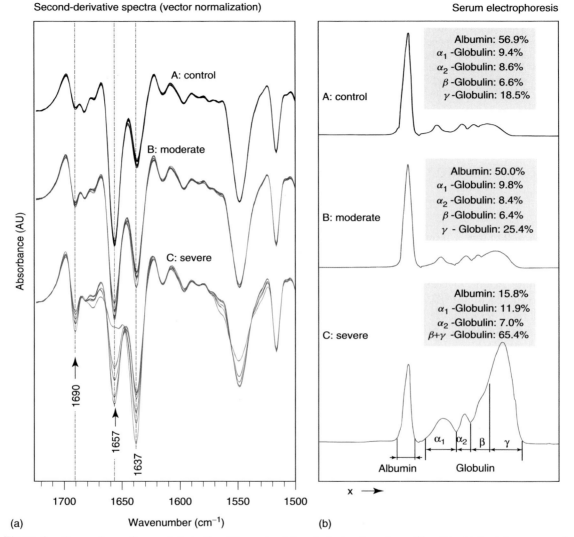

**Figure 9.** Comparison of second-derivative IR spectra (a) and electrophoresis profiles (b) obtained from sera of healthy control cattle (A) and cattle suffering from acute infections (B: moderate and C: severe infections). IR spectra were measured from liquid sera. Five spectra per class are given. Typical electrophoresis profile are shown in (b). The infrared spectral changes found in the amide I region are attributable to serum proteins and can be correlated with the albumin/globulin ratio detectable by serum electrophoresis (see text for details).

represent an intermediate state. The spectroscopic interpretation of these interesting findings is well known to spectroscopists: the amide I band of proteins is conformation sensitive, i.e., this band provides specific information on the secondary structure of proteins. For example, the amide I band component at $1657\,\text{cm}^{-1}$ is typically assigned to $\alpha$-helical structures and the two bands at 1637 and $1690\,\text{cm}^{-1}$ are characteristic for antiparallel $\beta$-sheet structures.[44,45] Obviously, many of the spectra are dominated by contributions from proteins that exhibit mostly $\alpha$-helical structures, while others, mainly from animals suffering from bacterial infections, display significant contributions originating from proteins rich in $\beta$-sheets.

In order to better understand the underlying molecular or compositional changes, sera of all types (classes A–C) were analyzed additionally by classical serum protein electrophoresis. A selection of typical electrophoresis profiles is shown in Figure 9(b). For example, the electrophoresis curve of a "class A" serum shows a high relative content of albumin and a low globulin concentration. In turn, the electrophoresis profiles of a "class C" serum samples exhibit a dramatic increase of the globulin, particularly of the $\gamma$-globulin fraction (see Figure 9b). These observations explain our spectroscopic findings quite well. Albumin is known to be a protein that contains almost completely $\alpha$-helical structures, while immunoglobulins as the major fraction of the $\gamma$-globulin fraction are rich in $\beta$-sheet structure elements.

The quantitative relationships between marker bands for $\alpha$-helical proteins ($1657\,cm^{-1}$) and $\beta$ sheets ($1637\,cm^{-1}$) and the fraction of albumins, or globulins, as determined by serum electrophoresis, are given in Figure 10. Both graphs demonstrate a high degree of linear correlation between the albumin/globulin ratio in serum and the intensities of these conformation-sensitive marker bands. Thus, the albumin/globulin ratio in serum can be precisely detected by IR spectroscopy from liquid serum samples. This fact could be of practical importance because the albumin/globulin ratio is an important parameter frequently determined in clinical routine diagnostics.

Furthermore, we found that a remarkably similar correlation also exists in human serum samples. Serial IR measurements on liquid serum samples from human donors displayed the same spectral pattern in the amide I region: spectra from individuals with acute bacterial infections and with confirmed high levels of $\gamma$-globulins exhibited significantly increased absorbance at 1637 and $1690\,cm^{-1}$ (see Figure 11 and Reference 46 for details).

In the following paragraphs, we wish to specify briefly the conditions that can alter the albumin/globulin ratio of serum. In clinical practice, much of the interest is focused on $\gamma$-globulins because immunoglobulins migrate to the $\gamma$-globulin region (cf. electrophoresis profiles of Figure 9b). Gammopathies, i.e., changed serum concentrations of immunoglobulins, can be monoclonal or polyclonal. Monoclonal gammopathies are clonal proliferations characterized by a spike-like pattern in the $\gamma$-globulin region. They can be found in some chronic processes (amyloidosis, cirrhosis)

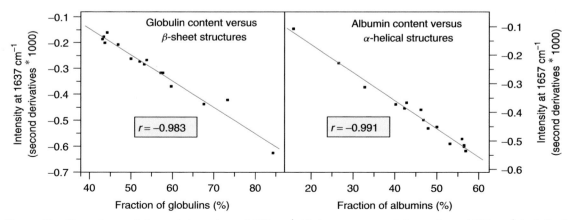

**Figure 10.** Dependence of the peak intensity at $1637\,cm^{-1}$ ($\beta$-sheet structures, left panel), or $1657\,cm^{-1}$ ($\alpha$-helical structures, right panel) on the relative globulin, or albumin content in the serum from cattle.

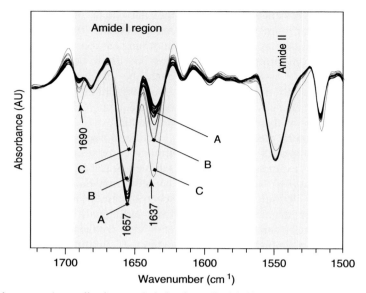

**Figure 11.** Infrared spectra (normalized second derivatives, liquid film technique) from sera of human donors with a normal albumin/globulin ratio (A), or moderate (B) and severe (C) cases of hyperglobulinemia. Band assignment: 1637 and 1690 cm$^{-1}$: antiparallel $\beta$-pleated sheet structures of proteins, 1657 cm$^{-1}$: $\alpha$-helical structures of proteins.

but are also typical for malignant conditions such as Hodgkin's disease, multiple myeloma, and chronic lymphatic leukemia (CLL). Polyclonal gammopathies are much more common and can be associated with any reactive or "classical" inflammation process.

Changes of the albumin concentration of serum are observed under circumstances in which there is either less production (liver diseases, starvation) or an increased loss or degradation of albumin such as in nephrotic syndrome, or in the protein-losing enteropathy of the bowel.

The results presented here illustrate how IR spectroscopy of serum can be used to obtain quantitative serum parameters, such as the concentration of particular serum constituents, or ratios between them. This approach differs conceptually from the DPR approach in that either a single or a set of IR parameters is obtained from serum. This once more highlights a particular strength of the IR technique, that both types of evaluation strategies are possible from one and the same spectrum.

## 6 OUTLOOK AND CONCLUSIONS

In the previous sections, we have presented experimental results that clearly demonstrate the potential of FT-IR spectroscopy and advanced methods of spectral data analysis for the diagnosis of TSE from serum. In an experimental scrapie study on terminally scrapie-infected hamsters, a sensitivity of 97% and a specificity of 100% were obtained. The hamster model was also used to test for the earliest time point at which a scrapie infection can be diagnosed. These investigations revealed subtle, but reproducible spectral changes in the sera of infected donors at a preclinical stage and permitted the identification of a scrapie infection at 100 dpi and later, but not at 70 dpi. Furthermore, in the context of an antemortem BSE field study, we attained classification results of up to 96% (sensitivity) and 92% (specificity). These findings demonstrated the presence of TSE-associated spectral markers in the sera of infected donors and suggested that further development of the method could lead toward a fully automatic, objective, and rapid technique for antemortem serum testing.

Within the scope of our studies, we have also attempted to reveal the molecular identity of the serum constituents that form the basis of the TSE serum tests. This is technically challenging but we have observed that exclusion of a few discriminative spectral features from classification analysis did not significantly alter the accuracy of classification. The latter finding indicates that the IR technique is not based on one unique serum marker alone and suggests that a plurality of serum constituents is involved. Thus, it seems more plausible that the IR serum technique is based on a number of yet unidentified TSE surrogate markers rather than on the misfolded pathological prion protein (PrP$^{sc}$).

In order to transfer the IR technique into a practical application, the classification accuracy needs to be improved. This could be achieved by extending the spectral databases by enlarging the number of training samples and, more importantly, by addressing covariates with potential impact on the IR spectral disease patterns. Particularly, the effects of alternative diseases, but also of age, breed, gender, and nutrition status, require further attention. These attempts should focus on an experimental proof to demonstrate the specificity of the surrogate markers (and of the IR features) for TSE.

On the other hand, we do not believe that improvements of the classification models will give significantly better classification results. In our experience, the sample's intrinsic compositional and structural properties form the basis of misclassifications (see discussion of Section 2), so that the presented metaclassifiers such as the decision tree[40] or the voting scheme used by the Roche group of Wolfgang Petrich[42] already constitute highly optimized classification models that offer only limited possibilities for further improvements. In this context, it might be interesting to discuss an alternative classification approach suggested by Menze *et al.*[47] The authors of this study reanalyzed the database of BSE spectra compiled by the Roche group by means of

a hierarchical classification scheme in which firstly binary subsets between samples originating from diseased and nondiseased cattle were defined along known covariates. Then, random forests,[48] which are combinations of tree predictors that vote together for the most popular class, were applied. Subsequently, binary discriminations within each subset were established by means of ridge regression and at the second stage the predictions from all linear classifiers were used as input to another random forest for final classification. Compared to the voting scheme applied in the original study,[42] the new classifier provided a nearly identical accuracy of classification; the authors reported that the hierarchical scheme performed as well as the neural network or the support vector machine subclassifiers.[42]

The largest potential to further increase the diagnostic sensitivity and specificity of the IR technique for TSE testing is expected from sample preconditioning. Researchers at the Institute for Biodiagnostics of the National Research Council in Canada recently presented an approach that allows sensitive IR spectroscopy of serum biomarkers of relatively low concentrations.[49,50] The technique is based on reproducible sample preconditioning by means of laminar fluid diffusion interfaces (LFDI) and allows detection of analytes below the IR detection thresholds of unprocessed serum.[49,50] Similar to techniques widely applied in proteomic mass spectroscopic analyses of serum, or plasma, the LFDI technique can be used for separation, i.e., for reproducible removal of high-abundance protein species of low diagnostic value such as albumin or immunoglobulins. The applicability of the technique to FT-IR serum analysis is expected not only to substantially broaden the diagnostic potential of the DPR fingerprint technique but also to sensitively detect diagnostically relevant low-abundance compounds from serum.

In summary, we have shown that a combination of FT-IR spectroscopy and advanced methods of data analysis can be employed for antemortem diagnosis of TSE infections from serum. In future studies, it will be important to identify the molecular nature of the underlying TSE-associated serum alterations by more molecule-specific analytical techniques, such as various hyphenated mass spectroscopy techniques, e.g., gas chromatography-MS or HPLC-MS. We have great confidence that the use of metabolomic and/or proteomic approaches is the way forward to identify TSE-specific surrogate markers from blood or its components.

## ABBREVIATIONS AND ACRONYMS

| | |
|---|---|
| ANN | Artificial Neural Network |
| AU | Absorbance Units |
| ATR | Attenuated Total Reflection |
| BSE | Bovine Spongiform Encephalopathy |
| CJD | Creutzfeldt–Jakob Disease |
| CLL | Chronic Lymphatic Leukemia |
| CNS | Central Nervous System |
| CWD | Chronic Wasting Disease |
| dpi | Days Post Infection |
| DPR | Disease Pattern Recognition |
| DTGS | Deuterated Triglycine Sulfate |
| FT-IR | Fourier Transform Infrared |
| GA | Genetic Algorithm |
| GC-MS | Gas Chromatography-MS |
| HPLC | High-Performance Liquid Chromatography |
| hpTLC | High-Performance Thin Layer Chromatography |
| HTS/XT | High Throughput Screening Extension |
| IR | Infrared |
| IRE | Internal Reflection Element |
| LDA | Linear Discriminant Analysis |
| LFDI | Laminar Fluid Diffusion Interfaces |
| MCT | Mercury Cadmium Telluride |
| MLP | Multilayer Perceptron |
| MS | Mass Spectroscopy |
| PrP | Prion Protein |
| QDA | Quadratic Discriminant Analysis |
| RKI | Robert Koch-Institut |
| ROC | Receiver Operating Characteristic |
| SNR | Signal-to-Noise Ratio |
| SVM | Support Vector Machine |
| TSE | Transmissible Spongiform Encephalopathy |
| vCJD | Variant Creutzfeldt–Jakob Disease |
| VLA | Veterinary Laboratories Agency |

## REFERENCES

1. D. Naumann, D. Helm and H. Labischinski, *Nature*, **351** (6321), 81–82 (1991).

2. H. Fabian, M. Jackson, L. Murphy, P. Watson, I. Fichtner and H.H. Mantsch, *Biospectroscopy*, **1**, 37–45 (1995).

3. C.O. Schultz, K.Z. Liu, J.B. Johnston and H.H. Mantsch, *Leuk. Res.*, **20** (8), 649–55 (1996).

4. N. Jamin, P. Dumas, J. Moncuit, W.H. Fridman, J.L. Teillaud, G.L. Carr and G.P. Williams, *Proc. Natl. Acad. Sci. U. S. A.*, **95** (9), 4837–4840 (1998).

5. M. Diem, S. Boydston-White and L. Chiriboga, *Appl. Spectrosc.*, **53** (4), 148A–161A (1999).

6. J.D. Kruse-Jarres, G. Janatsch, U. Gless, R. Marbach and H.M. Heise, *Clin. Chem.*, **36** (2), 401–402 (1990).

7. R.A. Shaw, S. Kotowich, H.H. Eysel, M. Jackson, G.T. Thomson and H.H. Mantsch, *Rheumatol. Int.*, **15** (4), 159–165 (1995).

8. R.A. Shaw, S. Kotowich, M. Leroux and H.H. Mantsch, *Ann. Clin. Biochem.*, **35**, 624–632 (1998).

9. A. Staib, B. Dolenko, D.J. Fink, J. Fruh, A.E. Nikulin, M. Otto, M.S. Pessin-Minsley, O. Quarder, E. Somorjai, U. Thienel, G. Werner and W. Petrich, *Clin. Chim. Acta*, **308** (1–2), 79–89 (2001).

10. E. Diessel, S. Willmann, P. Kamphaus, R. Kurte, U. Damm and H.M. Heise, *Appl. Spectrosc.*, **58** (4), 442–450 (2004).

11. E. Diessel, P. Kamphaus, K. Grothe, R. Kurte, U. Damm and H.M. Heise, *Appl. Spectrosc.*, **59** (4), 442–451 (2005).

12. L.P. Choo, D.L. Wetzel, W.C. Halliday, M. Jackson, S.M. LeVine and H.H. Mantsch, *Biophys. J.*, **71**, 1672–1679 (1996).

13. L.H. Kidder, V.F. Kalasinsky, J.L. Luke, I.W. Levin and E.N. Lewis, *Nat. Med.*, **3** (2), 235–237 (1997).

14. C. Marcott, R.C. Reeder, E.P. Paschalis, D.T. Tatakis, A.L. Boskey and R. Mendselsohn, *Cell Mol. Biol.*, **44** (1), 109–115 (1998).

15. P. Lasch and D. Naumann, *Cell Mol. Biol (Noisy-le-grand)*, **44** (1), 189–202 (1998).

16. K.Z. Liu, R.A. Shaw, A. Man, T.C. Dembinski and H.H. Mantsch, *Clin. Chem.*, **48** (3), 499–506 (2002).

17. K.Z. Liu and H.H. Mantsch, *Am. J. Obstet. Gynecol.*, **180**, 696–702 (1999).

18. M. Wang, M. Sowa, H.H. Mantsch, A. Bittner and H.M. Heise, *Trends Anal. Chem.*, **15**, 286–295 (1996).

19. V.R. Kondepati, U. Damm and H.M. Heise, *Appl. Spectrosc.*, **60** (8), 920–925 (2006).

20. D. Rohleder, G. Kocherscheidt, K. Gerber, W. Kiefer, W. Kohler, J. Mocks and W. Petrich, *J. Biomed. Opt.*, **10** (3), 031108 (2005).

21. W. Petrich, B. Dolenko, J. Frh, M. Ganz, H. Greger, S. Jacob, F. Keller, A.E. Nikulin, M. Otto, O. Quarder, R.L. Somorjai, A. Staib, G. Werner and H. Wielinger, *Appl. Opt.*, **39** (19), 3372–3379 (2000).

22. J. Schmitt, M. Beekes, A. Brauer, T. Udelhoven, P. Lasch and D. Naumann, *Anal. Chem.*, **74** (15), 3865–3868 (2002).

23. R.A. Shaw, H.H. Eysel, K.Z. Liu and H.H. Mantsch, *Anal. Biochem.*, **259** (2), 181–186 (1998).

24. J.N. Adkins, S.M. Varnum, K.J. Auberry, R.J. Moore, N.H. Angell, R.D. Smith, D.L. Springer and J.G. Pounds, *Mol. Cell Proteomics*, **1** (12), 947–955 (2002).

25. N.L. Anderson and N.G. Anderson, *Mol. Cell Proteomics*, **1** (11), 845–867 (2002).

26. S. Hu, J.A. Loo and D.T. Wong, *Proteomics*, **6** (23), 6326–6353 (2006).

27. S.B. Prusiner, *Proc. Natl. Acad. Sci. U.S.A.*, **95**, 13363–13383 (1998).

28. P. Brown, *Vox Sang.*, **89** (2), 63–70 (2005).

29. L. Cervenakova and P. Brown, *Expert Rev. Anti Infect. Ther.*, **2** (6), 873–880 (2004).

30. R. Masuch and D.A. Moss, *Appl. Spectrosc*, **57**, 1407–1418 (2003).

31. H. Fabian, P. Lasch and D. Naumann, *J. Biomed. Opt.*, **10** (3), 031103 (2005).

32. D. Naumann, 'Infrared Spectroscopy in Microbiology', in "Encyclopedia of Analytical Chemistry", ed R.A. Meyers, John Wiley & Sons Ltd., Chichester (2000).

33. D. Helm, H. Labischinski, G. Schallehn and D. Naumann, *J. Gen. Microbiol.*, **137** (1), 69–79 (1991).

34. C.M. Bishop, 'Neural Networks for Pattern Recognition', Oxford University Press, (1995).

35. J.P. Marques de Sa, 'Pattern Recognition: Concepts, Methods and Applications', Springer-Verlag, Berlin (2001).

36. P.A. McBride, W.J. Schulz-Schaeffer, M. Donaldson, M. Bruce, H. Diringer, H.A. Kretzschmar and M. Beekes, *J. Virol.*, **75** (19), 9320–9327 (2001).

37. P. Lasch, M. Beekes, J. Schmitt and D. Naumann, *Anal. Bioanal. Chem.*, **387** (5), 1791–1800 (2007).

38. T. Udelhoven, D. Naumann and J. Schmitt, *Appl. Spectrosc.*, **54** (10), 1471–1479 (2000).

39. J. Schmitt and T. Udelhoven, in 'Infrared and Raman Spectroscopy of Biological Materials', eds H.U. Gremlich and B. Yan, Marcel Dekker, New York (2000).

40. P. Lasch, J. Schmitt, M. Beekes, T. Udelhoven, M. Eiden, H. Fabian, W. Petrich and D. Naumann, *Anal. Chem.*, **75** (23), 6673–6678 (2003).

41. R. Masuch, A. Seidel, A. Wolf, K. Schuster and R. Jankowsky, A Novel Ante Mortem test System for the Diagnosis of Bovine Spongiform Encephalopathy Base on Direct FT-IR Measurements in Liquid Serum Samples, in Poster presented at the "Internatl. Prion Conference Munich 2003", München, Germany, Oct. 8–10, (2003).

42. T.C. Martin, J. Moecks, A. Belooussov, S. Cawthraw, B. Dolenko, M. Eiden, J. Von Frese, W. Kohler, J. Schmitt, R. Somorjai, T. Udelhoven, S. Verzakov and W. Petrich, *Analyst*, **9** (10), 897–901 (2004).

43. EFSA, **95** 1–14 (2006), http://www.efsa.europa.eu/en/science/tse_assessments/bse_tse/report_ej95_animal.html.

44. F.S. Parker, 'Infrared Spectroscopy in Biochemistry, Biology, and Medicine', Plenum Press, (1971).

45. A. Barth and C. Zscherp, *Q. Rev. Biophys.*, **35** (4), 369–430 (2002).

46. D. Naumann, P. Lasch and H. Fabian, 'Cells and Biofluids Analysed in Aqueous Environment by

Infrared Spectroscopy, Biomedical Vibrational Spectroscopy III: Advances in Research and Industry, eds A. Mahadevan-Jansen and W.H. Petrich, Proc. SPIE, vol. 6093, 1–12, (2006).

47. B.H. Menze, W. Petrich and F.A. Hamprecht, *Anal. Bioanal. Chem.*, **387** (5), 1801–1807 (2007).

48. L. Breiman, *Mach. Learn.*, **45** (1), 5–32 (2001).

49. C.D. Mansfield, A. Man, S. Low-Ying and R.A. Shaw, *Appl. Spectrosc.*, **59** (1), 10–15 (2005).

50. C.D. Mansfield, A. Man and R.A. Shaw, *IEE Proc. Nanobiotechnol.*, **3** (4), 74–80 (2006).

# Head and Neck Cancer: A Clinical Overview, and Observations from Synchrotron-sourced Mid-infrared Spectroscopy Investigations

**Sheila E. Fisher[1,2], Andrew T. Harris[1], John M. Chalmers[3] and Mark J. Tobin[4]**

[1] University of Leeds, Leeds, UK
[2] University of Bradford, Bradford, UK
[3] University of Nottingham, Nottingham, UK
[4] Australian Synchrotron, Clayton, Victoria, Australia

## 1 INTRODUCTION

In this chapter, we look at the role of vibrational spectroscopy in cancer diagnosis from the point of view of clinical need and biological diversity. We provide an introduction to clinical terminology and the basis of therapeutic decision making for those not familiar with these areas. Although we focus on head and neck (H&N) cancer, most of the material in this chapter is equally applicable to other cancers. Taking research in this field toward application requires close collaboration and understanding between spectroscopists and biologists, and, as the technology moves closer to patient use, doctors who have an active interest in clinical research. Variability in both patient characteristics and the disease itself has an impact on vibrational spectra taken from clinical settings, and these aspects form an integral part of testing prior to trials in patients.

H&N cancer represents an ideal model for clinical study as the disease is common globally and particularly so in the emerging economies. It is accessible for inspection and noninvasive or minimally invasive diagnostic surveillance, exhibits the full range of precancerous changes, carries a substantial risk of development of second primary tumors, and has a relatively short disease course, allowing robust outcome analysis. However, it is important for the basic scientist working in the field of biomedical diagnostic spectroscopy to understand the role of confounding factors that may have a significant influence on results: smoking, ingested material, drugs (medical or recreational), or even other disease states have the potential to impact on spectral features. Age, sex, ethnicity (especially skin or mucosal pigmentation), hormonal changes and mucosal thickness, and structure may all have an impact. Some treatment effects, especially the fibrotic changes seen

*Vibrational Spectroscopy for Medical Diagnosis*. Edited by Max Diem, Peter R. Griffiths and John M. Chalmers.
© 2008 John Wiley & Sons, Ltd. ISBN 978-0-470-01214-7.

after radiotherapy or chemoradiotherapy, may alter tissue architecture and composition. These factors are crucial for the development of diagnostic or therapeutic tools by strong research teams consisting of basic science/clinical research collaborators for cancer treatment, placing identification and analysis of biomarkers in context in terms of diagnosis, confounding variables, and outcome. The aims of this review are

- to provide an overview of clinical characteristics and management;
- to provide pointers that will guide translational research bringing ideas developed in basic science to clinical practice;
- to identify areas in which vibrational spectroscopy has potential to contribute to clinical care;
- to present some examples of tissue- and cell-based vibrational spectrocopy studies;
- to assist toward the development of new devices focusing on clinical need; and
- to consider future perspectives.

# 2 HEAD AND NECK CANCER: EPIDEMIOLOGICAL AND CLINICAL ASPECTS

## 2.1 Normal anatomy and descriptive terms

The term *head and neck* (H&N) cancer refers to those cancers arising from the mucosal lining of the upper aerodigestive tract and major specific structures within this anatomical area, especially the four major salivary glands and the thyroid gland. The area covered by this term is shown in Figure 1. Within the oral cavity, the sites described include lips, buccal mucosa, alveolus and gingiva (the tooth bearing segment of bone and the mucosa immediately surrounding the teeth), the tongue, hard and soft palate, and floor of mouth. The anatomy of the larynx is similarly complex but is generally separated into three key areas: supraglottis (above the vocal cords), glottis (at the level of the vocal cords), and subglottis

(below the level of the vocal cords). The pharynx is divided into the nasopharynx, the oropharynx, and the hypopharynx. In addition, the air sinuses (frontal, maxillary, ethmoid, and sphenoid) are considered as part of H&N as are the thyroid, parathyroid, and salivary glands.

It is important for the research scientist to be aware of this complexity as the normal mucosa varies in structure and thickness and epithelial type, a factor that must be considered when considering design of an "in vivo" probe. The ability to achieve penetration into the epithelial/connective tissue junction, where invasion—the biological event that defines cancer—occurs, is imperative. For example, the tongue, which is the most common site for oral cancer, has a thick epithelium with pronounced rete ridges overlying muscle fibers, whereas the floor of mouth, the next most common cancer site, has thin mucosa overlying loosely bound connective tissue.

## 2.2 Major aspects of H&N cancer

The main cancer that is discussed in this chapter is termed *squamous cell carcinoma* (SCCa). Carcinoma describes a malignant tumor that arises from epithelial elements. Over 90% of H&N cancers are SCCa. Other specialized tissues such as the thyroid and salivary glands have specific cancers, which are briefly addressed in a separate section. Cancers of connective tissue origin, sarcomas, are relatively rare but are briefly covered.

It is important for the research spectroscopist to know the derivation of the tissue to be assessed, the type of cancer, and the demographic details of the patient from whom it has been taken, as the characteristics of both normal and cancer tissues can vary considerably in appearance and genetic profile. Such variation must be considered in the generation of robust data sets for clinical use. As vibrational spectroscopic techniques are highly sensitive to biochemical change, it can be hypothesized that these may be reproducible within cancers and even between different cancers. Current work

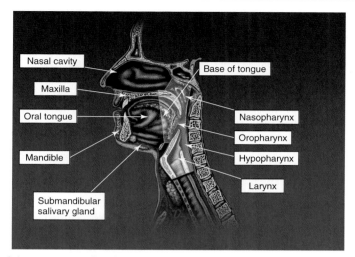

**Figure 1.** Anatomy of the upper aereodigestive tract. [Modified from diagram provided by Sanofi-Aventis.]

supports this view. For example, it has been shown that from a data set of 4120 high-quality point Fourier-transform infrared (FT-IR) spectra of colorectal adenocarcinoma derived from 28 patient samples and 12 different histological structures using an artificial neural network, a robust data set could be established and optimized for diagnostic sensitivity and specificity.[1] In the evaluation of Barrett's esophagus, Raman spectroscopy provided good sensitivity and specificity allied to a high level of agreement ($\kappa = 0.89$) with consensus pathology opinion.[2] Such studies provide promising evidence for the future efficacy of vibrational techniques in the clinical setting. (Note: $\kappa$, kappa, is a measure of agreement between two observers taking into account the agreement that could occur by chance (expected agreement). For a binary decision, the kappa value is calculated using the equation:

$$\kappa = \frac{\text{Observed agreement} - \text{Chance agreement (e.g., 50\%)}}{\text{Total observed (e.g., 100\%)} - \text{Chance agreement}} \quad (1)$$

A kappa value of less than 0 indicates no agreement; a value 0 to ca. 0.2 indicates poor agreement and a value of 0.8–1.0 indicates almost perfect

agreement. Note that kappa tells us nothing about the validity of the measurement.)

## 2.3 Epidemiology

Head and neck cancer is a global disease. Oral cancer alone accounts for nearly 220 000 cases in males and 90 000 cases in females each year, constituting 5 and 2%, respectively, of new cancer diagnoses globally.[3] It is particularly prevalent in the developing economies, India and South-East Asia, but subject to considerable regional variation. For example, in India the incidence ranges from 7.2/100 000 population in Bhopal to 2.4/100 000 population in Barshi.[4] In the United States, in 2003, 20 424 new cases of oral cavity and pharyngeal, 8981 laryngeal, 1091 nasal, and 5738 thyroid cancers were diagnosed.[5] In the United Kingdom, 7948 new cases, 5504 males and 2444 females, were registered in 2000, making it the eighth most common cancer (3% of total cancer diagnoses).[6] This level of variation shows the importance of racial origin, risk factors, environment, and also the need to understand underlying genetic susceptibilities.

In terms of sex distribution, more males than females are affected and it is generally believed that this relates to habits, particularly smoking and alcohol consumption. In recent years, reports of increased incidence in younger patients, perhaps related to human papillomavirus (HPV), are emerging.

## 2.4 Risk factors and screening

The main documented risk factors for head and neck SCCa are smoking and heavy alcohol consumption, which have an independent but also a synergistic effect. Substantial evidence indicates that these habits can cause important differences in both the population of patients affected by the disease and its biology. Such factors may act as confounders in vibrational spectroscopic analysis or, alternatively, prove that it is possible to investigate using this modality, yielding important clinically relevant information. The biological variation and its importance in clinical practice were considered by Lingen *et al.*[7] Smoking affects the expression of the important tumor suppressor gene, *p*53, often referred to as the *guardian of the genome*. It is possible that such basic genetic changes might be amenable to spectroscopic evaluation. Tobacco is associated with chewing habits either alone or as part of the betel quid. (The betel quid is made up of areca nut and mineralized slaked lime, wrapped in the leaf of the betel, a spice whose leaves have medicinal properties. This combination has been chewed in parts of India for thousands of years and is associated with the high rates of oral cancer seen in areas of mucosa in direct contact with the "quid".) A recent systematic review[8] concluded an odds ratio (OR) of 4.63 for oral cancer in heavy smokers, compared to people who have never smoked, and that of 2.14 for tobacco-free betel quid users.

Currently, there is more interest in the role of HPV, due to the proposed introduction of a vaccination program on a global scale. The intention is to vaccinate before any sexual activity. Recent publications[9,10] highlight the specific features and behavior of such cancers. HPV has been linked to oropharyngeal cancer in young patients who do not smoke or take alcohol in excess. One study,

although based on retrospective data, suggests a 60% reduction in 5-year mortality for HPV positive cancers.[11] The same study suggests that second primary tumors, i.e., a distinct new tumor, arising within the upper aerodigestive tract but distinct by site and behavior from the first or index cancer, and which represent a considerable problem in H&N cancer, were mainly seen in HPV negative or p53 mutation-containing tumors. Immunosuppression is an important risk factor and incidence increases considerably; rates as high as 20-fold more than the standard population have been reported in patients having transplant surgery. Higher rates are also seen not only in patients undergoing immunosuppressive therapy for other disease states, most notably human immunodeficiency virus (HIV), but also in patients with disorders such as rheumatoid arthritis. The effects, in terms of risk of induction of cancers, will require close monitoring as new biological therapies are introduced in the clinical practice for a range of chronic disorders.

Other factors, such as poor teeth, and/or chronic irritation, have been proposed but there is little definitive evidence to support their place as independent risk factors as they often coincide with major risk habits, such as smoking and alcohol consumption.

Given the immediate accessibility of the oral cavity to inspection, there has been much interest in screening, especially of those patients at particular risk (age over 40, male, smokers, especially where smoking is combined with alcohol consumption). In the areas of the world where the incidence is highest, screening has proved to be of some efficacy in reducing mortality.[12] However, screening is complicated by the variation in clinical appearance and the often long preinvasive phase, which are typical of these cancers.

## 2.5 Diagnosis and staging

Critical factors in diagnosis and management include complete resection of the primary tumor, adequate staging of the neck, and identification of metastatic disease that will impact on the modality of therapy.

Diagnosis is predominantly clinical as cancers in the oral cavity are readily visible and those further down the upper aerodigestive tract can be detected by flexible nasendoscopy. This is a simple procedure carried out in the clinic or office using a fine bore endoscopic device, introduced through the nasal cavity, to inspect the nasopharynx, oropharynx and hypopharynx (Figure 1). Visual diagnosis is far from accurate if widespread precancerous change, described as "field change",[13] is present (Figure 2). For many patients precancerous change may not progress or even regress, especially if they can reduce their risk factors. Where a cancer is clearly visible (Figure 3a and b), the diagnosis is simple to make. Where a widespread area of field change transforms to cancer, (e.g., Figure 3c), the exact area involved can be impossible to detect using the eye alone, and when a small tissue sample is taken for histopathological analysis (biopsy), this may not be representative or even taken from the biologically most abnormal area, thus limiting the ability of the pathologist to provide an accurate report and placing the patient at risk. Biopsy is a

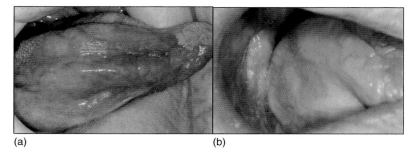

(a)                              (b)

**Figure 2.** Premalignant disease: (a) atrophic mucosa and (b) hyperkeratotic mucosa.

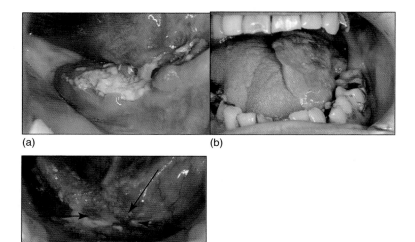

(a)                              (b)

(c)

**Figure 3.** Oral cancer: (a) ulcerated; (b) exophytic; and (c) arising in an area of field change (the area of malignant transformation is indicated by the arrows).

minor but invasive procedure that patients prefer to avoid, and for some sites, such as the larynx, it may require a general anesthetic with related risks as many of these patients have medical comorbidity. Serial biopsy causes distress to the patient through the physical discomfort experienced at the surgical site and the psychological stress of having to wait for a formal pathology report. It could be hypothesized that biochemical markers of progression or likely progression toward malignancy could be identified. If detection of such markers could be incorporated into a procedure, the clinical impact would be considerable. However, the challenges, not only of normal and pathological variation but also of managing the tissue probe interface, which may be air, saliva, food debris, necrotic ulcer slough (as in Figure 3a), inflammatory, hyperkeratotic, or atrophic must be considered. Stone *et al.*[14] considered the interface between probe and tissues and found that this was not important in terms of spectral readings, which supports the feasibility of development of future in vivo diagnostic probes.

The pathological diagnosis of SCCa is based on grading the biopsy, using standardized criteria including a description of morphology of individual cells and morphology of the tissue. Cancers vary in the degree to which they resemble the tissues from which they are derived, so a grading based on a small fragment of cancer may well not be representative of the whole cancer. The neck is assessed clinically by palpation and this can be supplemented by imaging. Metastatic disease to the nodes in the neck is termed *locoregional metastasis*. Where distant spread can be demonstrated, usually to the lungs, these are described as distant metastases.

In cancer treatment centers, information is exchanged using agreed systems, which indicate the sites and severity of disease. These are known as *TNM* (tumor-nodes-metastasis) *systems*, where T represents the primary tumor, N the nodal status, and M metastasis. An illustration of the way this system is derived from clinical or imaging information is shown in Table 1. Increasing numbers in each part of the TNM staging indicates a less good prognosis, so T4 compared to T1 indicates a large tumor, increasing N stage, the number of levels within the neck involved in disease and the size of nodes. $M^+$ indicates that distant metastases are present. These grades can be combined to give a single staging that correlates through to prognosis (Table 2). These descriptions derive from the two major published systems for staging, which have been developed by the International Union against Cancer (Union International Contre le Cancer, UICC)[15] and the American Joint Committee on Cancer (AJCC)[16] respectively. Both are periodically revised in terms of important prognostic factors, and information as to how this work is developing and about how to contribute to revisions can be accessed on the website for each organization.[17,18] Staging can be clinical, radiological, or pathological. The latter is most precise but only available for those patients having a surgical resection of their disease. Staging systems are available for all cancer sites and it is worthwhile, when placing research into a clinical context, for researchers to be familiar with the staging for their site of interest.

Decisions as to treatment are taken by a multidisciplinary team of clinicians who advise treatment based on known parameters of outcome, such as stage, presence, or absence of evidence of locoregional metastatic disease in the neck or of systemic metastatic disease. In principle, treatment aims to cure the patient. Where possible, efforts are made to spare the organ invaded by tumor, and this philosophy underpins the use of chemoradiotherapy rather than surgery. The requirement is that the organ preserved must be functional after completion of therapy. Surgery is indicated where it is considered possible to remove the whole cancer with a margin and to achieve a functional reconstruction of the area in which the cancer was located. Radiotherapy preferentially targets malignant rather than nonmalignant cells on the basis of the increased cell turnover and the decreased ability for repair after biological injury exhibited by the former. In recent times chemotherapy, usually combined with radiotherapy, has played a prominent role at some disease sites, after publication of a

**Table 1.** TNM staging of H&N cancer: this table gives the staging for oral cavity as a representative illustration.

| TNM | Stage | Description |
|---|---|---|
| T status | T0 | Primary tumor cannot be detected |
| | Tis | Cancer "in situ", the histological appearance is that of a cancer but actual invasion cannot be demonstrated. |
| | T1 | Tumor less than 2 cm in its greatest measurable dimension |
| | T2 | Tumor between 2 and 4 cm in its greatest measurable dimension |
| | T3 | Tumor more than 4 cm in its greatest measurable dimension |
| | T4 | Tumor invades adjacent structures; e.g., cortical bone, extrinsic muscles of tongue, external skin |
| | TX | Primary tumor cannot be assessed |
| N status | N0 | No evidence of regional lymph node metastasis |
| | N1 | Metastasis in a single ipsilateral lymph node, 3 cm or less in greatest dimension |
| | N2a | Metastasis in a single ipsilateral lymph node, more than 3 but less than 6 cm in greatest dimension |
| | N2b | Metastasis in multiple ipsilateral lymph nodes, none more than 6 cm in greatest dimension |
| | N2c | Metastasis in bilateral or contralateral lymph nodes, none more than 6 cm in greatest dimension |
| | N3 | Metastasis in a lymph node more than 6 cm in greatest dimension |
| | NX | Regional lymph nodes cannot be assessed |
| M status | M0 | No evidence of distant metastasis |
| | M1 | Distant metastasis |
| | MX | Distant metastasis cannot be assessed |

In this illustration, the nodes being assessed are those of the neck.
In this classification, T indicates the staging of the primary Tumor; N indicates the staging of the lymph Nodes; M indicates the presence of Metastatic Disease.
For any of these: the addition of the suffix "X", e.g., TX, NX, means that aspect of the staging cannot be accurately performed.

**Table 2.** Correlation between TNM stage and stage grouping. This system allows the more detailed staging developed through TNM to be reported as a single figure, which correlates with prognosis. A patient with Stage III or IV disease will fare significantly worse than someone with stage I or II disease.

| Stage 0 | Tis (no invasive element identified) | N0 | M0 |
|---|---|---|---|
| Stage I | T1 | N0 | M0 |
| Stage II | T2 | N0 | M0 |
| Stage III | T1, T2 | N1 | M0 |
| | T3 | N0, N1 | M0 |
| Stage IV | T1, T2, T3 | N2 | M0 |
| | T4 | Any N | M0 |
| | Any T | Any N | M1 |

Tis = Cancer in situ.

meta-analysis showing a survival advantage for this management strategy.[19] For reasons yet to be elucidated, oral cavity cancers appear to have less sensitivity to chemoradiotherapy, and surgery remains the preferred option. In principle, small cancers are amenable to single modality treatment, either surgery or (chemo) radiotherapy, whereas larger cancers may well require a multimodality approach, which has traditionally been surgery followed by radiotherapy or surgery followed by chemoradiotherapy. Sites for which primary chemoradiotherapy is employed preferentially in many centers include the nasopharynx and oropharynx. Most of this chapter relates to the prediction of surgical parameters, so surgical management of the cancer in the context of challenges in diagnosis or management is discussed in the next section.

# 3 HEAD AND NECK CANCER: TREATMENT AND PATHOLOGY

Understanding the treatment—and hence the ways in which tissue has been handled, preserved, and prepared for analysis—is important in the development of spectroscopy-based diagnostic systems and also to ensure that material gathered for spectroscopy research is harvested in the most appropriate way. The major issues relating to standardization were the focus of a recent international workshop[20] and it is proposed that recommendations to assist scientific and clinical researchers in this field will be posted on a website. This section considers some of the challenges in day-to-day clinical research involving tissue, from the perspective of the primary cancer, locoregional spread, and distant metastatic disease.

## 3.1 The primary tumor

The most important factor in improving outcomes in terms of disease-free survival and avoidance of local recurrence is the achievement of clear resection margins by complete removal of the cancer, which can be difficult because microscopic invasion is not apparent on visual inspection or on palpation, and also because of proximity to key anatomical structures.[21] A detailed review of 301 radical resection specimens showed that 70 (23%) had involved margins, some in a single area around the primary cancer and the remainder at more than one site. The site distribution of involved margins was mucosal in 11, bone in 10 and deep soft tissue in 61. Residual disease at the bone and deep margins is likely to present late when clinical examination alone is relied on for surveillance and potentially at a stage when curative therapy is no longer possible.[22] The more "functional" the resection and the more normal tissue that can be retained, the better will be the outcome in terms of ability to eat, chew, and speak successfully.[23] In the case of soft tissue, detection of margins requires a frozen section, in which a small piece of tissue at the periphery of the resection is excised, snap frozen, stained, and examined histologically. This takes time and the sample reported may not be representative. Where hard tissue is involved (e.g., bone in oral or cartilage in laryngeal resections), standard histological reports can only be issued after decalcification, which may take several weeks. In situations where margins are close or involved, further surgery would be ideal but is often precluded by the presence of advanced reconstructions. In this situation, removal or revision of the reconstruction would cause too much morbidity to be justifiable. In practice postoperative radiotherapy is advocated and has efficacy in eliminating small volume residual disease. However, a decision about whether or not to offer radiotherapy is required before, using currently available techniques, the definitive report is obtained on the status of the bone margins.

An additional prognostic factor is tumor differentiation. This is graded from "well" to "poor", depending on individual cell morphology and architectural features of the epithelium, based on the resemblance of the tumor to the parent tissue (Figure 4). Even in experienced hands this has an element of subjectivity and our work shows a change in differentiation between biopsy and definitive resection in 50% of cases.[24] A reliable way to characterize tissue and to immediately report margins is very much required and has the potential to contribute to both locoregional control and survival.

## 3.2 Locoregional metastasis

The first point of metastasis in H&N disease is the neck and this reduces survival on a stage-by-stage basis by 50%.[25] The likelihood of nodal metastasis is determined by tumor size and site, depth of infiltration, differentiation and pattern of infiltration, including vascular and perineural invasion. A comprehensive review of pathological features and their clinical significance in relation to oral and oropharyngeal cancers has recently been published by Woolgar[26] and the

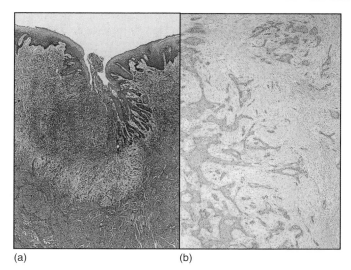

(a)                                         (b)

**Figure 4.** (a) Well-differentiated SCCa (low power view), the tissue resembles normal architecture and keratin is formed; (b) poorly differentiated SCCa (high power view) with widespread foci of cancer.

same parameters apply to all H&N sites. Patients with neck disease are at greater risk of distant metastasis especially if more than three nodes are positive on histological examination or show metastasis at multiple levels and are associated with macroscopic or microscopic extracapsular spread (ECS).[25,27,28] As the majority of patients in the United Kingdom present with late (stage III or IV) disease (Table 2), the need to assess the neck thoroughly is clear. Disease-specific survival correlates well with stage and the presence of nodal disease confers a change to a less favorable stage in this internationally accepted classification.

In clinical practice a subset of patients present to the clinic with a visible and palpable neck mass, indicating the presence of cancer; however, on inspection no cancer is visible. In this case imaging is essential to try to locate the primary cancer. Even where the cancer can be seen, imaging of both primary site and neck by computerized tomography (CT) or magnetic resonance imaging (MRI) can improve staging and decision making for treatment. Figure 5 shows a cancer imaged using MRI. Criteria have been developed to predict the likelihood of imaged changes

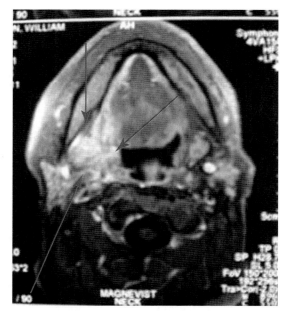

**Figure 5.** MRI image of oropharyngeal cancer. The cancer is highlighted using the arrows.

representing true metastatic disease in the neck. Positron emission tomography (PET) is currently being widely used. PET has been evaluated in major studies and shows most promise when combined with CT. A comprehensive review of imaging has recently been published by Rumbolt et al.[29] However, as discussed later in this section, the main parameters for outcome relate to the number and pattern of nodal metastases, to the spread of cancer beyond the lymph node capsule ECS and to the deposits of cancer outside lymph nodes (soft tissue disease (STD)).

In case of evidence of neck disease, it is usual to carry out a neck dissection (ND). The standard approach to the neck and the structures seen during ND is shown in Figure 6. The aim is to remove all the lymph nodes, which might carry cancer cells. Different types of ND exist: "radical", which includes major vascular and neural structures with considerable morbidity to the patient, and "modified radical" in which some structures can be preserved. The neck is divided into "zones" or levels, and the levels excised depend on the site of the primary cancer. Histology reports should state the level as this will help define where disease has been present, thus guiding the planning of postsurgical radiotherapy, usually indicated in node positive disease. These decisions about the extent and radicality of ND have a marked impact on the quality of life of the patient in the posttreatment phase but failure to

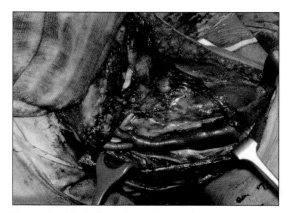

**Figure 6.** Neck dissection: clinical view of structures in the neck. This is a "selective" neck dissection in which the carotid artery, the jugular vein, and the accessory nerve (which moves the shoulder) are all preserved.

fully excise the cancerous tissue places the patient at a considerable risk of recurrence and death. A tool that could assist in peroperative surveillance would be of considerable benefit. To attempt to reduce unnecessary surgery and the morbidity of surgery, research is in progress to determine whether it is possible to assess to what degree the cancer has spread to the lymph nodes and to which nodes it has spread. This involves the use of a dye technique with or without a radioactive tracer and removal of positive nodes (sentinel node biopsy). This has gained a routine place in some cancer sites, most notably melanoma and axillary dissections for breast cancer, but remains experimental for H&N. A spectroscopy-based probe with sufficient penetration could allow imaging of nodes during ND. As shown in the illustration, a significant amount of tissue is removed and there would be a requirement for any in vivo surveillance to address the field of resection.

Once the ND is completed the excised tissue is usually fixed in formalin, then embedded in paraffin wax, sectioned, stained with hematoxylin and eosin (H&E) and mounted on glass slides and carefully assessed by a specialist histopathologist (Figure 7). However, in practice this can be difficult in terms of workload for the pathologist with a requirement to visually examine many sections of neck and nodal tissue. The extent of tissue to be examined can be estimated from the ND and the fact that sections are about $25\,\mu m$ thick. The exclusion of areas from analysis carries the risk of missing a cancer deposit and wrongly staging the neck. If a neck is incorrectly downstaged, the risk is recurrence, and successful salvage of the patient by further therapy once recurrence has become clinically manifest is unlikely.

Concern about neck staging and its direct clinical relevance has led to the publication of an agreed minimum pathology data set for H&N tumors.[30] In terms of treatment decisions, success in treatment, and the need to detect disease, which may not be apparent by clinical or imaging examination, understanding of occult metastasis, ECS, and soft tissue deposits are critically important. ECS is defined as the presence of cancer beyond the capsule that encases the lymph node, and

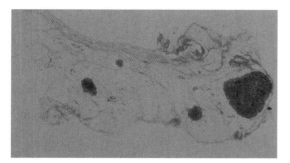

**Figure 7.** Histological section from a neck dissection: macroscopic view. The lymph nodes appear dark blue after fixing in formalin and staining with hematoxylin and eosin. Over 100 lymph nodes may be removed in a neck dissection. [Reproduced by permission of Professor K. A. MacLennan, Professor of Tumour Pathology, University of Leeds, UK.]

the term soft tissue disease describes foci of cancer lying in the neck without apparent association with lymph nodes. Both are of considerable importance in prognosis and, as deposits may be less than 3 mm in maximum diameter, can be hard to locate by traditional methods of analysis. This aspect is therefore summarized in detail. The variability of material, which requires analysis by the pathologist, is shown in Figure 8. Metastases are often occult and it is worrying that in necks that are staged as having no cancer in the lymph nodes (N0) Coatesworth and MacLennan found a prevalence of 19% microscopic ECS, with 7.9% showing STD.[31] They concluded that ECS and STD could occur even at an early stage in metastasis. Jose *et al.*[32] examined 237 NDs

from patients with upper aerodigestive tract SCCa and found that 25% of clinically N0 necks and 44% of N1 necks had ECS and/or soft tissue deposits. The same team found that distribution patterns of ECS and STD follow those of cervical lymphadenopathy, with the parts of the neck that are most likely to be involved by nodal disease also being most likely to have ECS and STD.[33] In their retrospective data analysis, Myers *et al.*[34] found that the 5-year overall survival rate for patients with positive nodes on pathological staging with ECS was significantly lower than that of patients with a positive pathological nodal staging but no ECS—29 and 51%, respectively. Similarly, although local recurrence rates for pathologically staged N0 necks and node positive but ECS negative and node positive with ECS were similar, the regional recurrence and distant metastasis were significantly higher in the patients with ECS. The number of nodes with ECS impact negatively on survival and metastasis. Patients whose pathological staging indicates the presence of multiple nodes with ECS have a poorer prognosis with "higher odds ratio of death from disease... overall death... and increased recurrence".[35] Woolgar *et al.*[36] followed up 173 patients with oral or oropharyngeal SCC. They also concluded that ECS, whether macro- or microscopic, was an important predictive factor for survival and recommended that it should be included into the pathological staging system. Jose *et al.*[37] carried this work through to impact on survival, looking at soft tissue

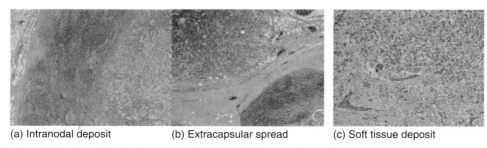

(a) Intranodal deposit     (b) Extracapsular spread     (c) Soft tissue deposit

**Figure 8.** Histological variation in the pattern of metastatic disease in the neck: (a) intranodal deposit; (b) extracapsular spread; and (c) soft tissue deposit. [Reproduced by permission of Professor K. A. MacLennan, Professor of Tumour Pathology, University of Leeds, UK.]

deposits specifically. Sixty-five percent (101/155) had evidence of metastatic disease. Seventy-one percent of this population had ECS and/or STD. Of the entire population 24% (37/155) had STD, with 43% of them having STD alone. They found that STD had a significant adverse effect on the patient's survival when compared with patients with pathologically N0 and pathologically $N^+$ necks but with no ECS. When compared with ECS alone, there was no significant difference and this led them to conclude that both ECS and STD behave in a similar manner, and that both are indicative of aggressive disease.

MacLennan et al.[38] concluded that SCCas that metastasize to cervical nodes range from the less aggressive ones that metastasize without ECS or STD to those that show these adverse characteristics. The survival of patients with the less aggressive disease does not differ overtly from that of patients with a pathologically negative (pN0) neck, whereas the presence of ECS or STD is indicative of increased chances of developing distant metastasis.[39] Therefore, it is the detection of ECS and STD that is important in precise staging and which in turn is important for survival.

This summary of current evidence emphasizes the need for meticulous examination of the neck at a level that cannot currently be achieved in routine practice. The development of a laboratory-based screening device to scan all sections and focus the attention of the pathologist on those slides that require his/her detailed assessment would have a radical impact on both quality of assessment of neck disease and risk of metastatic disease. Such a system, linked to advanced learning tools, such as neural networks, will assist with quality control and production of predictive data sets, which can, in turn, inform the use of new, emerging, and expensive chemotherapeutic regimens.

### 3.3 Distant metastasis

Until recently, when advanced reconstructive surgery and advances in chemo- and radiotherapy have secured gains in locoregional control, it was argued by many authorities that distant metastatic disease was not important in H&N cancer. However, now it has been seen that this disease behaves in a similar way to cancers at other sites. Recognition of aggressive disease by careful staging of the neck will be important in deciding who would be offered additional treatment, probably using chemotherapy.

Markers of aggressive disease may well be possible through analysis of peripheral blood or even salivary markers, both fields that are promising for exploration. We have examined bone marrow for markers of metastasis with a high correlation with ECS and STD but meticulous preparation of the specimens is necessary.[40] Work toward a spectral diagnostic system is one of our current aims.

## 4 BIOLOGICAL FACTORS

The mapping of the human genome has opened radically new avenues to both diagnosis and therapy. Targeted biological agents hold much promise but also are expensive and do have side effects. About 2000 new agents were noted in a recent report to be close to clinical trials.[41] If it were possible to predict which patients will and which will not respond to a particular therapeutic regimen, this would be beneficial in terms of patient care and health economics.

A number of genes and pathways have been explored in detail. The p53 gene has been particularly extensively studied and, as already noted, there are alterations in terms of environmental factors such as smoking. In terms of therapeutic strategies, these must focus on pathways, which are not required in health, and the most extensively investigated target in H&N and some other cancers is the epidermal growth factor receptor (EGFR). Anti-EGFR agents have shown promising results in clinical trials[42] and are entering routine practice in some geographical areas.

A detailed discussion on the complex field of cancer molecular medicine is outside the scope of this chapter; the interested reader is referred to the excellent book by Knowles and Selby.[43] Our own work, Tobin *et al.*,[44] see later, showed key changes in the mid-IR spectrum relating to the activation of the growth factor signaling mechanism. This, and work at other cancer sites, shows promise for vibrational spectroscopy in terms of correlation with known molecular events of importance. This is important in terms of current research in that there are two approaches. The first is to produce robust data sets depending on reproducibility alone but without consideration of the underlying biological mechanisms. This may well lead to diagnostic applications but is not likely to establish sensitivities to biological agents. The second is to relate spectral changes to cell lines of known characteristics and to track, ideally through live cell work, the correlation between biologically important pathways, and spectral changes. This may well have important implications for future cancer therapy.

# 5 RARER HEAD AND NECK CANCERS

Although SCCa is by far the most common H&N cancer, there are research groups working actively in some of the other areas, so a brief clinically related summary is presented here.

## 5.1 Thyroid cancer

Approximately 1550 new cases of thyroid cancer are diagnosed annually in the United Kingdom,[6] representing 0.6% of all new cancer diagnoses. Vibrational spectroscopy study of this disease is discussed in detail in **Introduction to Spectral Imaging, and Applications to Diagnosis of Lymph Nodes** of this volume and this short comment highlights only the clinical context. The biological interest lies in the fact that these cancers behave in a completely different way from H&N SCCa and the patients are often young and predominantly female, without known risk factors. The thyroid gland is important in

metabolism producing T3 (triiodothyronine) and T4 (throxine) under the control of stimulation by the pituitary gland, which is found in the area of the cranial cavity called the *sella turcica*. Calcitonin is another hormone made by the thyroid glands that works in partnership with the parathyroid glands, which, as indicated by their name, lie very close to the thyroid gland and which can be affected by or even removed during thyroid surgery. The core activity of this pathway is to control calcium metabolism, whose disturbances can have significant consequences on important bodily functions.

Cancer of the thyroid gland is interesting because of its biological diversity. There are four main types:

- Papillary cancer
- Follicular cancer
- Medullary cancer
- Anaplastic cancer.

Of these, papillary cancer is the most common, accounting for 60% of new diagnoses. It is more common in women, especially younger patients. It is slow growing and associated with a favorable prognosis. It can metastasize to the lymph glands of the neck and may be diagnosed incidentally during ND for other pathology.

Follicular cancer is usually seen in young or middle aged people and accounts for 15% of new diagnoses. It is associated with systemic metastasis, most often to the lungs or bones.

Medullary thyroid cancer is the variant of greatest biological interest accounting for between 5 and 10% of new diagnoses. The interest stems from its origin, in about 25% of cases, through the presence of an inherited familial genetic change.

Anaplastic cancer is the most aggressive cancer and is associated with a very poor prognosis being refractory to treatment in most cases. It is usually a disease of older patients (75% of cases older than 60 years).

Treatment of thyroid cancer is a complex subject, beyond the remit of this chapter. It consists of surgery, radiotherapy usually in the form of radio-iodine and supplementation of thyroid function by thyroxine.

The clinical dilemma lies in the diagnosis between the main types, as this has an impact on treatment. Diagnosis involves fine needle aspiration (FNA) of palpable lumps, which, as already noted for SCCa, may not sample a representative area. A probe that could identify the type of thyroid cancer at surgery would contribute to care.

## 5.2  Salivary tumors

Each of us possesses four major salivary glands: the parotid glands, which are situated on each side of the face just in front of the ear and extending below the earlobe, and the submandibular glands, which lie just below and medial to the lower border of the mandible (lower jaw). There are also sublingual glands situated bilaterally in the floor of the mouth and numerous small glands in the lip (especially the lower lip) and palate.

Salivary tumors are highly diverse in their biology and behavior and the researcher interested in working in this field will find a definitive classification and description in the World Health Organization textbook, which gives a definitive account of the pathology and genetics of all H&N tumors[45] and the definitive review by Zarbo.[46]

By far the most common salivary tumor is the pleomorphic adenoma, a benign neoplasm, which presents as a lump, usually in the parotid gland, accounting for 85% of tumors at that site. This is best treated by careful and complete surgical excision, usually as a superficial parotidectomy. When incompletely excised, recurrence is common, usually as numerous small nodules in the tumor bed or overlying skin. Tumors close to or breaching the excision margin are routinely managed by postoperative radiotherapy to prevent or delay this unpleasant outcome. When left untreated the tumors can transform into the malignant variant, cancer expleomorphic adenoma, a highly aggressive cancer, which carries a very poor prognosis despite radical surgery and radiotherapy.

Other important cancers at this site include mucoepidermoid cancer where the behavior depends on a classification of low grade, which carries a favorable prognosis, or high grade, which does not. Markers to indicate which cancer falls into each subtype would be a welcome addition to current clinical and pathological assessment.

Two further cancers are Warthin's tumor, or adenolymphoma, which, although associated with lymphadenopathy, carries an excellent prognosis, leading to some authorities questioning its inclusion in the salivary cancer classification and polymorphous low grade adenocarcinoma, which is associated not only with lymphadenopathy but also with good outcomes.

The most difficult cancer to manage is the adenoid cystic carcinoma, which typically infiltrates widely especially along the trunks of nerves, with skip lesions as far as 2 cm apart. Confirmation of complete excision is almost impossible given these characteristics and radiotherapy is usually given to prevent or delay the almost inevitable local recurrence, which is managed by serial resection until such a time as the cancer becomes inoperable. The course is indolent but relentless and progression is associated with the development of lung metastases, which can be present in the patients in apparently good health for many years.

## 5.3  H&N sarcomas

By definition, a sarcoma derives from nonepithelial tissue and comprises cancers of muscle, bone, cartilage, blood vessels, and fat, termed respectively *rhabdomyosarcomas, osteosarcomas, chondrosarcomas, angiosarcomas*, and *liposarcomas*. These tend to affect younger patients, including infants and children, and are rare. Management is often by chemo- and radiotherapy with excision of residual abnormal areas, either on clinical examination or on imaging. Prognosis is variable depending on the type and characteristics of the primary cancer.

## 6  VIBRATIONAL SPECTROSCOPY FOR THE DETECTION OF H&N CANCER

Work toward in vivo and laboratory assessments has included other modalities as well as vibrational spectroscopy, especially fluorescence and elastic scattering spectroscopy (elastic scattering spectroscopy is a real-time, in vivo optical technique that detects changes in the physical properties of cells). Such work falls outside the remit of this work but should be acknowledged in that, for some applications, tools other than those based on vibrational spectroscopy may play a role and for others a multimodal approach, bringing different technologies together, may be effective.

Our work has involved both mid-infrared (mid-IR) and Raman spectroscopy, taking an approach by which a data set is evolved from single cell work in either cancer tissue or cell lines of known provenance toward evaluation of tissue samples, blood and potentially saliva, and, if feasible, bone marrow.

For the majority of our mid-IR studies, mostly undertaken prior to 2002, we used synchrotron-sourced mid-IR radiation coupled into a commercial FT-IR microscope to differentiate between cell types present in normal and malignant oral mucosa, producing tissue maps that allowed discrimination between normal and cancer tissue.[44,47] The FT-IR spectrometer was a Nicolet model 730 (Nicolet Instruments, Inc., Madison, USA) interfaced to a Nic-Plan® IR microscope, fitted with a 32× objective and equipped with a narrow-band (cutoff ca. 700 cm$^{-1}$) mercury–cadmium–telluride detector. The spectra were recorded at the synchrotron radiation source (SRS), Daresbury Laboratory, UK The high brightness (low divergence) of mid-IR radiation emanating from a synchrotron makes it eminently appropriate for high (<10 μm diameter) lateral spatial single-point-mapping resolution studies using FT-IR microspectroscopy; this is exemplified in Figure 9. Since our studies, recent developments in both FT-IR instrumentation and synchrotron beamline design have significantly improved the speed of collection of high-quality data at high spatial resolution.

Cryo-microtomed tissue sections, nominal thickness 5 μm, were mounted on 0.5-mm-thick $BaF_2$ windows and their spectra were recorded in transmission mode. Parallel sections were stained conventionally (with H&E) to facilitate identifying regions of particular interest. Some of the sections used for infrared examination were similarly stained after they had been studied spectroscopically. Cell culture samples were deposited on low-e glass slides (Kevley Technologies, USA) and examined in transflection mode; full details

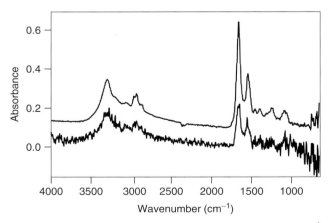

**Figure 9.** Comparison of SRS (upper) and globar-source (lower) FT-IR spectra, 8 cm$^{-1}$ spectral resolution, 1024 scans, from a 10 μm × 10 μm area on a tissue section. The upper spectrum has been offset for clarity.

can be found in Reference 44. The sampling area interrogated was $10\,\mu m \times 10\,\mu m$ for tissue section studies (unless stated otherwise) and $15\,\mu m \times 15\,\mu m$ for cultured cell samples, the latter allowing spectra to be recorded from single cells without interference from adjacent cells. Spectra were collected at $8\,cm^{-1}$ spectral resolution; 512 or 1024 scans were coadded for tissue section samples, 128 scans were coadded for cell culture samples. These gave adequate signal-to-noise ratio (SNR) for subsequent multivariate data analyses.

Because of the time requirements associated with single-point mapping measurements, it was only possible, compared to what can now be achieved using imaging mid-IR spectroscopy (see examples in other chapters in this book), to interrogate small data sets; nevertheless, studies such as these paved the way for highlighting the potential of mid-IR spectroscopy for medical diagnoses.

The potential of mid-IR spectroscopy to discriminate between oral squamous carcinoma cells (SCCa), normal gingival tissue and normal subgingival tissue, using key spectral bands had previously been demonstrated including those associated with the structure of nucleic acids and intracellular collagen.[48] Using the SRS beamline we were able to achieve spectral mapping at a lateral spatial resolution of $20\,\mu m$ (or less), allowing single cell analysis.[44,47]

## 6.1 Mid-IR synchrotron radiation FT-IR studies of oral tissue sections

Using fresh, unstained, air-dried sectioned tissue material mounted on a $BaF_2$ window, we identified regions of interest. A parallel H&E stained section was used for comparison, where necessary. Oral tissues containing normal and cancerous cells were used to test the capability of multivariate analysis techniques to discriminate between them. Preprocessing corrections were made to all tissue spectra for variations in sample thickness (normalization to the amide II band), baseline correction, and water vapor absorption subtraction using routines within the

Nicolet Omnic32™ software supplied with the FT-IR spectrometer. Sets of spectra between 1806 and $938\,cm^{-1}$ were then subjected to multivariate analysis using the Pirouette® version 3 software package (Infometrix, Inc., Woodinville, USA). This software included routines for hierarchical cluster analysis (HCA), principal component analysis (PCA), and soft independent modeling of class analogies (SIMCA). Empirically, it was decided that all data sets would be also preprocessed, within the Pirouette® software package as first-derivative (15 points) spectra with variance scaling; cross-validation (leave one out) was used for PCA evaluations.

Figure 10(a) shows a $4\times$ magnification visual image from an H&E stained oral tissue section; the stroma and tumor regions are clearly discriminated by their light and dark purple stains, respectively. Figure 10(b) shows a $32\times$ magnified visual image from a portion of a parallel, unstained section; the superimposed dashed white line separates the visually different morphologies. In sequence, five spectra were recorded from each of the three distinct regions using an aperture of $10\,\mu m \times 10\,\mu m$. The locations of these are marked by a $+$ on Figure 10(b) and numbered as 1–5 for the upper tumor region, 6–10 for the central stroma layer, and 11–15 for the lower tumor region. The 15 synchrotron-sourced FT-IR transmission spectra as recorded from these positions are shown as absorbance plots overlaid in Figure 11. Very clearly the interspectral differences associated with pathlength, baseline, and water vapor intrusion outweigh any interclass absorbance variations. These spectra following preprocessing, and over the region selected for further interrogation by multivariate analysis procedures, are shown in Figure 12 (the abscissa scale inversion is a consequence of the way the data set was input into the Pirouette® software package). The region contained 901 data points. The interclass spectral differences are still not obviously apparent over any intraclass variations. Figure 13 shows a dendrogram, obtained from HCA, using a Euclidean distance metric and group average linkage. This simple example

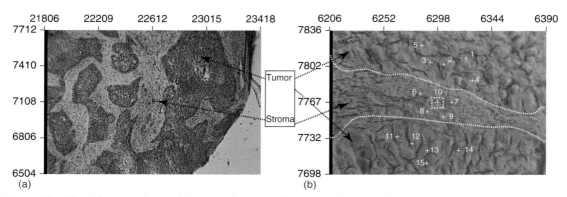

**Figure 10.** Visual images of an oral tissue section: (a) H&E stained, 4× magnification; (b) unstained, 32× magnification, showing the centers, marked with a +, of the $10\,\mu m \times 10\,\mu m$ areas from which FT-IR microscopy spectra were recorded. The white dotted lines in (b) have been superimposed to highlight the boundaries between the tumor and stroma regions. The white dashed square around point 10 delineates the aperture size.

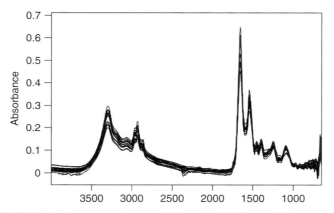

**Figure 11.** Overlaid SRS FT-IR absorbance spectra recorded from the 15 $10\,\mu m \times 10\,\mu m$ areas marked in Figure 10(b).

shows that the unsupervised classification method clearly separates the set of spectra into two distinct classes that represent stroma and tumor.

Overlays of the average of the normalized absorbance spectra from each of three regions shown in Figure 10(b) are presented in Figure 14. The two average spectra of the tumor regions (the arithmetic means of spectra 1–5 and 11–15) closely overlay, but are readily distinguished from the average spectrum of the stroma region, determined as the average of spectra numbers 6–11. From these normalized amide II band intensity data, some general observations may be made. There is an apparent decrease in the relative intensity of the amide I band in connective tissue spectra compared with tumor. This is accompanied by a general increase in relative absorbance over the range ca. $1500–1130\,cm^{-1}$ in stroma over tumor, and a reduction in relative intensity of the band at ca. $1090\,cm^{-1}$ in stroma regions. However, it must be remembered that these differences could have arisen as a consequence of the normalization procedure employed.

Two other areas from the tissue section were similarly examined, and individually realized similar results. In total 44 single-point spectra were recorded from the three examinations of separate areas and combined into one preprocessed

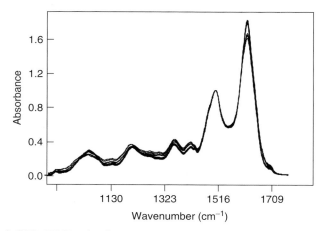

**Figure 12.** Preprocessed SRS FT-IR absorbance spectra over the range 1806–938 cm$^{-1}$ recorded from the 15 10 μm × 10 μm areas marked in Figure 10(b), and used as input data for multivariate analysis using Pirouette® software. (The abscissa scale reversal is a consequence of loading the data into the Pirouette software for multivariate analysis.)

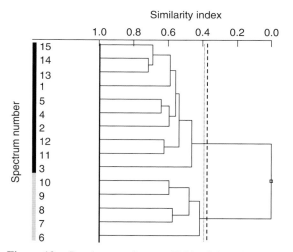

**Figure 13.** Dendrogram from an HCA of the 15 spectra shown in Figure 12.

data set for subsequent HCA, the dendrogram from which is shown in Figure 15. This essentially classified the samples into two main groups comprising those purporting from the tumor regions and those purporting from the stroma regions, excepting #37, which was misclassified, according to our visual perception, as stroma. (Although by no means clear, with hindsight,

closer visual inspection of the morphology of the tissue section region concerned indicated some suggestion of "mixing" of stroma and tumor types in the regions around measurement point #37, which lay close to a boundary between two visually different morphologies.)

A white-light image of a fourth area of this tissue section is shown in Figure 16. Superimposed on this picture is a grid highlighting the region studied, and showing the individual area elements from which spectra were recorded. A numbered grid is shown below, indicating the numbers allocated to individual spectra. This area of the tissue section is more complex than the previous three regions examined; a preliminary histopathological study showed it to contain primarily three regions. These are indicated on the figure as follows: tumor, stroma, and an area of early keratinization. Keratin is the protein that covers the surface of mucosa or skin. Keratinization is seen where a cancer attempts to adopt the morphology of its parent tissue. As this represents a favorable prognostic morphology, the ability to identify its presence by IR microspectroscopy is important. Preliminary multivariate analysis very clearly showed that the spectrum recorded from area element #46 was atypical of the remainder

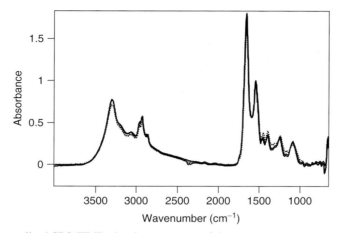

**Figure 14.** Overlaid normalized SRS FT-IR absorbance spectra of the average spectrum recorded within each of the three regions shown in Figure 10(b). The two average spectra of the tumor regions, solid lines, (the means of spectra 1–5 and spectra 11–15) closely overlay, but the average spectrum (mean of spectra 6–10) from the stroma region, dashed line, shows significant differences.

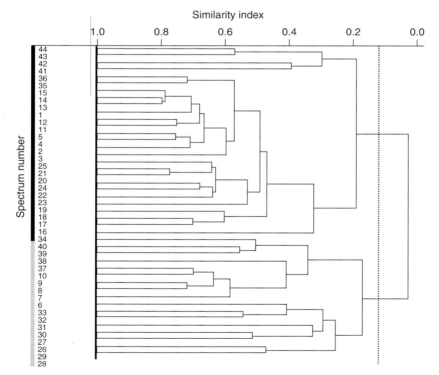

**Figure 15.** Dendrogram from a HCA of preprocessed mid-IR absorbance spectra for 44 spectra over the range 1810–900 cm$^{-1}$ from the same tissue sample, see text for details.

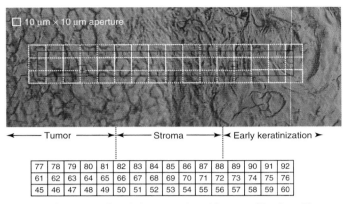

**Figure 16.** White-light image of an area of oral tissue section, 32× magnification. The superimposed dashed white line grid on the image shows the areas (10 μm × 10 μm) from which SRS FT-IR spectra were recorded. These were numbered as shown in the grid below the image.

in the set and an outlier. When overlaid on the others in the set, this spectrum, even after the preprocessing steps, clearly contained a more distorted baseline than the rest, and was therefore excluded from the data set and any further analyses. Preliminary PCA and SIMCA analyses of the data set, including a SIMCA prediction based on a model generated from the spectra #1–44, highlighted that the spectra recorded from within the region marked as early keratinization were spectrally similar but very different from others in the data set. A simple HCA interrogation indicated the presence of three clusters. The more separated cluster contained all those spectra recorded from the region indicated as early keratinization excluding number 88, but including number 55. Both spectra numbers 88 and 55 lie in the border between stroma and early keratinization, and close inspection of the white-light image of Figure 16 clearly shows that the boundary, indicated by differing morphologies, meanders through area element numbers 88, 72, 56, and 55. (The distinct gray-scale contrast between the left half and right half of the image of Figure 16 is artificial; it is a consequence of the image being a composite of two independent pictures corresponding to each half.) In a similar manner the boundary between the marked stroma and tumor regions does not follow a vertical line as indicated, but rather appears to meander

somewhere through the area contained within the area elements 50–52, 65–67, and 80–82. Hence, there was some swap-over in the HCA of these boundary spectra from the cluster anticipated from the preliminary histopathological survey. Closer histopathological inspection highlighted that there had been invasion of the stroma region by tumor within the vicinity of the boundary between the two layers. Since, we were primarily concerned at this stage of our study with ascertaining spectral characteristic of essentially distinct classes of tissue cells, rather than gradation processes or mixed types, those within the two boundary regions, that is spectra corresponding to area elements 50, 51, 55, 56, 65, 66, 72, 81, 82, and 88, were excluded, along with the outlier 46, from the final multivariate analyses of the data. Figure 17 shows a similar PC1-2 projection from each of a PCA and SIMCA analysis of the residual set of 37 spectra. Those from the early keratinization region clearly form a distinct cluster well separated from the others; the spectra from the tumor are also well clustered and separated from those of the stroma region. The spectra from the central region of the stroma, although forming a cluster, are more dispersed, probably indicating that they are less "pure", being possibly invaded from each side by characteristics of tumor and early keratinization. In particular 71 and 87, which lie close to the stroma/early keratinization boundary, seem to

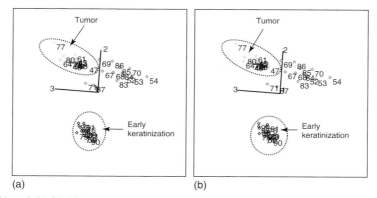

(a)  (b)

**Figure 17.** (a) PCA and (b) SIMCA scores plot output projections for Factor 2 versus Factor 3 from the preprocessed absorbance spectra, excluding those recorded from the boundary regions, see text for details.

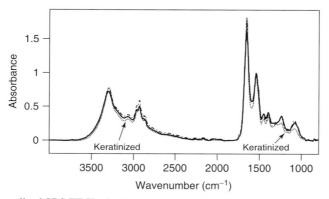

**Figure 18.** Overlaid, normalized SRS FT-IR absorbance spectra of the average spectrum recorded from within in each of the three regions shown in Figure 16, see text for details. The solid line (———) represents the average spectrum of the tumor region spectra; the dotted ( · · · · ) line is the average spectrum of the early keratinization region spectra; the average spectrum of the stroma region spectra is denoted by the dot-dash line (− · − · −).

suggest perhaps some invasion of early keratinization characteristics. On the basis of the evidence mentioned above, three subgroups of spectra were selected to generate the average spectra characteristic of each region in Figure 18. The differences between spectra from tumor and stroma regions noted in Figure 14 are essentially repeated in the comparison in Figure 18, while the average spectrum characteristic of early keratinization is readily distinguishable in this comparison from the other two types.

Figure 19 is a visual image of an oral tissue section from another patient. The boundary between the two regions identified as stroma and tumor is visibly distinct. An area of this section encompassing both regions and the boundary was mapped point-by-point by synchrotron-sourced FT-IR microscopy, using a $12\,\mu\text{m} \times 15\,\mu\text{m}$ aperture. In total, 30 spectra were recorded, in 3 rows of 10, as depicted by the grid superimposed on the visual image of Figure 19. Below the tissue image shown in this figure is a duplicate grid in which the numbers in the area elements correspond to the numbers allocated to the recorded spectra. A HCA of these spectra did not resolve them into three sets; spectra from the boundary region were mixed in with those from the tumor classification. While both preliminary PCA and SIMCA

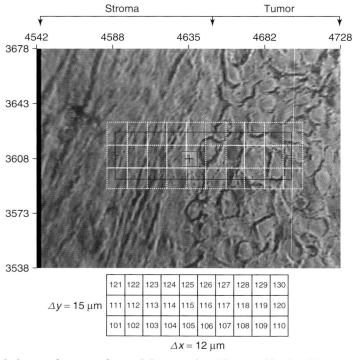

**Figure 19.**   White-light image of an area of an oral tissue section, 32× magnification. The superimposed dashed white line grid on the image shows the areas (15 μm × 12 μm) from which SRS FT-IR spectra were recorded. These were numbered as shown in the grid below the image. See text for details.

3D plots of Factors 1–3 could be oriented to indicate that the spectra recorded from aperturedareas overlaying the boundary (spectra numbers 105, 155, and 125) were similar and had more bias toward tumor than other spectra within the tumor classification, such depictions were overly subjective. More objective pictures were obtained from the SIMCA outputs of "class projection" and "class distances". The class projection provides a visual evaluation of the degree of class separation and is generated from a three-factor SIMCA of the spectra training set. Figure 20 is a bar-chart representation of the "class projection" output for Factor 2 from a SIMCA, in which the modulus of the class distinctions is 10 or greater for all spectra except numbers 105, 115, and 125. The values for these spectra are about 7, 4, and 1 respectively, suggesting much less distinction (particularly for spectrum number 125), as might be expected from

spectra recorded along a boundary region. A class distance plot for the two main categories is shown in Figure 21. The transition from stroma to tumor along a row of spectra is clearly evident in the regions of spectra numbers 105, 115, and 125; the other crossovers occur at the end of each row of spectra in the grid-map, i.e., at spectra numbers 111 and 121. Figure 22 is one plot projection of the three SIMCA factors, in which spectra 105, 115, and 125 lay nearer the tumor classification boundary spectra than others within the tumor cluster. As a consequence, we have categorized these three spectra separately from both tumor and stroma in our false-color coding of the mapped area shown in Figure 23. This simple example of "spectroscopic staining" accords well with that from conventional histopathology H&E staining.

As a final example of some of our SRS FT-IR microscopy studies on oral tissue sections,

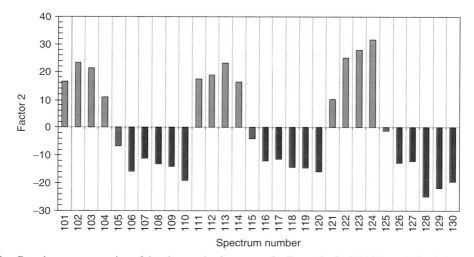

**Figure 20.** Bar-chart representation of the class projection output for Factor 2 of a SIMCA analysis of the preprocessed absorbance spectra numbered 101–130, see text for details.

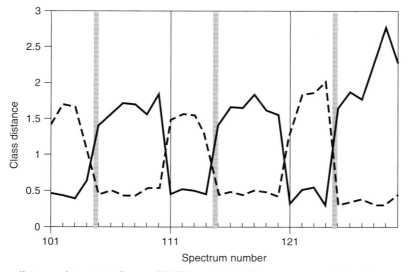

**Figure 21.** Class distance plot output from a SIMCA analysis of the preprocessed absorbance spectra numbered 101–130, see text for details.

Figure 24 shows a set of five white-light images taken from an oral tissue section from a patient. Histopathological examination showed this to be a complex region containing stroma, tumor, and necrotic tissue; the extent of each of these is indicated in Figure 24. A line profile of infrared spectra, consisting of consecutive points, as depicted in Figure 24, was recorded across this region. The spectra recorded are numbered as 201–258; there is some overlap between the consecutive images shown in Figure 24. Preliminary multivariate analyses clearly indicated that the spectrum number 246 was an outlier. Close inspection of the white-light image shows that the

SIMCA

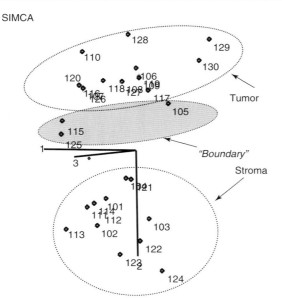

**Figure 22.** A SIMCA plot projection for Factors 1, 2, and 3 from the analysis of the preprocessed absorbance spectra numbered 101–130, see text for details.

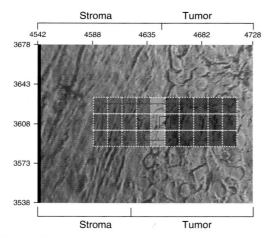

**Figure 23.** "Spectroscopic staining" of the SRS FT-IR mapped area of the white-light image shown in Figure 19. The tumor region is colored in red; the stroma region is colored blue; the boundary region is highlighted in purple. See text for details.

position from which this spectrum was recorded lay close to a tear (parting) in the tissue section, so it is possible that this spectrum was affected

by some stray-light intrusion. Spectrum number 255 was also considered an outlier, although the underlying reason for this was not so clear, except that it appears to lie on what is probably a ridge in the tissue section. Since borders between the stroma, tumor, and necrotic regions were not distinct, four spectra encompassing each boundary region were excluded from further multivariate analyses. Thus in total 13 spectra were eliminated from the set of 58 spectra before further data processing. The spectra left out were numbered 211–214, 230–233, 244–247, and 225. Figure 25 shows a HCA output from the remaining 45 spectra data set. The extreme spectra (numbers 252–258) from the necrotic region form a tight cluster, although those nearer the boundary (numbers 249–251) are clustered with the spectra associated with the tumor region. The spectra from the stroma region form a separate cluster, as depicted in Figure 25. Figure 26 is a PCA analysis of this 45 spectra data set, and yields a similar prediction as the HCA.

The examples described above serve to highlight the potential of mid-IR spectroscopy for the diagnosis of oral tissue sections, but emphasize its present reliance on the "gold standard" of H&E histopathology staining. In addition, the difficulties that are associated with collecting spectra from complex samples with variable contour boundaries using a definite geometrically shaped aperture to delineate sample microregions for spectral analysis can be seen.

This work showed promise in the discrimination of cancer tissue.[44,47] On one occasion, the clinician's analysis of the section indicated noncancer tissue; however, the spectral profile was suggestive of cancer, a finding confirmed by subsequent inspection of the stained section. This work proved the feasibility of producing a robust data set for cancer/noncancer evaluation at single cell level using the high lateral spatial resolution available through the use of synchrotron mid-IR radiation. Continuing improvements in bench-top spectrometers and sampling instrumentation now make translation of such initiatives to the clinic and clinically laboratory a feasible option.

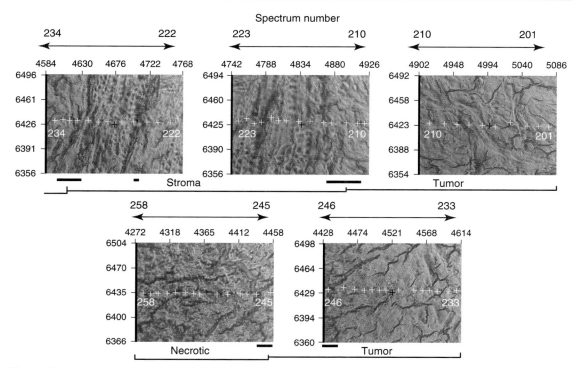

**Figure 24.** Set of overlapping white-light images taken from an oral tissue section from a patient, 32× magnification, showing the centers, marked with a +, of the $10\,\mu m \times 10\,\mu m$ areas from which SRS FT-IR microscopy spectra were recorded. The spectra were numbered 201–258. The histopathological diagnosis is indicated. Spectra excluded from the detailed multivariate analysis occur in the regions marked by the solid black rectangles beneath the images, see text for details.

## 6.2 Mid-infrared synchrotron radiation FT-IR studies of cultured cells

Mid-IR spectroscopy has also been used to explore cell changes, as described by Tobin et al.[44,47] in the study related to epidermal growth factor (EGF). EGF is known to be critically important in a number of epithelial cancers including H&N and cervix.[49] These two pathologies share many similarities, and to further explore the place of EGF a carcinoma cell model (in this case derived from cervix) was selected. These cultured carcinoma cell studies were undertaken in parallel with the oral tissue section studies discussed above with the goal of gaining both a better understanding of the inherent variability in the mid-IR spectra of such epithelial cells

and further demonstrating the capability of synchrotron-sourced FT-IR microscopy to make informative measurements at the single cell level.[44] The cell culture samples were A431 cultured cervical cells, which were stimulated with the growth-stimulating hormone EGF. (Full details about the cells, their preparation, and sampling, and storage can be found in Reference 44.) As stated above, the cells were examined using the transflection mode. Figure 27 shows an example of the mid-IR spectra of A431 cells cultured on a low-e glass slide after drying where cell nuclei, subcellular organelles, and fibers attached to the substrate can be seen. The crosses overlaid on the figure indicate the positions (including cells) from which spectra were recorded. The cell spectra, see example

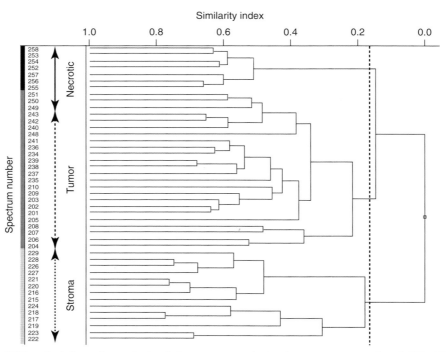

**Figure 25.** HCA output from the analysis of the preprocessed absorbance spectra numbered 201–210, 215–229 (excluding 225), 234–243, and 248–258, see text for details.

in Figure 28, were broadly similar in overall appearance to those recorded from the oral tissue sections.

In the cell study, prepared samples were exposed to EGF for periods of time between 1 min and 3 h by incubation in phosphate-buffered saline (PBS) solutions containing EGF at concentrations in the range $1 \times 10^{-7}$–$1 \times 10^{-9}$ M. For each EGF concentration and each time interval, spectra were recorded from up to 30 individual cells from each sample. As a preliminary assessment of the spectral differences occurring as a consequence of the treatment, several spectral features were measured. These were as follows: the measurement of the amide I and amide II peak positions; the amide I and amide II absorption band areas (baseline corrected between 1800 and 790 cm$^{-1}$); the $\nu$ C=O (1730 cm$^{-1}$) peak area (baseline 1760–1723 cm$^{-1}$) ratioed (normalized) against the amide I band area; the 970 cm$^{-1}$ band area (baseline 980–940 cm$^{-1}$) ratioed against the

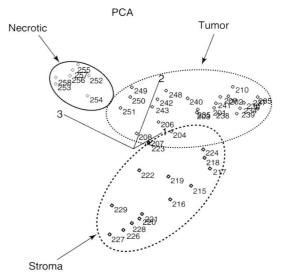

**Figure 26.** A PCA scores plot output from the analysis of the preprocessed absorbance spectra numbered 201–210, 215–229 (excluding 225), 234–243, and 248–258, see text for details.

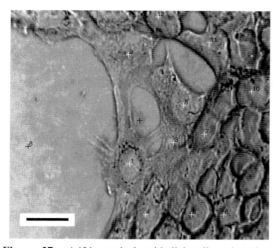

**Figure 27.** A431 cervical epithelial cells cultured on infrared reflective glass slides. Crosses marked selected data collection points. Bar represents 30 μm. [Reproduced from Tobin *et al.*[44] with permission from The Royal Society of Chemistry, © 2003.]

amide I area; and the 1170/1155 cm$^{-1}$ band area ratio (baseline 1182–1142 cm$^{-1}$). Although no changes in these spectral measurements individually correlated well with EGF concentration or time after stimulation with EGF, cross correlation of some pairs did show EGF concentration and time dependences.[44] The most prominent of these was the one between the area of the 970 cm$^{-1}$ absorption band area and the amide I peak maximum wavenumber position,

see for example Figure 29. After an interval of 1 min, similar correlations were observed for both cells treated with $1 \times 10^{-8}$ M EGF (Figure 29a) and untreated cells. After 10 min, half of the cells at this concentration showed both an increase in absorption area at 970 cm$^{-1}$ and a shift toward higher wavenumber of the amide I band maximum (Figure 29b). A smaller fraction showed these effects at 45 min (Figure 29c). This correlated change was still evident in some cells after 3 h (Figure 29d). At the lower concentration ($1 \times 10^{-9}$ M EGF), which is close to physiological levels of EGF, after only 1 min (Figure 29e) the spectral correlation distribution was similar to that observed at 10 min for the $1 \times 10^{-8}$ M EGF solution (Figure 29b) and persisted to varying degrees after 10 min (Figure 29f) and 45 min (Figure 29g). However, after 3 h exposure (Figure 29h) it had mostly returned to unstimulated distribution. At $1 \times 10^{-7}$ M EGF concentration a slight change in the correlation distribution was observed after 45 min but the effect was weaker than that shown for the two lower concentrations.

Tobin *et al.*[44] discussed some possible explanations for their empirically derived observed correlation. It is known that the amide I absorption band in such biological samples occurs primarily from the C=O stretching vibration of proteins within the sample. Changes in its peak maximum position are indicative of changes in the average

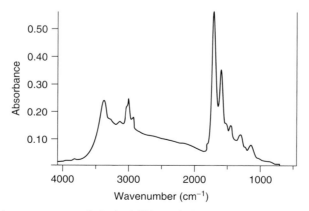

**Figure 28.** Infrared absorbance spectrum of single A431 cervical carcinoma cell. [Reproduced from Tobin *et al.*[44] with permission from The Royal Society of Chemistry, © 2003.]

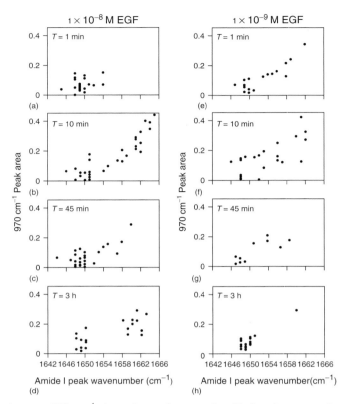

**Figure 29.** Correlation between $970\,cm^{-1}$ absorption peak area and amide I peak wavenumber position for A431 cells at time after addition of $1 \times 10^{-8}\,M$ (left column) and $1 \times 10^{-9}\,M$ EGF (right column). Each point represents one cell. [Reproduced from Tobin *et al.*[44] with permission from The Royal Society of Chemistry, © 2003.]

secondary structure of proteins. The shift in this band maximum position in the mid-IR spectra recorded from stimulated cells from $1648\,cm^{-1}$ toward $1664\,cm^{-1}$ was considered characteristic of a shift in protein composition from mostly $\alpha$-helical structures toward a higher level of turns and bends. The shift observed in the EGF stimulated cells from around $1648\,cm^{-1}$ toward $1664\,cm^{-1}$ is characteristic of a shift from highly $\alpha$-helical structure toward a higher level of turns and bends. This may be spectroscopic evidence of increased protein translation in response to the mitogenic signal of the EGF, resulting in elevated levels of differently folded protein within the cells.

Such changes are investigated in molecular medicine as epigenetic phenomena and

relationship to cancer behavior is a matter of much interest. If such molecular changes can be explored by spectroscopy, this may well elucidate important changes in cell metabolism. Bruni *et al.*[50] suggested changes perhaps related to high levels of DNA, collagen, carbonates, and lipids in infiltrating cancers. They emphasized the complexity of biological systems but pointed to the promise of perhaps predicting aspects of cancer behavior based on single cell analysis. Holman *et al.*[51] provided evidence of a shift in amide I and amide II bands corresponding to changing protein morphologies, characteristic of dying cells, together with a significant increase in the intensity of an ester carbonyl $C{=}O$ peak at $1743\,cm^{-1}$. Taken together, these papers indicate the promise of mid-IR spectroscopy in predicting biochemical events at the single-cell level.

One possible interpretation discussed for the simultaneous correlated increase in intensity of the weak absorption band at $970\,cm^{-1}$ was that it was also related to triggering of the protein phosphorylation cascade following EGF stimulation.[44] A similar increase in absorption was observed by Sanchez-Ruiz and Martinez-Carrion[52] in the mid-IR spectra recorded from phosvitin and ovalbumin after the phosphorylation of the proteins at specific residues. An alternative explanation, based on the work of Diem *et al.*[53] and Boydston-White *et al.*,[54] (*see also* **Infrared and Raman Microspectroscopic Studies of Individual Human Cells** in this book), points to bulk changes in the DNA absorption properties being responsible for changes in the relative intensity of the absorption band at $970\,cm^{-1}$. Whatever the reason(s), it is likely that high lateral spatial resolution spectroscopy at the subcellular level will be required to determine whether the observed spectral changes are the consequence of cytoplasmic events, such as protein phosphorylation and translation, and/or, nuclear events, such as DNA replication.[44] The high SNR of mid-IR spectra attainable at diffraction-limited spatial resolution, ca. $4\,\mu m$ at $1650\,cm^{-1}$, using synchrotron-sourced FT-IR microscopy will undoubtedly remain a key tool in these continuing types of study.

FT-IR and Raman microspectroscopy, this group has also recently described differences between cell lines that were and were not resistant to chemotherapeutic agents.[56] Prediction of response would be valuable in the clinical environment, ensuring targeting of cancers by an agent to which they are known to be responsive, while avoiding futile and expensive therapies that inevitably carry morbidity for the patient.

Much of the in vivo work in H&N as well as other sites has been led by Stone (a coauthor of **Raman Spectroscopy as a Potential Tool for Early Diagnosis of Malignancies in Esophageal and Bladder Tissues** in this book). His preliminary work, on laryngeal tissue, published in 2000[57] used linear discriminant analysis to separate spectra with the most visible differences being seen at 850, 950, 1200, and $1350\,cm^{-1}$. He also concluded that the relative intensity of the nucleic acid peak increases with progression to malignancy. Not only do several publications indicate efficacy in the differentiation of nonmalignant and malignant cell phenotypes but they also demonstrate an evolving consensus as to what they represent in terms of cell behavior. It is our belief that attempts should be continued to relate spectral changes to known characterization and manipulation, and we are currently undertaking studies of thyroid and oral cell lines, using Raman spectroscopy and a neural network–based mathematical model with encouraging early results.

## 6.3   Raman studies of H&N samples

In view of its potential for in vivo probes, the feasibility of studying many cancer sites by Raman spectroscopy has been investigated. Within H&N, Krishna *et al.*[55] studied the material from biopsies of five malignant and five "normal" tissues. They observed distinct differences through PCA, which was especially encouraging as the "normal" tissue was taken from the same patients as the cancers. In the possible presence of field change, there may not be any truly normal tissue, yet the distinction from cancer remained clearly definable. In a study combining

## 7   SUMMARY

The primary aim of this chapter was to place the application of vibrational spectroscopy to H&N cancer in particular, in a clinical context and to indicate its promise for future clinical use. To ensure progression from the science laboratory to the clinic requires both a mutual understanding of spectroscopy and medicine and that the translational element be driven by clinical need. Work toward a clinical setting must take account of biological variation in both patients and cancers, and clear recommendations exist for the development of data sets for clinical trials taking into account patient variables. A good model to achieve this is the phased approach used in the pharmaceutical industries. Evaluation must commence with a Phase 1 trial, in which the safety of the protocols and the establishment of an appropriate candidate dosage schedule are established in a volunteer population. Any obvious confounding factors in a candidate data set are identified and removed in the Phase 1 trial. The study is continued through a Phase 2 trial, which is similar to Phase 1 but involving a small patient population. Phase 2 trials usually require 30–40 patients. For an initial assessment of a proposed spectroscopic diagnostic tool or intervention, a trial of similar size would allow any less than robust spectroscopic markers to be identified and removed and a final refined data set would be carried forward; this methodology has been followed for colorectal adenocarcinoma with significant promise.[1] The Phase 2 trial also allows a sample calculation to be performed for a major (Phase 3) evaluation and premarketing trial after which formal submission for regulatory approval is completed.

Vibrational spectroscopy carries much potential for cancer diagnosis and the development of predictive and prognostic models. Much of this consideration of H&N cancers can be generalized to other sites. The key message is that a basic understanding of the disease and its therapy facilitates scientific advances. Recent developments in technology make laboratory-based and even hand-held probes a feasible development within the next 5–10 years. If this promise is confirmed, vibrational spectroscopy will make a substantial contribution to clinical medicine. For this reason, careful collaboration between clinicians and spectroscopists forms the ideal base for robust research leading to results that are likely to form the basis for a dependable and reproducible data set and results in a direct clinical impact.

## ACKNOWLEDGMENTS

The spectroscopic studies that are reported in this chapter were undertaken primarily at the Synchrotron Radiation Source (SRS), Daresbury Laboratory, UK, under a UK Engineering and Physical Sciences Research Council (EPSRC) funded project (GR/M62778) as part of its Physics for Healthcare program. The collaborative project brought together researchers from the University of Nottingham, Leeds University, Derby Royal Infirmary, and Daresbury Laboratory. The authors are indebted to many members of these institutions that formed part of the research team of spectroscopists, clinicians, and histopathogists, in particular: Professor Michael A. Chesters and Dr Frank J.M. Rutten from the School of Chemistry, Nottingham University; Mr Ian M. Symonds from the Department of Obstetrics and Gynaecology, University of Nottingham and Derby City General Hospital; Dr Richard Allibone from Queen's Medical Centre, University of Nottingham; and, Dr Andy Hitchcock from the Derbyshire Royal Infirmary, Derby.

Our recent work, in Leeds, has been funded by Cancer Research UK, which funded the research undertaken by Andrew Harris through a Clinical Research Training Fellowship. Professor Ken MacLennan, Professor of Tumor Pathology at the University of Leeds, has provided illustrations for this chapter and much support to our work in this field.

## ABBREVIATIONS AND ACRONYMS

| | |
|---|---|
| AJCC | American Joint Committee on Cancer |
| CT | Computerized Tomography |
| ECS | Extracapsular Spread |
| EGF | Epidermal Growth Factor |
| EGFR | Epidermal Growth Factor Receptor |
| EPSRC | Engineering and Physical Sciences Research Council |
| FNA | Fine Needle Aspiration |
| FT-IR | Fourier-Transform Infrared |
| HCA | Hierarchical Cluster Analysis |
| HIV | Human Immunodeficiency Virus |
| H&E | Hematoxylin and Eosin |
| H&N | Head and Neck |
| HPV | Human Papillomavirus |

| | |
|---|---|
| mid-IR | Mid-Infrared |
| MRI | Magnetic Resonance Imaging |
| ND | Neck Dissection |
| OR | Odds Ratio |
| PBS | Phosphate-Buffered Saline |
| PCA | Principal Component Analysis |
| PET | Positron Emission Tomography |
| SCCa | *Squamous Cell Carcinoma* |
| SIMCA | Soft Independent Modeling of Class Analogies |
| SNR | Signal-to-Noise Ratio |
| SRS | Synchrotron Radiation Source |
| STD | Soft Tissue Disease |
| TNM | Tumor-Nodes-Metastasis |
| UICC | Union International Contre le Cancer |

# REFERENCES

1. P. Lasch, M. Diem, H. Wolfgang and D. Naumann, *J. Chemom.*, **20**, 209–220 (2006).

2. C. Kendall, N. Stone, N. Shepherd, K. Geboes, B. Warren, R. Bennett and H. Barr, *J. Pathol.*, **200**, 602–609 (2003).

3. D.M. Parkin, P. Pisani and J. Ferlay, *Int. J. Cancer*, **80**, 827–841 (1999).

4. Indian Council of Medical Research, 'Summary of Specific Sites of Cancer', National Cancer Registry Programme, Chapter 6 (2002), Available at: http://www.canceratlasindia.org (accessed Dec 2007).

5. U.S. Cancer Statistics Working Group, 'United States Cancer Statistics: 1999–2003 Incidence and Mortality Web-based Report', U.S. Department of Health and Human Services, Centres for Disease Control and Prevention and National Cancer Institute, Atlanta (2007), Available at: http://www.cdc.gov/uscs (accessed Oct 2007).

6. Cancer Research UK, Cancer Statistics (2004) Available at: http://www.cancerresearchuk.org.

7. M. Lingen, E.M. Sturgis and M.S. Kies, *Curr. Opin. Oncol.*, **13** (3), 176–182 (2001).

8. S. Thomas, C.J. Bain, D. Batistutta, A.R. Ness, D. Paissat and R. Maclennan, *Int. J. Cancer*, **120**, 1318–1323 (2007).

9. C. Fakhry and M.L. Gillison, *J. Clin. Oncol.*, **24** (17), 2606–2111 (2006).

10. M.L. Gillison, *J. Clin. Oncol.*, **24** (36), 5623–5625 (2006).

11. L. Licitra, F. Perrone, P. Bossi, S. Suardi, L. Mariani, R. Artusi, M. Oggionni, C. Rossini, G. Cantu, M. Squadrelli, P. Quattrone, L.D. Locati, C. Bergamini, P. Olmi, M.A. Pierotti and S. Pilotti, *J. Clin. Oncol.*, **24** (36), 5630–5636 (2006).

12. R. Sankaranarayanan, K. Ramadas, G. Thomas, R. Muwonge, S. Thara, B. Mathew and B. Rajan, *Lancet*, **365** (9475), 1927–1933 (2005).

13. D.P. Slaughter, *Surgery*, **20** (1), 133–146 (1946).

14. N. Stone, C. Kendall, N. Shepherd, P. Crow and H. Barr, *J. Raman Spectrosc.*, **33**, 564–573 (2002).

15. L.H. Sobin and C. Wittekind, 'TNM Classification of Malignant Tumours', ISBN: 0-471-22288-7, 6th edition, John Wiley & Sons, Hoboken (2002).

16. American Joint Committee on Cancer (AJCC), 'Cancer Staging Manual', 6th edition, Springer-Verlag, New York (2002).

17. International Union Against Cancer, TNM and Prognostic Factors (2007) http://www.uicc.org/index (accessed Nov 2007).

18. Collaborative Staging. http://www.cancerstaging.org/ (accessed Dec 2007).

19. J.P. Pignon, J. Bourhis, C. Domenge and L. Designe, *Lancet*, **355** (9208), 949–955 (2000).

20. M. Romeo and M. Diem, Spectral Diagnosis Workshop (2007) Available at: http://www.spectraldiagnosis.com/ (accessed Oct 2007).

21. D.N. Sutton, J.S. Brown, S.N. Rogers, E.D. Vaughan and J.A. Woolgar, *Int. J. Oral Maxillofac. Surg.*, **32**, 30–34 (2003).

22. J.A. Woolgar and A. Triantafyllou, *Oral Oncol.*, **41** (10), 1034–1043 (2005).

23. S.N. Rogers, S.E. Fisher, J.S. Brown and E.D. Vaughan, *Br. J. Oral Maxillofac. Surg.*, **40**, 11–18 (2002).

24. A.W. Njiru, S.E. Fisher, T.K. Ong, D.A. Mitchell, K.A. MacLennan, A.S. High and W.J. Hume, *Br. J. Oral Maxillofac. Surg.*, **41** (4), 31 (2003).

25. A. Ferlito, A. Rinaldo, K.T. Robbins, C.R. Leemans, J.P. Shah, A.R. Shaha, P.E. Andersen, L.P. Kowalski, P.K. Pellitteri, G.L. Clayman, S.N. Rogers, J.E. Medina and R.M. Byers, *Oral Oncol.*, **39**, 429–435 (2003).

26. J.A. Woolgar, *Oral Oncol.*, **42** (3), 229–239 (2006).

27. E. Genden, A. Ferlito, J. Bradley, A. Rinaldo and C. Scully, *Oral Oncol.*, **39**, 207–212 (2003).

28. B. Vikram, E.W. Strong, J.P. Shah and R. Spiro, *Head Neck Surg.*, **6** (3), 730–733 (1984).

29. Z. Rumbolt, T.A. Day and M. Michel, *Oral Oncol.*, **42**, 854–865 (2006).

30. T. Helliwell and J.A. Woolgar, 'Datasets for Histopathology Reports on Head and Neck and Salivary Neoplasms', Royal College of Pathologists, London (2005).

31. A.P. Coatesworth and K. MacLennan, *Head Neck*, **24** (3), 258–261 (2002).

32. J. Jose, A.P. Coatesworth and K. MacLennan, *Head Neck*, **25**, 194–197 (2003).

33. J.W. Moor, J. Jose, C. Johnston, A.P. Coatesworth and K. MacLennan, *Acta otolaryngol.*, **124**, 97–101 (2004).

34. J.N. Myers, J.S. Greenberg, V. Mo and D. Roberts, *Cancer*, **92** (12), 3030–3036 (2001).

35. J.S. Greenberg, R. Fowler, J. Gomez, V. Mo, D. Roberts, A.K. Naggar and J.N. Myers, *Cancer*, **97** (12), 3030–3036 (2003).

36. J.A. Woolgar, S.N. Rogers, D. Lowe, J.S. Brown and E.D. Vaughan, *Oral Oncol.*, **39**, 130–137 (2003).

37. J. Jose, J.W. Moor, A.P. Coatesworth, C. Johnston and K.A. MacLennan, *Acta Otolaryngol.*, **123** (3), 336–339 (2004).

38. K. MacLennan, J. Jose, A. Ferlito, K.O. Devaney, K.T. Robbins, J. Moor and A. Rinaldo, *Acta Otolaryngol.*, **123** (3), 336–339 (2003).

39. J. Jose, J.W. Moor, A.P. Coatesworth, C. Johnston and K.A. MacLennan, *Arch. Otolaryngol. Head Neck Surg.*, **130** (2), 157–160 (2004).

40. A.W. Njiru, 'Bone Marrow Micrometastasis in Patients with Head and Neck Cancer', University of Leeds, Leeds (2004).

41. R. Rosen, A. Smith and A. Harrison, 'Future Trends and Challenges for Cancer Services in England: A Review of Literature and Policy', ISBN 978 1 85717 549 3, King's Fund (2006).

42. J.A. Bonner, P.M. Harari, J. Giralt, N. Azarnia, D.M. Shin, R.B. Cohen, C.U. Jones, R. Sur, D. Raben, J. Jassem, R. Ove, M.S. Kies, J. Baselga, H. Youssoufian, N. Amellal, E.K. Rowinsky and K.K. Ang, *N. Engl. J. Med.*, **354**, 567–578 (2006).

43. M. Knowles and P. Selby, 'Introduction to the Cellular and Molecular Biology of Cancer', Oxford University Press (2005).

44. M.J. Tobin, M.A. Chesters, J.M. Chalmers, F.J.M. Rutten, S.E. Fisher, I.M. Symonds, A. Hitchcock, R. Allibone and S. Dias-Gunasekara, *Faraday Discuss.*, **126**, 27–39 (2004).

45. L. Barnes, J.W. Eveson, P. Reichart and D. Sidransky, 'World Health Organisation Classification of Tumours; Pathology and Genetics: Head and Neck Tumours', IARC Press, Lyon (2005).

46. R.J. Zarbo, *Mod. Pathol.*, **15** (3), 298–323 (2002).

47. M.J. Tobin, F. Rutten, M. Chesters, J. Chalmers, I. Symonds, S. Fisher, R. Allibone and A. Hitchcock, *Eur. Clin. Lab.*, **21**, 20–22 (2002).

48. Y. Fukuyama, S. Yoshida, S. Yanagisawa and M. Shimizu, *Biospectroscopy*, **5**, 117–126 (1999).

49. K.K. Ang, B.A. Berkey, X. Tu, H.-Z. Zhang, R. Katz, E.H. Hammond, K.K. Fu and L. Milas, *Cancer Res.*, **62**, 7350–7356 (2002).

50. P. Bruni, C. Conti, E. Giorgini, M. Pisani, C. Rubini and G. Tosi, *Faraday Discuss.*, **126**, 19–26 (2004).

51. H.Y. Holman, M.C. Martin, E.A. Blakely, K. Bjornstad and W.R. McKinney, *Biopolymers (Biospectroscopy)*, **57**, 329–335 (2000).

52. J.M. Sanchez-Ruiz and M. Martinez-Carrion, *Biochemistry*, **27**, 3338 (1988).

53. M. Diem, S. Boydston-White and L. Chiriboga, *Appl. Spectrosc.*, **53**, 148A–161A (1999).

54. S. Boydston-White, T. Gopen, S. Houser, J. Bargonetti and M. Diem, *Biospectroscopy*, **5**, 219–227 (1999).

55. M.C. Krishna, G.D. Sockalingum, J. Kurian, L. Rao, L. Venteo, M. Pluot, M. Manfait and V.B. Kartha, *Appl. Spectrosc.*, **58** (9), 1128–1135 (2004).

56. M.C. Krishna, G. Kegelaur, I. Adt, S. Rubin, V.B. Kartha, M. Manfait and G.D. Sockalingum, *Biopolymers*, **82**, 462–470 (2006).

57. N. Stone, P. Stavroulaki, C. Kendall, M. Birchall and H. Barr, *Laryngoscope*, **110** (10), 1756–1763 (2000).

# Infrared Spectroscopic Imaging Protocols for High-throughput Histopathology

## Rohit Bhargava[1] and Ira W. Levin[2]

[1] *University of Illinois at Urbana-Champaign, Urbana, IL, USA*
[2] *National Institutes of Health, Bethesda, MD, USA*

## 1 INTRODUCTION

### 1.1 Prostate cancer and enhancements needed in histopathology

Prostate cancer is one of the leading causes of death in the USA and is widespread with an incidence rate of over 234 000 new cases estimated in 2006. Consequently, population screening is being increasingly employed to detect the disease. The emphasis in screening is to use a high sensitivity, but simple, diagnostic tests. For example, the prostate-specific antigen (PSA) assay[1] acts to triage individuals at risk for prostate cancer. A cutoff level (typically $4 \, \text{ng} \, \text{mL}^{-1}$) or the rate at which the PSA level increases (PSA velocity) implies that the screened person should be at heightened surveillance. A manual examination for prostate growth, namely, a digital rectal examination (DRE) may be employed. A DRE is less sensitive in the diagnosis of tumors; hence, it is often used in conjunction with PSA screening. If PSA[2] or DRE screens are abnormal,[3] a biopsy is essential for detecting or ruling out cancer. In general, careful surveillance not only leads to a large majority of cancers being detected early but also results in large numbers of men being biopsied. Morphologic structures in biopsied tissue, as diagnosed by a pathologist, are the only definitive indicator of disease and form the gold standard of diagnosis.[4] Along with clinical history, stage and PSA values, pathologic diagnoses form a cornerstone of clinical therapy and serve as a basis for a vast majority of research activity.[5]

The large number of biopsy samples generated in the management of prostate cancer point to several undesirable issues. First, widespread screening results in a large number of men having biopsy samples taken (estimated >1 million annually).[6] This, in turn, places an increasing demand on the clinical services (typically, 6–23 needle biopsies per patient are acquired). Operator fatigue and associated guidelines limit the workload and rate of sample examination. Further, newly detected cancers are increasingly moderate grade tumors, in which inter- and intrapathologist diagnostic variations complicate disease management. Consistency in determining the primary characteristic of disease severity (Gleason score) is

*Vibrational Spectroscopy for Medical Diagnosis*. Edited by Max Diem, Peter R. Griffiths and John M. Chalmers.
© 2008 John Wiley & Sons, Ltd. ISBN 978-0-470-01214-7.

often not present. Intraobserver measurements in various studies indicate that pathologists confirm their own Gleason scores only 50% of the time and ±1 of their score in no more than 80% of the cases.[7] Hence, the diagnoses for ~50% of the cases studied may change and may be significantly altered for ~20% of cases, which ultimately leads to changes in the recommended mode of therapy for a patient subset.[8] Second opinions[9] improve assessment and are cost-effective,[10] and often mitigate the effects of increased healthcare costs, lost wages, morbidity, or potential litigation. Thus, a reduction in variability by providing an immediate second opinion, preferably using a completely different methodology, would be beneficial. Lastly, the issues will exacerbate in aging populations, as age is the single largest risk factor in prostate cancer.[11]

For the reasons outlined above, there is an urgent need for high-throughput, automated, and objective pathology tools in prostate cancer management and for research. Our overarching hypothesis is that these requirements are fulfilled through spectroscopic imaging approaches that are both compatible with, and add substantially to, current pathology practice. Hence, we have developed tools and methodologies to integrate spectroscopic imaging data with current clinical protocols and workflow. Vibrational spectroscopy offers the most direct measurement of the molecular content of any material, including that of tissue. Since vibrational spectroscopic techniques involve noninvasive radiation and exclude perturbing probes, they are completely compatible with downstream tissue processing and overall clinical work flow. At the initiation of this work, mid-infrared (mid-IR) spectroscopic imaging was chosen over Raman spectroscopic imaging techniques since the requirement for scanning large tissue areas could not be met at that time by Raman imaging methods. Further, it is unclear whether the higher spatial resolution of Raman spectral imaging will prove useful in prostate pathology as a majority of diagnostic features are of a length scale that can be addressed by mid-IR imaging. An analysis of the spatial resolution advantage of Raman spectroscopy definitely needs

to be studied at some future time, however, as there may be unknown diagnostic indices that are not apparent at present. In this chapter, we focus only on mid-IR spectroscopy and describe the results and development of a vibrational infrared (IR) spectroscopic imaging protocol for prostate histopathology. We emphasize specifically the integrative nature of the spectral data acquisition, numerical processing, and statistical analysis of the results. We describe in detail the role of spatial resolution, signal-to-noise ratio (SNR), classification parameters and statistical considerations in prostate histology. Lastly, we discuss the potential of and outlook for this approach.

## 1.2 The promise and role of FT-IR imaging in tissue histopathology

In pathology practice, multiple samples are withdrawn from the prostate during biopsy. Each biopsy sample is fixed and embedded in a sectioning medium for processing. A section, typically of 3–7 μm in thickness, is deposited onto a glass slide for review by a pathologist. Since tissue does not have a useful contrast in optical brightfield microscopy, the sectioned tissue is stained. A mixture of hematoxylin and eosin (H&E) is commonly employed, staining protein-rich regions pink and nucleic acid–rich regions of the tissue blue. An example is shown in Figure 1(a) and (b), where different structural units of prostate tissue are clearly delineated only after staining. In stained tissue, a pathologist can easily recognize specific cell types and alterations in local tissue morphology that specifically indicate disease. In prostate tissue, for example, the cells of interest are epithelial cells, as more than 95% of cancers arise in this one cell type. Epithelial cells line three-dimensional ducts in the prostate, forming its functional units. In two-dimensional thin sections, the cells lining the ducts appear to demarcate empty circular regions (lumen). Distortions in the normal lumen appearance provide evidence of cancer and characterize its severity (Gleason grade) (Figure 1c and d). A catalog of the distortions of epithelial patterns

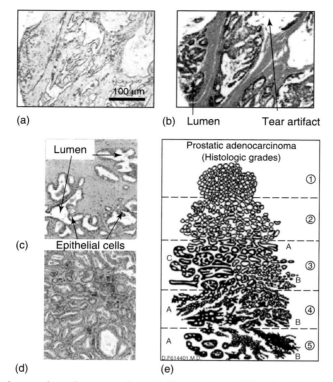

**Figure 1.** Prostate histology and carcinoma grading. (a) Unstained and (b) stained tissue. Histologic structures are only apparent upon staining sectioned tissue (a). Grading for prostate cancer is based on patterns of epithelial cells around lumen as, for example, shown in (c) and (d). If lumen is nearly regular and well-spaced, tissue is benign. Increasing disruption of this structure leads to higher grade tumors on the Gleason Scale (e). The circled numbers indicate Gleason grades. The letters A, B and C refer to sub-classes of indicated grades. [Based on figures reproduced with permission from References 12–14.]

forms the key reference for diagnosing disease (Figure 1e). In engineering terms, the process of disease recognition is a manual pattern recognition that seeks to match unknown observations to known healthy or diseased morphologies.[12]

Visual examination of biopsies is quite powerful in that an investigator not only can recognize disease but can also overcome confounding preparation artifacts (e.g., shown in Figure 1b), detect unusual cases, and recognize deficiencies in diagnostic quality. Manual examinations, unfortunately, are time consuming and frequently lead to a variation in subjective judgments concerning the disease grade.[15] As an alternative, computer-based pattern recognition approaches to diagnose disease promise more accurate, reproducible, and automated approaches that could reduce the variance in diagnosis, while proving economically favorable. The potential integration of automated approaches in current clinical practice can be seen in Figure 2. Automated approaches could be used to relieve workload by identifying cases that are clearly not cancer, by providing input at initial diagnosis, and by serving as quality control tools for final diagnosis. Hence, attempts have been made to characterize morphology using H&E image analysis[16] and to perform spectroscopic imaging on both stained and unstained tissues. While spectroscopic approaches and other automated strategies have been proposed and are reasonably accurate, the ability to provide robust diagnoses is the key failing of these methods.

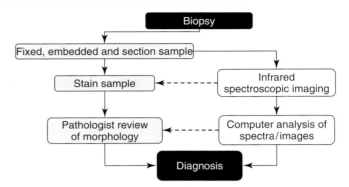

**Figure 2.** Potential application of FT-IR imaging for pathology. The current paradigm of cancer diagnosis and grading upon biopsy involves sample processing, staining, and pathologist review (left, shaded boxes). To implement the paradigm of automated analysis (right, unshaded boxes), IR chemical imaging is followed by computer analysis for diagnosis. Since IR imaging is label free and nonperturbing, the sample can be stained, providing the pathologist with both IR chemical and conventional stained images. [Reproduced from Bhargava.[13] © Springer, 2006.]

A potential solution is emerging in the form of chemical imaging and microscopy.[17] As opposed to conventional probe-assisted molecular imaging, chemical imaging[18] seeks to measure directly the identity and/or concentration of chemical species in a sample using spectroscopic methods. Instead of molecular probes, computer algorithms are used to extract information from the data and statistical methods are used to provide confidence intervals. The approach is limited only by the ability of the technology to sense specific types of molecules or otherwise resolve chemical species and morphologic structures. Among the prominent approaches are vibrational spectroscopic imaging, both Raman and IR, as well as mass spectroscopic imaging (MSI)[19,20] and magnetic resonance spectroscopic imaging (MRSI).[21] While each technology promises a specific measurement (e.g., proteins or metabolic products) for specific situations (e.g., in vivo or ex vivo), IR spectroscopic imaging[22] is particularly attractive for the analysis of tissue biopsies in that it permits a rapid and simultaneous "fingerprinting" of inherent biologic content and metabolic state.[23–26] Recent reviews address biomedical applications of Fourier transform infrared (FT-IR) spectroscopy and imaging,[27–32] especially related to diseases and cancer. Other chapters in this

book (**Antemortem Identification of Transmissible Spongiform Encephalopathy (TSE) from Serum by Mid-infrared Spectroscopy; Head and Neck Cancer: A Clinical Overview, and Observations from Synchrotron-sourced Mid-infrared Spectroscopy Investigations; Spectral Histopathology of the Human Cervix; and Raman Spectroscopy as a Potential Tool for Early Diagnosis of Malignancies in Esophageal and Bladder Tissues**) describe applications to various diseases involving other organs. Similarly, the technology for IR spectrometry, microscopy, and imaging is described in a recent book.[33]

The commercial availability of high-fidelity FT-IR imaging instruments and advances in computer technology and data analysis algorithms generate an increasing number of studies reporting automated disease diagnoses. At the same time, there is considerable debate emerging on various aspects of this subject. Reports have focused on different sample acquisition and processing techniques, on differing instrumentation, on data acquisition or handling protocols, and on the application of a variety of decision-making algorithms. Hence, direct comparisons between studies are difficult and various facets such as resolution (both spatial and spectroscopic), biological diversity, and chemometric or statistical methods are not universally accepted.

Although universal standards for treating biological data do not exist, the field continues to develop rapidly. Indeed, there may be no universally superior method and a single standard for all organs and diseases may not be possible or desirable. We believe that progress in this area can be accelerated by carefully obtaining the results that seem promising and subjecting them to the usual optimization needed for routine spectroscopic analyses. Once a protocol has been developed, it must be rigorously tested not only for validity of results but also for its sensitivity to various experimental parameters. The limitations and key performance measures of the protocol can thus be readily established and compared with other protocols applied to the same organ. Further, many excellent studies have generally focused on one aspect, e.g., spatial resolution. Their role in overall classification, however, is less understood. We believe that the development of clinical protocols is necessarily integrative and

the understanding of each factor in the classification of tissue must be in the context of the disease. In this chapter, we present recent efforts at understanding various experimental variables in the context of prostate tissue, but emphasize that the approach is applicable to other tissue types.[34] Other groups have recently reported exciting results in the diagnosis, grading, and classification of prostate cancer,[35–38] including the effects of zonal anatomy[39] and cytokinetic activity on spectra.[40] The discussion here is aimed at providing insight for formulating better protocols and for understanding the performance of classifiers, rather on the specific results obtained in our laboratory or elsewhere.

## 1.3 Advances in prostate histopathology using FT-IR imaging

The application of FT-IR imaging for pathology is illustrated in Figure 3. Conventional H&E stained

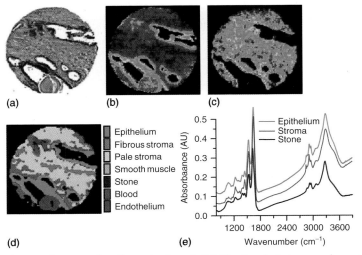

**Figure 3.** Correspondence of conventionally stained and FT-IR chemical images for pathology applications. (a) Hematoxylin and eosin (H&E) stained image of prostate tissue section. Hematoxylin stains negatively charged nucleic acids (nuclei and ribosomes) blue, while eosin stains protein-rich regions pink. The diameter of the sample is ~500 μm. Simple univariate plots of specific vibrational modes provides for enhancement or suppression of specific cell types. (b) Absorption at 1080 cm$^{-1}$, commonly attributed to nucleic acids, highlights nuclei-rich epithelial cells in the manner of hematoxylin. (c) Spatial distribution of a protein-specific peak (~1245 cm$^{-1}$) highlights differences in the manner of eosin. The entire spectrum can be analyzed for a series of markers that provide more information than H&E or univariate images, as shown in (d) where specific cells are color coded based on their spectral features (e). [Reproduced from Bhargava.[13] © Springer, 2006.]

images require human interpretation (Figure 3a), while plots of absorbance values for every detector pixel (Figure 3b and c) provide similar visualizations for specific vibrational modes of species in the IR. The IR image, further, contains a full spectrum at every pixel with numerous features indicative of compositional differences between cells and those characteristic of disease. The challenge is then to utilize the spectral data from various cell types, patients, and clinical settings to provide robustly a color-coded image that is, at least, equivalent to H&E images. As histologic classes contain mostly identical chemical components (proteins, nucleic acids, carbohydrates, and lipids), their characteristic IR vibrational spectra only differ marginally. Univariate analyses are generally unable to distinguish all histologic constituents, thus creating a need for multivariate descriptions. Although the IR data only demonstrate univariate representations in the images, automated mathematical algorithms can utilize multivariate techniques to determine the cell types and their locations within the image. These methods can, additionally, be transparent in their assignments. The probing of methods for numerical characteristics of the classification process provides quantitative measures of accuracy and the basis for statistical confidence in results.[41] The image resulting from a composite multivariate analysis (Figure 3d) would automatically label each cell or disease state with specific colors, thus providing easy interpretation and obviating the need for manual image analysis. The classification task becomes particularly challenging as spectral differences between histologic entities and pathologic conditions may be small and diagnostic differences may be masked by measurement noise, biologic variability, and limited scope of studies due to various hurdles.

The approach can be extended to samples from many patients and from different clinical venues. For example, we have examined samples from several hundred patients obtained from 30 different clinical settings using tissue microarray (TMA) technology. The application of TMAs for spectroscopic analyses was a critical approach that enabled the large sampling diversity.[42] Briefly,

TMA slides consist of multiple tissue samples placed on a single substrate. The production of these slides requires a master block in which multiple biopsy samples are placed in a grid format. The block is subsequently sectioned with a microtome to provide numerous TMA sections, providing a convenient production route as well as correspondence between consecutive sections. Automated array construction, image acquisition, and correlations between different instruments are now routinely possible. Details on the use of TMAs and their potential for spectroscopic analyses are provided elsewhere.[17] The major developments needed for our studies were the construction of TMAs specifically to include a large number of cell types from many patients and software to reconstruct large TMA images from acquired data sets. Here, we note that TMAs are an ideal platform for both discovery of spectral biomarkers and for validating vibrational spectroscopic studies for large populations of patients.

We developed protocols for automated prostate histology and applied them to a number of TMA sets.[43] An example of a TMA is shown in Figure 4. The advantages of IR-based histologic classification over conventional H&E stained images are obvious. No human interpretation is required, cell types can be readily tabulated and quantified, no stains are required, and the process can be readily adapted in clinical settings. Furthermore, multiple steps are not required to obtain information for more than one cell type. For example, multiple immunostains are required for the identification of various stromal compartments. Fibroblasts are characterized by expression of vimentin, which forms a mesenchymal cell intermediate filament, and the absence of smooth muscle markers. Early- and late-stage muscle markers are characterized by smooth muscle (sm)−actin microfilaments and associated calponin, respectively. Myofibroblasts are considered intermediate between fibroblasts and smooth muscle and, thus, express both vimentin and sm−actin but not calponin. These identifications can now be accomplished in a single step.

Pathologic assessment of tissues has proved to be more challenging. One of the major drawbacks

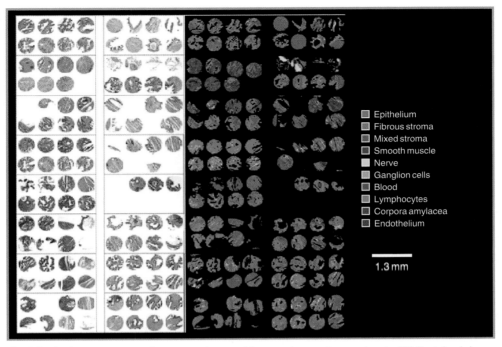

**Figure 4.** Histologic assessment of prostate tissue without stains or human input. H&E stained images (left) provide histologic identification that has to be manually inferred. Application of multivariate analysis methods lead to color-coded images (right) denoting various histologic units. [Reproduced from Fernandez *et al.*[43] with permission from Macmillan Publishers, © 2005.]

has been the absence of universally accepted standard measures of disease severity. A complicating concern has been that the morphologic and chemical clusters of disease may not be closely related. Hence, many efforts have been focused on distinguishing malignant from benign samples, while other studies have been attempted to find direct correlation with disease grade.[44] We have, again, followed the path of clinical compatibility. Hence, we first segment tissue into histologic classes. Since prostatic adenocarcinoma (>95% of prostate cancers) are epithelial in origin, we focused directly on the epithelium. We have adopted three broad strategies to describe cancer diagnoses: first, we examined the spectral content of epithelial cells; second, we combined spectral and spatial information to determine cancer probability; and third, we used comparative analyses between benign and malignant tissues of the same patient as a baseline for statistical analyses

and formulation of protocols. Our first attempt at determining the possibility of disease diagnoses for prostate cancer using the histology-based approach was to compare normalized spectra from the same patients.

In a study of 50 patients, multivariate analysis was able to consistently separate malignant from benign epithelium for each patient by normalizing the spectroscopic metric values that provided optimal tissue segmentation. In this chapter, we use the term *metrics* to denote a set of spectral and spatial features that provide a quantitative assessment of the classification. The term *feature* is reserved for what is seen directly in the spectrum or image, whereas a metric may be, for example, the ratio of the absorbance at one wavenumber to the absorbance at another. The fractions of total epithelial pixels classified as carcinoma for the benign and malignant sample of each patient demonstrated that there was

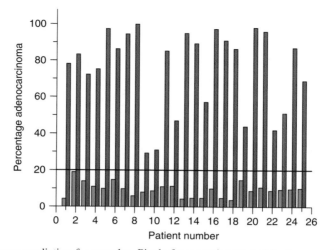

**Figure 5.** Adenocarcinoma prediction for samples. Pixels from carcinoma and benign samples from 25 patients are classified as belonging to benign or carcinoma subgroups. The fraction of pixels classified erroneously as carcinoma for benign samples, indicated by green bars, is consistently below 20%, whereas, the fraction classified correctly as malignant for carcinoma samples, indicated by red bars, is significantly higher. The solid line at 20% provides a convenient threshold to segment epithelial foci. [Reproduced from Fernandez *et al.*[43] with permission from Macmillan Publishers, © 2005.]

significantly higher misclassification for disease diagnoses than there was for histologic segmentation. Figure 5 provides a patient-by-patient breakdown of the classification of both benign and malignant samples. While benign samples contained, at most, ~16% adenocarcinoma spectral pixels, tissue that was classified as malignant usually demonstrated a significantly larger number of pixels classified as benign. While the error in classifying pixels from benign tissues may be taken to be the limits of the classification process, the high fraction of benign pixels in malignant tissue deserves further examination. There may be a biological basis to this diversity. It must be noted that the above analyses are for pixel-level classification. They do not imply accuracy in detecting disease, but do provide a sense of the error rates and the possible sensitivity and specificity, if single cells were to be examined. A diagnosis is not made on the basis of single cell features in prostate tissue, but on the appearance of a large number of epithelial foci. Hence, the approach we took for classifying the entire tissue was to term it malignant if more than ~20% pixels were classified as malignant. In small data sets,

as used above, we do not recommend assigning a specificity or sensitivity value to the results due to limited statistical sampling. The number of pixels, however, is significantly larger, allowing for accuracy measures for each patient. Again, the data should be used only to infer the variation of the model's performance in a set of patients and not to suggest that the entire population will behave similarly. More studies to address the issues of overall diagnostic accuracy are underway.

For prostate tissue, the spatial distribution of cytoplasm and nuclear-rich pixels was additionally measured for epithelial cells, as shown in Figure 6.[42] A morphologic index, constructed from spectral identification of these two subclasses of epithelium, correlates with malignancy and appears to be an important parameter in predicting disease in the prostate. Specifically, decreased cytoplasmic-to-nuclear ratios are common in malignant cells in comparison to benign cells. The data, however, did not consistently follow for different grades of malignant prostate cells. Instead, a common feature is an increased variability with increasing Gleason grade. Analogous to spectral parameters, in which no single peak

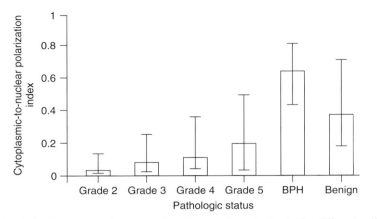

**Figure 6.** Epithelial polarization measured as a cytoplasmic-to-nuclear pixel ratio for different pathologies in prostate tissue. The ratio represents an example of a spatial metric. The significant variability in measurements presents a challenge for automated recognition algorithms. [Reproduced from Bhargava *et al.*[42] with permission from Elsevier, © 2006.]

predicts malignancy, this index does not predict the pathologic state of the tissue by itself and must not be misconstrued as such. In our classification paradigm, it becomes an additional metric for differentiating benign and malignant tissues using multivariate methods and can be treated like any spectral feature. As with spectral metrics, this spatial metric is simply a number that can help distinguish classes of pathology. Hence, FT-IR imaging provides the possibility of utilizing effectively both the morphologic and the spectral content as metrics in disease diagnosis. A detailed discussion of the sources of error and potential confounding variables for spatial indices is available elsewhere.[42]

## 2  APPROACH AND ESSENTIALS

The key elements of our approach are displayed in Figure 7. The issues of sampling, imaging, analysis, and validation are interconnected. TMAs and principles of statistical validation are described elsewhere. We focus here, however, on detailing the role of pattern recognition and FT-IR imaging. To better understand the role of imaging, it is instructive to first examine our data analysis protocols and the unique insights that they offer.

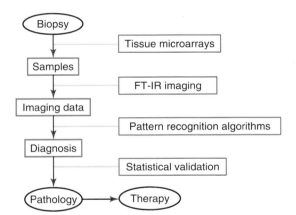

**Figure 7.** Key elements of an automated approach (green rectangles) to clinical practice (blue ellipses). To develop and validate this approach, four key areas were employed, as indicated in the red rectangles.

### 2.1  Pattern recognition algorithms

Several methods have been reported to employ spectral data for histopathologic information. Broadly, all methods can be divided into supervised and unsupervised approaches. We believe that unsupervised methods are more suited to research and discovery, as they ultimately involve some level of human interpretation in the final assignment of model for histopathology (for example, the number of classes in clustering defines the universe of possible cell types in histology). Algorithm training and subsequent predictions for such methods do not need prior pathology knowledge and can be widely applied. A potential drawback may be the limited significance to clinical practice for some of the discovered classes. Supervised methods are preferred when the data need to be related to known conditions, as for example the universe of histologic classes or clinical diagnosis that each pixel must belong to is denoted prior to the training and application of the algorithm. Thus, these approaches need carefully annotated data. Conceptually, a protocol based on a supervised classification of IR spectral imaging data for histopathology is fairly straightforward,[32] as shown in Figure 8.

First, a model for classification that includes all possible outcomes for any pixel in the image is selected. We term each histopathologic constituent of the model a *class*. A class may not necessarily correspond to specific cell types or entities corresponding to morphology-based pathology, but should have chemical features that are distinct from other classes. Hence, the method presents opportunities to not only reproduce morphology-based pathology but also include additional classes discovered by chemistry-based imaging. Uncoupling class definitions from conventional histology also allows both simplification of the model to be used for classification and the user to selectively focus on specific cells relevant in disease. For example, a simple two-class model of epithelium and stroma may be useful in rapidly segmenting epithelial from all other cells in the spectroscopic analysis of adenocarcinoma. While conventional pathology is bound by morphological differences, chemical definition of classes is

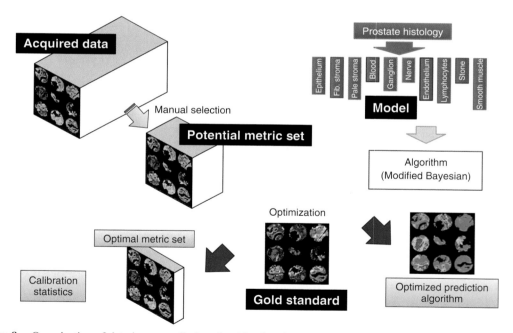

**Figure 8.** Organization of data into a prediction algorithm involves several steps. Acquired FT-IR imaging data (top, left) are reduced by manual selection to a set of features that capture the essential elements of spectra from all tissue types. A model (top, right) is selected for the data and employed to develop an algorithm. The algorithm is applied to the entire metric set and prediction capabilities are optimized. Results of the optimization provide an optimal metric set for validation studies, the parameters of the algorithm to be applied, and statistical measures of the accuracy of the protocol applied to calibration data. The optimized algorithm is applied to acquired data without supervision. [Reproduced from Srinivasan and Bhargava.[32] © 2007, Advanstar Communications.]

also likely to prove useful in the discovery of different entities that may appear morphologically similar. For example, we have observed layers of compositionally different materials in prostate stones, whereas the clinical description of the entire entity is that of corpora amylacea. One of the exciting areas to emerge from this chemical imaging approach to pathology is likely to be the interpretation of heterogeneity observed chemically in homogeneous classes defined morphologically in conventional histopathology.

Next, data from a large number of tissue samples are recorded using TMAs. Pixels are identified and specifically marked (gold standard) by different colors to correspond to known regions of tissue. The gold standard set of pixels provides the true identity of the tissue for comparison to that predicted by the algorithm. This is usually accomplished not only by comparison with an H&E stained image or with immunohistochemically stained images,[34] but also by using clustering methods on the spectral data. Spectral data typically consist of several thousand resolution elements per pixel, including many features with redundant information. Hence, it is desirable to reduce the size of the data by recognizing a (lower) inherent dimensionality. For our purposes, the spectral data are reduced to a smaller set of measures that capture the classification potential of an entire spectrum. These characteristics are termed *spectral metrics* to denote that they measure the most important features of the spectrum for classification purposes (vide supra). A metric is a general term for any feature that may be useful in classification. This should not imply that all metrics will be used or are useful for tissue segmentation. Metrics must be defined in a manner, however, that no important ones are excluded. Hence, the metric set is a universal set that will not only necessarily include all metrics used in classification but also include some metrics that are not found to be useful. While there are no formal terms for features used in previous studies, the metric data set typically consists of a manual selection of large spectral regions, factors from principal components analysis, or a grouping of regions from

genetic algorithms. We have also proposed a sequential forward selection (SFS) algorithm that is integrated with the discriminant step. Briefly, in this method, metrics are added one at a time to create a new classification model. The optimal model is the one with the lowest number of metrics that provide acceptable accuracy.

After a metric set is obtained, a numerical classification algorithm is then chosen, as for example, linear discriminant analysis, neural network, soft independent modeling of class analogies (SIMCA), or modified Bayesian classifier. Since specific combinations of features provide different performances in different classifier systems, the metric selection step is necessarily integrated with the classification step and is optimized in a similar manner. In our scheme, the classifier is optimized iteratively to predict the training data set by including or eliminating metrics based on their ability to improve the classification. Hence, the large number of metrics selected above is whittled down by training the algorithm to a significantly smaller fraction to be used in the classification protocol. Subsequently, the algorithm is applied to a second data set for an independent validation. The validation data set is independently associated with a specific class to yield a second gold standard. A comparison of the gold standard marking with the computationally predicted class provides a measure of the accuracy of the procedure. The accuracy is typically characterized by the fraction of positive cases that were correctly predicted, or the sensitivity of the test. For histopathology, it is equally important to know how many of the cases found negative upon the test were in fact, indeed, negative. It is also important to know what and where the misclassifications are occurring to remove any possible bias. Hence, we have examined accuracy in three forms: receiver operating characteristic (ROC) curves,[42] confusion matrices, and images. ROC curves plot the sensitivity (true positive) of the test against the complement of the specificity (false positive), illustrating the sensitivity and specificity trade-off of the classifier. ROC curves provide information in addition to simple summary measures, such as overall accuracy. The area under the receiver operating characteristic curve (AUC) is a summary measure that is well-known to characterize the performance of the classifier. The AUC can be understood to be the probability that the classifier will assign a higher score to the positive example than to the negative one if one positive and one negative sample are randomly picked. An area of 1 represents a perfect classifier, whereas an area of 0.5 represents a test with no ability to correctly segment the tissue. Confusion matrices provide the fraction of pixels classified for each class, thereby providing measures of both accuracy and source of errors. Classified images that can be compared pixel-for-pixel to other images yield qualitative measures of accuracy, as well as a spatial distribution of the error, as illustrated, for example, in Figure 9.

A unique feature of the classification method that we have developed is an intermediate step in which the probability of a pixel belonging to each class is explicitly determined. The graph in Figure 10(a) displays the fraction of epithelial pixels that are correctly classified (labeled correct) and the fraction of other cell types that are classified as epithelium (labeled incorrect). The optimally classified image is shown at the bottom (Figure 10b) for reference. The image panel to the right (Figure 10c) demarcates pixels that are recognized as epithelium (white) at specific threshold values, illustrating an increase of pixels that are classified according to a given class when the probability threshold is lowered. At high thresholds, not all epithelial pixels are recognized, leading to a limited sensitivity of the computation. At low thresholds, the number of pixels erroneously classified as epithelium increases, which enhances sensitivity but leads to a lower specificity value. The sensitivity—specificity trade-off is always implicit in classifiers, but in this case, a user may directly select the detection capability or "purity" of the result. The bounds to selection are described by the correct—incorrect plot (Figure 10) for each class. An important caveat must be mentioned: the procedure described above is a simplified representation. The actual class in a complete

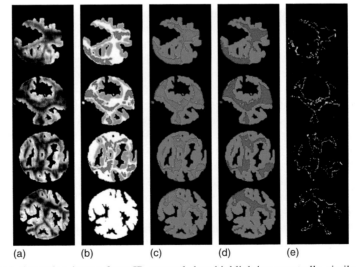

(a)      (b)      (c)      (d)      (e)

**Figure 9.** Normalized absorption image from IR spectral data highlighting spectrally similar regions. Using H&E images as a guide for metric images (a), epithelial pixels (green) and stromal pixels (magenta) were marked manually, as shown in (b). (c) All pixels were color coded based on a careful matching of H&E and IR images by an operator. (d) The classified pixels are color coded for epithelium (green) and stroma (magenta), using algorithms described later in this chapter (e). White pixels indicate a difference between the automatically classified image in (d) and the one classified by hand in (c). [Reproduced from Fernandez *et al.*[43] with permission from Macmillan Publishers, © 2005.]

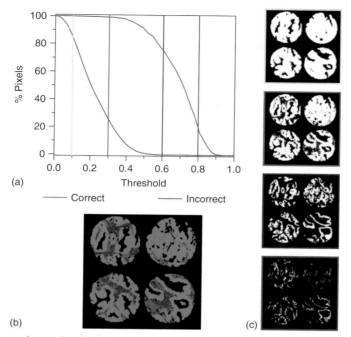

**Figure 10.** Explicit control over the classification process is possible using the classification approach described. A classified image (b), in which epithelial pixels are denoted by green color. The correct and incorrect fraction of pixels as a function of the adjustable threshold for the epithelial class is given in the panel at (a). Panel (c) indicates the pixels that are classified as epithelium (white) and those that are not (black). An increasing number of pixels are classified as epithelium when the threshold is progressively lowered (bottom to top). [Reproduced from Bhargava *et al.*[42] with permission from Elsevier, © 2006.]

classification involves a trade-off between the threshold corrected values of the discriminant for multiple class and is very difficult to optimize without explicit models. In any case, the curves in Figure 10 represent the limits of performance and are useful for estimating the bounds of multi-class classifier performance. The curves are also used to calculate the ROC curves for histologic accuracy.

Once the classification process is optimized using Figure 10, the entire data handling protocol for providing a single classified image is then processed without human supervision. The data analysis algorithm also incorporates filters to eliminate pixels with insufficient SNR, pixels demonstrating significant optical edge effects,[45] pixels that do not exhibit absorption features characteristic of biological materials (as, for example, dust particles), and pixels that may be outliers based on the failure to fall within the distribution

of most metrics. We rigorously examine outliers in every class for each metric by extensive, visual examination of scale-expanded histograms; generally, only a small number of pixels (less than ~0.025% pixels in a data set) appear to be outliers. The outlier pixels may arise from natural variations, from sampling artifacts, optical effects or other, yet unknown, causes. Since we follow routine clinical practice in preparing tissue, this pixel impurity level likely reflects coagulated fixing and embedding media, dust, or other laboratory contaminants (particulate matter).

We evaluate the classification accuracy using ROC curves; hence, we use the same measures to optimize the prediction model.[46] Once the metrics have been defined in our model, they are used to determine, individually, their ability to classify a given class. A histogram of the metric values for two cases is plotted—one for the class under consideration and one for all other classes. The overlap between the two curves determines the pairwise error. The lowest pairwise error for any class, thus, represents the metric with the best ability to distinguish that class from the rest of the tissue. We order metrics based on pairwise errors. While the use of metrics reduces data dimensionality and pairwise error ordering determines the importance of specific metrics, the net effect of using a specific set of metrics for classification needs to be determined.

A metric with the lowest pairwise error is first chosen. The probability of each class for every pixel in the calibration data set is calculated and the accuracy of the prediction is evaluated using the AUC of the ROC plots. The AUC represents a particularly powerful measure of the "overall" segmentation capability and the average sensitivity. We evaluate the classification accuracy and the contribution of each potential member of the final metric data set based on the AUC increase that arises from the metric by sequentially adding one metric at a time to the classification process. ROC curves are calculated for a single class at a time, thereby eliminating any bias that may arise from the discriminant function. Metrics are added and the AUC value after each addition to the classification set is recorded. The AUC as a function of the number of metrics increases rapidly at first and stabilizes. A metric set at the point of stable AUC values represents the optimal metric set to be used in developing the classifier. A second step is conducted in which a metric is left out one at a time ("leave-one-out" procedure) and the classifier accuracy redetermined using the AUC values. Metrics that demonstrate little improvement or degrade results can be eliminated at this step. Finally, an optimal set of metrics is obtained. Hence, although metrics were defined using spectroscopy knowledge, the selection of the final metric set for classification is completely objective.

The major difference between our method and those methods that seek to recognize spectral features of importance using computer algorithms is that our metric definitions are based on spectroscopic knowledge without regard to the specific features that may help segment the tissue. In contrast, some methods seek to optimize the spectral regions chosen by directly using them for classification of the training data. The integrated optimization of metric selection and classifier training then objectively determines the metric set. Hence, we do not anticipate that the initial definition of metrics using spectroscopic knowledge has any deleterious effect on the final results of the classifier.

The AUC for the ROC of each class is shown in Figure 11(c) and (d) for the training data set and an independent sample set (validation), respectively, while part (b) displays the average ROC curve. The classification accuracy achieves a stable value for the ~20 metrics used for the calibration data set and then decreases further for the validation data, which arises from modeling the spectral noise. The small decrease in the validation data and its agreement with the performance of the calibration data set based upon a small number of features is surprising as most classification methods are significantly more complicated. The efficacy of this procedure arises from the robust sampling of a large population of samples afforded by TMAs, which provides excellent definition of the priors for the Bayesian classification. The agreement between calibration and validation data sets also requires a careful selection, rather than random inclusion of large spectral regions, of spectral metrics. The results are further verified by the results of the "leave-one-out" procedure after a small set of metrics is selected. For example, one metric is removed at a time from the original set of 20 metrics. Classification accuracy was further improved, however, by a few percent when 2 metrics were omitted. Hence, a final prediction model was based on 18 metrics and is used routinely in our laboratories for other studies.

# 3 FT-IR IMAGING

## 3.1 Need for spatially resolved data

The need for spatially resolved data has been recognized to avoid contaminating effects of

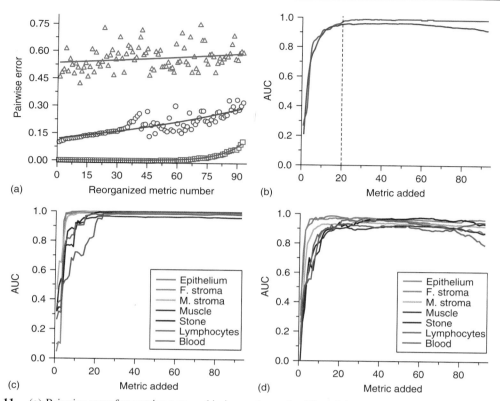

**Figure 11.** (a) Pairwise error for metrics arranged in increasing order. The minimum error (red squares), average error (blue circles), and maximum error (green triangles) for all classes indicate that there are a number of metrics useful for segmenting at least two classes. (b) Area under the ROC curve (AUC) is obtained for each class by building successive prediction algorithms based on the number of spectral metrics. The AUC values are calculated for classification on the calibration data. (c) The same exercise as in (b) is repeated for an independent calibration data set. (d) The average AUC values for all classes as a function of the number of metrics. The red curve indicates the values calculated for calibration data while the blue curve denotes the average value of curves for validation data. [Reproduced from Bhargava *et al.*[42] with permission from Elsevier, © 2006.]

neighboring cell types,[47] but the effect of limited lateral spatial resolution on the classification procedure has received little attention. The primary drawback of coarse spatial resolution is the production of "impure" boundary pixels, which are defined as pixels that are assigned to one class, but that would likely be assigned to other or additional classes if a higher resolution image was obtained. Thus, it is important to assess the spectral content of boundary pixels and to determine whether mixing leads to errors in the classification.[48] As image resolution becomes coarser, the fraction of pixels in

an image belonging to boundary pixels increases; hence, we sought to determine the increase in pixels that are likely contaminated at different spatial resolutions by simulating data acquired at 6.25 μm pixel size from 148 samples to 10, 15, 20, 30, and 50 μm pixel sizes. It must be noted that there is an important distinction between pixel size and achieved spatial resolution. The pixel size denotes the best possible optical resolution subject to the limits of the array detector. The achieved resolution may be limited to other factors, such as spectral wavelength and optical effects due to the sample itself.[45,47,49–52]

The number of neighbors of a class determines whether the pixel may possibly be contaminated. The probability of a pixel surrounded by eight of the same class being a mixed pixel is very low; here, we assume it to be zero. The neighbors of each pixel are assigned to a given class to provide a semi-quantitative probability of it being a mixed pixel. The number of neighbors for epithelial pixels for different spatial resolutions was obtained after a classification of the freshly binned data, as shown in Figure 12. The first observation is that, in this example, a large majority of pixels have the same class as all eight neighbors. It is highly likely that different tissue types will have different spectral features of interest and different sensitivities to limited spatial resolution. Even within the same tissue, different pathologies may affect such results differently. Given that prostate-type morphologies are fairly common and that we employed a large number of samples, however, the results should be generally applicable. The fraction of pixels with all neighbors of the same class decreases rapidly with decreasing spatial resolution and stabilizes at ~20 μm. Hence, a spatial resolution coarser than a certain value (here, 20 μm) is unlikely to have an effect on the classification result, but is expected to lead to about 25% more epithelial pixels being contaminated compared to 6.25 μm pixel sizes. The precise effect on a specific sample is dependent on sample morphology and is generally associated weakly with pathologic state. This statistical measure does not imply that results from coarser resolution studies would be invalid; however practitioners must recognize that error rates may be higher and that this contribution may be mitigated by using commonly available imaging systems.

One danger of classifying mixed composition pixels is whether they may be assigned to an entirely different class or discarded from the data set as belonging to no class. Hence, we conducted a second simulation to determine the performance of our classifier. Pixels of composition ranging from 0 to 100% for pairs of each class were simulated and noise was added to simulate different data acquisition conditions. An

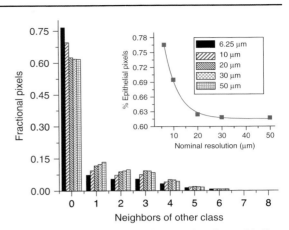

**Figure 12.** Neighbors of cell types other than epithelium or empty space for different spatial resolutions. The inset shows the decrease in percent epithelial pixels that do not have any other cell types as neighbors. [Reproduced from Bhargava.[13] © Springer, 2006.]

example of the data can be seen in Figure 13. The mosaics in Figure 13(a) and 13(c) consists of squares with the number of squares on any row or column being equal to the number of classes. Figure 13(a) is an example of the data values, as a function of composition, at a specific spectral feature (1080 cm$^{-1}$). Figure 13(c) is the classification result of all compositions in the square. Each square represents a mixture model of two classes. Hence, a square in the first column and second row (from the top) will represent mixtures of class 1 and class 2. The top of the square is 100% of class 1 while the bottom of the square is 100% of class 2. Any position in the middle of the square varies linearly in concentration. From left to right, the composition of the square is the same, but noise is added to simulate different experimental conditions. An average spectrum from each of the two classes was baseline corrected and used as the spectral data. Figure 13(b) demonstrates the classification of this gradient data set. In general, the classification works well, favoring the class with higher pixel concentration. The classifier is also stable at the examined noise levels.

A surprising result was that pixels contiguous to epithelium and fibroblast-rich stroma were classified as mixed stroma. This drawback, however,

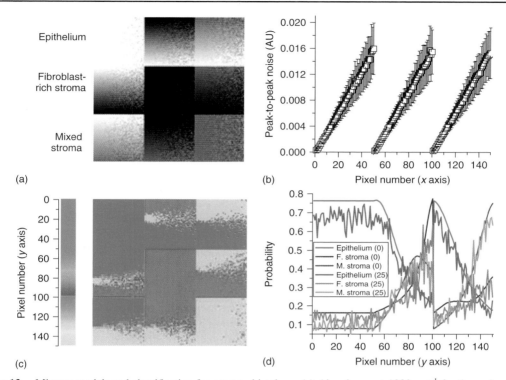

**Figure 13.** Mixture models and classification for prostate histology. (a) Absorbance at $1080\,\mathrm{cm}^{-1}$ for three classes and their mixtures. The first column contains mixtures of epithelial cell spectra with the average spectrum from fibroblast-rich stroma and mixed stroma. The second and third columns contain mixtures of fibroblast-rich stroma (F. stroma) and mixed stroma (M. stroma), respectively. The concentration changes from 0 to 100% linearly along the $y$ direction as indicated by the color bar in (c). (b) Along the $x$ axis of the composite image, the noise in each pixel increases linearly. Error bars are standard deviations of noise in the spectra. (c) Classified image for the data, demonstrating the effect of composition and noise on classification. (d) Probability profiles of the three cell types at columns 1 and 25, demonstrating the effect of noise. [Reproduced from Bhargava.[13] © Springer, 2006.]

was the only example of two classes mixing to yield an entirely different one. This difficulty stems, in part, from the definition of the mixed stroma class. While the mixed stroma class was designed specifically to assign stromal cells that were not clearly fibroblasts or smooth muscle in origin, but appeared mixed, the mix of epithelium and fibroblast-type stroma also leads to a classification of mixed stroma. Noise levels considered here appear to have little effect on this behavior.

The full simulation of all classes (not shown) reveals that mixed pixels are largely classified as the constituent class with the higher concentration. Clearly, boundary pixels at epithelial-fibroblast-rich regions must be handled with

care. The increase in boundary pixels at lower spatial resolution conditions also implies that this type of systematic misassignment may arise more frequently for coarser resolutions. The rate of occurrence of boundary pixels may be less for synchrotron source–based mid-IR imaging that is conducted at high pixel densities or in emerging approaches that utilize synchrotron-IR source-based interferometers and array detectors. The simulated example above, however, demonstrates that simply oversampling a spatial region to increase pixel density may allow for better definition of the interface and assignment of pixels, although it will not address spectral purity. Hence, for analyses based on spectral discrimination, mixture models will have to be developed based on entire spectra.

A further complication arises in using data from histologic classification for pathologic diagnoses. For example, the boundary epithelial pixels classified above may disproportionately contribute to classification errors. We have found evidence for this in studies of both cancer pathology and histology in tissue from different organs. For example, the boundary pixels in benign tissue may be misclassified as cancerous, leading to a major source of error in applying this approach to pathology. At this time, the evidence is anecdotal and requires further investigation to quantify the extent of the error and its mitigation by advanced numerical processing.

The last interesting aspect of low spatial resolution conditions is that it tends to overpredict certain classes. For example, Table 1 demonstrates the regression results of each sample's composition against that obtained at 6.25 µm for three classes. While the regression coefficient

is high, it is clear that epithelial and mixed stroma fractions are overestimated and fibroblast-rich stroma is underestimated with a decreasing pixel size. There are also differences based on the underlying pathology. For example, normal epithelium is generally encountered in 10- to 40-µm-wide strips, while high-grade tumors may be hundreds of micrometers to millimeters in size. Individual sample variability reflected in the regression coefficient decreases with increasing pixel size. In spectroscopic models for predicting diseases that include morphological units but are based on average spectra, mixed pixels may lead to estimates with large errors. For example, a 1:1 mixed region of epithelial and fibroblast pixels at 6.25 µm pixel size increases to ~1.19:1 for a 50 µm pixel size. Hence, the use of histologic mixture models at limited spatial resolution may not be estimated correctly, providing evidence that the percent content of cell types in a limited field of view is likely to be a less robust measure of tissue histopathology.

**Table 1.** Correlation of composition for samples between 6.25 µm pixel sizes and other pixel sizes. The first row in each section of the table denotes the composition factor for that pixel size and class. For example, for every 100 µm², the area of epithelial pixels at 10 µm pixel size is 99.13% of that at 6.25 µm pixel size. This implies a composition factor of 0.9913× in the column. Increasing/decreasing numbers represent pixels being increasingly/decreasingly classified as that class. The ratios are not uniform for every sample, and the regression coefficient of the best-fit line passing through the origin is provided (in brackets) in the second row of the each table cell. Decreasing coefficients reflect greater variance from the fit line.

| Pixel size (µm) | Epithelium | Fibroblast-rich stroma | Mixed stroma |
|---|---|---|---|
| 10 | 0.9913× (0.9976) | 0.9847× (0.9923) | 1.0300× (0.9957) |
| 20 | 1.0156× (0.9906) | 0.9671× (0.9775) | 1.0473× (0.9787) |
| 25 | 1.0404× (0.9896) | 0.9768× (0.9624) | 1.0262× (0.9617) |
| 30 | 1.0720× (0.9773) | 0.9683× (0.9507) | 1.0175× (0.9363) |
| 50 | 1.1180× (0.9459) | 0.9410× (0.8947) | 1.0390× (0.8723) |

Reproduced from Bhargava.[13] © Springer, 2006.

### 3.1.1 Evolution and capabilities of current instrumentation

It is now clear that the measurement of single spectra from large tissue samples is unlikely to provide accurate histopathologic information; hence, microspectroscopy is mandatory for such applications.[53] Single spectra (initially measured by dispersive spectrometers and later by FT-IR spectrometers) have been recorded from microscopic samples for over 50 years by restricting light incident on the sample through an aperture. More than 85% of cancer arises in epithelial cells, which often form surface layers that are $10-100\,\mu m$ wide. As we demonstrated in the previous section, however, a finer spatial resolution than $\sim 10\,\mu m \times 10\,\mu m$ is preferable. Consequently, as the illuminated spot at the sample is made smaller, throughput decreases proportionally, which in turn decreases the SNR of the acquired spectra. Orders of magnitude brighter sources, e.g., synchrotrons, may be employed to recover the lost SNR. Unfortunately, synchrotron or free electron lasers[54,55] are prohibitively expensive, and no laboratory lasers exist for a wide spectral region. An alternative is to average successive measurements (coadding) to statistically increase the SNR. Since the SNR increases only as the square root of the number of averaged spectra, long averaging periods are required. This situation may be eased by using higher condensing optics, sources at higher temperatures, slightly faster scanning times,[a] gain ranging,[56] or ultrasensitive detectors.[57] Even if a hypothetical instrument with all these advances was constructed, only a $\sim 10$- to 20-fold reduction in time would be obtained. We note that this estimate underestimates the time required since losses due to diffraction or stage movement were not considered.

In prostate tissue, for example, the situation is similar to Figure 1. Epithelial cells form 10- to 35-$\mu$m-wide foci around the cross-sections of ducts. Ducts appear as white circles in Figure 1(b), surrounded by epithelial cells that are depicted in blue. To analyze this morphology, aperture dimensions of $\sim 6\,\mu m \times 6\,\mu m$ ($\sim$cell size) were proposed;[42] for this case, the point mapping approach would require $\sim 1028$ hours for a $500\,\mu m \times 500\,\mu m$ sample. Under these conditions, mapping is not a viable option. In contrast to point mapping using apertures, large fields of view are measured in FT-IR imaging. Depending on the microscopy configuration, thousands of moderate resolution spectra can be acquired at near diffraction limited spatial resolution in minutes.[22,58] The time advantage over mapping is nominally the number of pixels in the array detector (16- to 65 000-fold) but the noise characteristics of focal-plane array (FPA) detectors are poorer than sensitive single point detectors.[59] Hence, the SNR-normalized advantage is lower.[60] Faster detectors are being used for imaging and promise significantly higher SNR at the same time. For example, we have employed a $128 \times 128$ element mercury cadmium telluride (MCT) array operating at $\sim 16\,kHz$ to acquire a full data set in $\sim 0.07\,s$. The data set consists of 16 384 pixels, with each pixel containing an interferogram of 1024 elements. These rates of data acquisition are approximately a factor of 10 higher than commercially available, but are required for practical data acquisition times.

The emergence of these fast detectors, with integration times that approach frame times, is likely to provide practical speeds for clinical translation. The limited data acquisition speed currently remains a bottleneck for applications of IR imaging to routine clinical studies. Coupled with the complexity and cost of instrumentation, present technology provides preliminary capabilities but is unlikely to prove practical for clinical translation. If both the classification and instrument development technology that is now the subject of research proves to be successful, there is incentive for most hospitals to purchase these instruments. With the large number of hospitals generating a volume of business, cost will drop and the user friendliness of instruments will increase substantially. In our opinion, however, substantial development is needed before widespread clinical acceptance.

The first step of demonstrating the robust capabilities of FT-IR imaging for many tissue types is well underway. The next step of integrating these developments for large-scale clinical trials to demonstrate patient benefit is likely to commence for some organs soon.

# 4 HIGH-THROUGHPUT SAMPLING AND STATISTICAL PITFALLS

## 4.1 Quantitative analyses of results

The best imaging instruments (which employ sensitive array detectors and therefore have a significant multichannel advantage) acquire data in about 0.1% of the time required for mapping for equivalent parameters. The time advantage arises from both the multichannel advantage as well as from a small detector size that is customized to the spatial resolution of the imaging spectrometer. Thus, imaging instruments typically allow for an order of magnitude larger number of samples to be analyzed with modern systems. For example, in a recent report ~10 million spectra from ~1000 samples were analyzed at a spatial resolution of 6.25 μm.[43] This quantitative validation is necessary for any automated biomarker approach[61] (vide infra). Studies are underway in our and other laboratories to correlate spectral patterns with other physiologic and pathologic conditions. Published studies and other chapters in this book provide results that verify robustness and potentially wide applicability of FT-IR microscopy.[62,63]

## 4.2 Sample size

Although numerous studies have demonstrated the capability of FT-IR imaging for histopathology, considerable debate exists on reproducibility and accuracy measures. Indeed, the first response of many practitioners to new data is to question its validity based on limited statistical confidence intervals. A detailed understanding is emerging from the work of several groups regarding appropriate sample control[64,65] and confounding factors due to biology.[66] Inherent differences between patient cohorts, effects of sample preparation, and measurement noise are topics that can be addressed with the available imaging technology but are yet to be fully explored. Hence, validating robust spectral markers for large sample populations[67,68] is exceptionally challenging since the opportunity exists for chance and bias to influence the results. The first line of defense is to validate the results on large numbers of samples. The second strategy is to model the results and demonstrate that statistical power is sufficient for the claims made in the study.

More importantly, the fundamental question of predicting a required sample size ab initio remains open. There are two major concerns: first, the optimal sample size in forming calibration sets and a prediction algorithm must be determined; and second, investigators must support the results by means of statistical considerations. While the first problem is essentially that of optimizing a model and its prediction algorithm, the second aspect impacts directly the quality of the results and claims of applicability. Results are awaited from our studies to address the first issue; hence, in this chapter, we examine only the second aspect. The statistical validity of obtained results and dependence on data acquisition parameters is discussed later in this chapter. Specifically, we estimate sample size based on the standard error for the area under an ROC curve.

## 4.3 Gold standard

The selection of spectra measured at individual pixels (subsequently referred to simply as pixels) as gold standards for calibration requires great care. It must be performed independently of any classifier training or validation, thus ensuring a blinded study design. Once the gold standard set is determined, it must not be changed. This will ensure that there is no bias in the process. Care must be taken to avoid pixels that do not lie on the tissue or those that are at the boundaries of tissue classes, as these may artificially inflate the error. The use of all pixels in an image has been suggested and specific exclusions have been proposed to contribute to selection bias.[69] Selection bias, however, cannot arise if pixels are chosen independent of validation algorithms. The exclusion of boundary pixels is necessary in both training (to avoid spurious probability distribution functions) and validation (to prevent introduction of errors). There are major technological difficulties in relating stained pathology samples to IR images acquired from unstained tissue since chemical effects during the staining process often leads to errors. Hence, it is proposed that the exclusion of boundary pixels is analogous to the performance of a classifier containing a reject option for boundary pixels.[43]

## 4.4 Sampling, archiving, and consistency

While a choice for an optimal sample size is unclear, it is certain that a large number of tissue samples is required for effective validation. While it may theoretically be possible to train on a single sample, protocol validation requires a significant sample size to attain statistical validity. We recognized that one does not need to observe the entire surgically resected tumor for validating IR protocols, as a representative small section would suffice. Hence, we employed TMAs[70] as a platform for high-throughput sampling. TMAs consist of a large number of small tissue samples arranged in a grid and deposited on a substrate. Sample processing times are easily increased 100-fold in TMA sections compared to that in surgically resected tissue. Additionally, valuable tissues are optimally utilized and consecutive TMA sections can be used for correlating with staining results. The construction and analysis of TMAs has been automated, further increasing the throughput. For spectroscopists, TMAs provide a ready source of tissue to test hypotheses and to develop prediction models. Hence, we suggest its use as a platform for developing diagnostic tests.

The validity of employing TMAs for prostate cancer research and, especially, for cancer grading has been addressed by a number of authors.[71] For example, in a study surveying genitourinary pathologists[15] using images from TMA cores, ~90% of the respondents considered this approach useful for pathology teaching and for resident training. Further, a Gleason score was easily assigned to each TMA spot of a 0.6-mm-diameter prostate cancer sample. The sample handling technique from pathology can also prove quite useful for data handling in spectroscopy. Large tissue sections can all lead to a very large data size for a given study, hence precluding visualization of all samples at the same time. Virtual TMAs could be constructed from different areas of large samples, thus providing many subsamples for within-patient and among-patient comparisons. This approach has not yet been reported but is likely a useful extension of the TMA concept for spectroscopic imaging.

## 5 PREDICTION ALGORITHMS AND HIGH-THROUGHPUT DATA ANALYSIS

### 5.1 Univariate algorithms

The major technological advances of rapid FT-IR microscopy and high-throughput tissue sampling have been addressed by imaging and TMAs, respectively. There is still some confusion and widespread disagreement, however, about the "best" approach to extract histopathologic information from FT-IR imaging data. Several early manuscripts employed univariate correlations to disease states. While the results were exciting, it is now realized that they were statistically flawed and were not necessarily fundamentally based on cancer biology. To our knowledge, there is no manuscript that has expressly demonstrated, using statistical arguments, the reason univariate analyses are likely to fail. There is widespread consensus and anecdotal evidence, however, among practitioners that argues against the approach. Consider the distributions for a univariate measure (absorbance at $1080\,\text{cm}^{-1}$

that is normalized to the amide I peak height) for benign and malignant cases as shown in Figure 14.

The normalized histograms reveal that for specific, single samples the distribution of pixel absorbances at the indicated wavenumber clearly indicates that the metric is sufficient for cancer discrimination. When the distribution from all samples is considered, however, there is little difference in the distributions. Hence, many univariate measures described in the literature do not hold up in wide population testing. A TMA-based, high-throughput validation can easily prove that the measure is generally unsatisfactory but does discriminate for some samples. In Figure 14, it can be seen that a cutoff value can generally be found that distinguishes disease for an individual, leading to the erroneous conclusion that the feature is universally indicative of a diseased state. Since a typical mid-IR spectrum has numerous absorption bands and perhaps nonchemically specific features that can provide disease discrimination, a small number of samples

increases the probability of finding such discrimination by chance alone. Univariate measures that apparently provide discrimination when none exists can be equated to the false discovery rate (FDR)[72] of metrics. The FDR is very different from the *p* value for determining that a metric separates two distributions; a much higher FDR can be tolerated than can a *p* value. Similarly, a false negative rate has been proposed,[73] which is not critical for our case, as we have observed high accuracy without use of any erroneously left out metrics. While detailed calculations and their underlying concepts are too lengthy to reproduce here, for the sake of completeness, it suffices to say that for the expected number of metrics demonstrating discrimination, the FDR tends to zero for larger than ∼30 samples. While correlations due to chance can be minimized by this approach, there is the potential for unknown bias or error in the predictions for small number of samples; that is, the algorithm must be integrated with sampling considerations.

## 5.2 Multivariate algorithms

It was argued in the previous section that univariate analysis might not provide a good measure of a population distribution. Alternatively, it can be argued that the individual differences in univariate measures are masked when population statistics of the same measure. Similarly, multivariate techniques may mask the individual measures in population testing. Our philosophy has been to employ a multivariate, supervised classification in which the measures are univariate. This enables us to carefully examine each measure for both population as well as individual sample relevance. While unsupervised clustering approaches provide good insight into spectral similarity, a supervised method forces a relation to common clinical knowledge. For example, as shown in Figure 8 for prostate tissue, we consider a 10-class model to determine histology. The drawback is that the sensitivity of the approach to individual samples is lost at the expense of generality. One could potentially combine clustering and

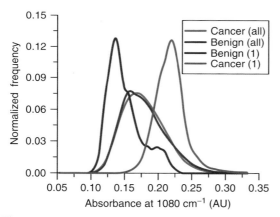

**Figure 14.** Distribution of absorbance for individual spots and all pixels from each class, normalized by the total number of pixels in the class, demonstrates that the examination at patient level and at a global level may not correspond. All denotes the absorbance values from ∼40 patients, (1) denotes values from two selected patients. While the single univariate measure provides discrimination based on two patients, it does not prove to be reliable for a large population. [Reproduced from Bhargava.[13] © Springer, 2006.]

supervised classification. Clustering information in the training data set would emphasize individual sample distributions, which could allow for supervised classification tailored to each cluster type. Such an approach has not yet been implemented, but is being attempted in our laboratories to classify samples optimally.

## 5.3 Dimensionality reduction

It is well recognized that the spectrum at each pixel should be reduced to a smaller set of useful descriptors for capturing the essential information inherent in the spectrum. The reduction of full spectral information to essential measures (metrics) helps eliminate from consideration those spectral features that have no information (wavenumbers at which there is minimal absorption), little biochemical significance (e.g., apparent absorption at nonchemically specific wavenumbers), inconsistent measures that may degrade classification, and those with redundant information. The number of useful measures is significantly smaller than the number of spectral resolution elements and, hence, the process is also termed dimensionality reduction. Dimensionality reduction and further refinement (vide infra) also helps reduce the incidence of prediction by chance alone, as well as reducing computation time and storage requirements.

It may be argued that the metrics are not selected in an objective manner due to human intervention and computer routines must necessarily be employed. While the use of an automated computer program is most certainly objective and reproducible, the algorithm that drives such a program is generated from detailed spectroscopic knowledge. A well-trained spectroscopist can recognize spectral features and assign them to a biochemical basis. While a computer algorithm may be able to enhance subtle features in the spectrum, automated peak picking algorithms run the risk of substantial error as they are based on very specific criteria that may not be universally valid. We believe that computer algorithms are more suited in finding correlations and patterns because

of the size and complexity of data. The process of determining which spectral features to consider, however, is entirely manual in our approach. It must be emphasized that a universal set of metrics is selected manually, but the data reduction step to a set of metrics to be used in further algorithms is based on objective determinations. Manual refinement of metrics for a classification is, obviously, not recommended since there exists a possibility of overlooking specific features that leads to a bias in the final set of metrics. Dimensionality reduction is also intimately linked to the data quality and the employed classification algorithm. We note, further, that the selection of metrics in this manner obviates the need to remove $CO_2$ contributions or perform baseline corrections. We do not use second derivatives as a preprocessing step, even though they have shown to be useful in increasing sensitivity for segmentation by many groups. We believe that mathematical operations on a large data set must be kept to a minimum to ensure fast classification using routine computing power.

## 5.4 Classification algorithm

A number of supervised algorithms have been applied to dimensionally reduced data, including those based on linear discriminant analysis, neural networks, decision trees, and modified Bayesian classifiers. An intermediate step in some of these algorithms provides a fuzzy result in which the spectra measured at every pixel have a finite probability of belonging to every class. For example, in our approach the spectrum measured at each pixel can lead to a probability (between 0 and 1) of belonging to each class. A discriminant function then assigns each pixel to a class based on a decision rule. The prediscriminant data set, termed the *rule imaging set*, contains important information. In our algorithm, it is a direct measure of the probability of the pixel belonging to a specific class. The probability value may then be used to compare the potential of two protocols in distinguishing a cell type or in quantifying confidence limits.

### 5.5 Measures of accuracy and optimization

Quantitative measures of performance and accuracy are perhaps the weakest portion of reports using IR spectroscopy for cancer pathology. Typically, sensitivity and specificity have been employed as summary measures. While these are indeed relevant, we demonstrate that they are insufficient and classification analyses must utilize additional measures to understand the process. We specifically recommend the use of ROC curves and AUC values for the curves as better measures of classification potential. The AUC further provides both a summary measure for a quantitative understanding of the discrimination potential of the model and a convenient measure to compare multiple classification models. The third tool we introduced was the confusion matrix. While ROC curves provide the potential for correct classification of a binary rule, confusion matrices correspond to a particular point on the ROC curve under the constraints of the accuracy measures for other classes. These also directly correspond to the final segmentation of the rule image under an optimization condition. The optimization condition may simply be either the maximization of the accuracy or the minimization of certain types of errors.

We prefer the use of the AUC for both optimizing algorithms and for validating results. Confidence in the value of the AUC is the primary test for the validity of developed algorithms and is characterized by the standard error of the value. For example, Figure 15 shows the cumulative distribution of AUC in a TMA, which is obtained from validating the discrimination of epithelial from stromal pixels in a blinded validation set. More than 20% of the TMA samples had an AUC $>0.95$ and no AUC value below 0.8 was recorded. One drawback of using ROC curves and AUC values is that the results are valid for one-at-a-time classification in which the global optimization of all classes cannot be undertaken. In Figure 15 we have analyzed the segmentation of epithelium from all other cell types. The tissue is classified into 10 classes as before, but the

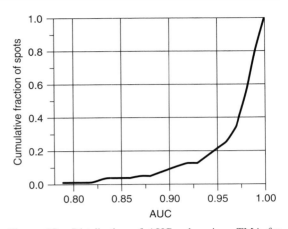

**Figure 15.** Distribution of AUC values in a TMA for discriminating epithelium from stroma using the 10-class model. [Reproduced from Bhargava.[13] © Springer, 2006.]

results are lumped into epithelial and nonepithelial pixels. Although not all TMA cores have all cell types, the two-class model allows us to examine a large number of samples. Additionally, we excluded TMA samples that did not contain at least 100 pixels of each class, leaving 103 cores for the analysis.

### 5.6 Discriminant and class assignment

In a multiclass analysis, our approach in evaluating ROC curves for a class is one-at-a-time, i.e., all other classes are essentially lumped in the rule data and the highest probability of the lumped ensemble is compared to the class whose ROC curve is being built. Hence, the AUC values must be regarded as a potential for classification. They are best suited to answer the binary question of whether a pixel is correctly identified or not when considering a single class. This method is ideally suited to a cascaded classifier in which one pixel at a time is assessed. Such a classifier has not been reported yet for this type of work but would provide a means to explicitly determine the error for any given classification scheme.

# 6 EXPERIMENTAL PARAMETERS AND CLASSIFICATION

Here, we take advantage of the trading rules of FT-IR spectroscopy and imaging to model the effects of the experimental parameters on the classification process. While the SNR and resolution are generally arbitrarily fixed in most studies, we demonstrate their importance in classification.

## 6.1 Effect of signal-to-noise ratio

There are two issues: first, what is the "best" SNR to formulate algorithms and second, for a given algorithm, what is the lowest SNR that would provide adequate classification? Only the latter issue is examined here. As with conventional FT-IR spectrometers, imaging spectrometers obey the trading rules of IR spectroscopy; that is, for an $n$-fold reduction in SNR with the same results, the data acquisition will be $n^2$-fold faster. Thus, in addition to an interesting fundamental behavior of the classifier, the role of SNR has a direct bearing on the speed at which data is acquired.

We examined classification accuracy as a function of average spectral noise. To strictly examine the effect of noise, data must be acquired for different numbers of coadded spectra. Since the time required for imaging an array multiple times, is prohibitive, we computationally added random, Gaussian noise to the original spectral data. Peak-to-peak and root-mean-square (rms) noise were measured in the 1950–2150 cm$^{-1}$ region adjacent to the amide I peak.[b] Representative single-pixel spectra from the data sets are shown, as a function of noise, in Figure 16(a). We additionally plotted the observed noise levels against the added noise to verify linearity (plot not shown). The linear relationship conforms to the expected result and provides a scaling factor to express the equivalent reduction in data acquisition time (coaddition) that would be realized at that noise level. For example, the addition of noise with a peak-to-peak amplitude of 0.005 absorbance units (AU) raises the peak-to-peak noise level from 0.0013 to 0.015 AU, corresponding to a decrease in data acquisition time by a factor of ~100 for this data set. In addition to increasing noise, we employed a minimum noise fraction (MNF) transform[75,76] based algorithm to mathematically reduce noise. Since the observed peak-to-peak noise was 0.00017 AU, corresponding to an increase in data acquisition time by a factor greater than ~100, the data examined span about 5 orders in magnitude of collection time.

The average height of the amide I peak was 0.42 AU in all cases, providing a SNR of 2500

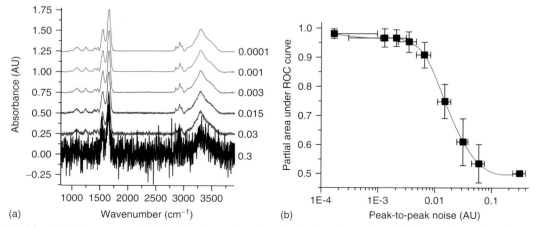

**Figure 16.** (a) Noise observed in the data set as a function of added random noise. (b) Effect of spectral noise on the accuracy of classification as measured by AUC values. [Reproduced from Bhargava.[13] © Springer, 2006.]

(MNF-corrected data) to 1.5 for the data sets. Accuracy as a function of the noise level is shown in Figure 16(b). While the abscissa error bars indicate the standard deviation of noise levels in pixels, the ordinate error bars indicate the standard deviation in AUC values of all 10 classes. As a general rule, the classification improves with decreasing noise levels. We first note that the classification does not become perfect for any noise level and that there is a significantly diminishing return in increasing the SNR beyond a given level. At the other extreme, the ability to distinguish classes is entirely lost when the peak-to-peak noise exceeds ~0.1 AU Performance across multiple data sets observed using our prediction model indicates that the increases demonstrated at noise levels lower than ~0.003 AU are within the variance, indicating that there is little benefit in decreasing the noise levels below ~0.003 AU for this data set, or to increasing the SNR of the amide I band beyond ~150. It must be emphasized that the model, prediction algorithm, and discriminant function are intimately linked in a nonlinear manner. While this makes

it impossible to predict the behavior generally for all classification approaches, this simple exercise may be conducted to determine the optimal data acquisition parameters. For our selected metrics and model, it appears that the data acquisition time can be decreased by a factor of ~3 without significant degradation in accuracy.

## 6.2 Spectral resolution

We next examined the effect of spectral resolution on the results that would be obtained using the developed algorithm. As in the previous section, the data were not rerun but were acquired using a neighbor binning procedure to reduce the number of data points in the ratio of resolution. Spectra from the same epithelial class pixel, at different resolutions (Figure 17a), demonstrate the effect of degrading the spectral resolution on feature definition. Figure 17(b) demonstrates, first, that the peak-to-peak noise levels over the region remain the same with spectral resolution, $\Delta \tilde{\nu}$. As previously observed, noise is an important control in comparing spectra; the peak-to-peak

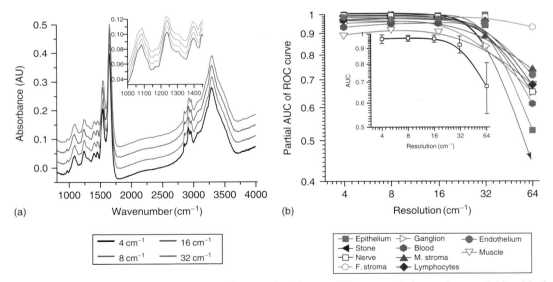

**Figure 17.** (a) Spectra obtained by downsampling acquired data to different resolutions using a neighbor binning procedure. The inset demonstrates the effect of resolution on narrower features in the spectrum. (b) AUC values for each class and average AUC values as a function of spectral resolution demonstrate a decrease only for coarse spectral resolution. [Reproduced from Bhargava.[13] © Springer, 2006.]

noise over the same number of data points was preserved by neighbor binning. In practice, for an FT-IR spectrometer operated at constant optical throughput (which is the case for all imaging instruments), the noise level is inversely proportional to $\Delta\tilde{v}$. Second, the performance of the classifier is very nearly the same for finer spectral resolutions and degrades only significantly at $32\,\text{cm}^{-1}$. While the results may appear to be surprising, a closer analysis of the basis of the algorithms provides insight into the trends. During the preparation of this chapter, one of the editors (PRG) suggested that the use of spectra measured at very low resolution but with higher zero-filling to reduce the data point spacing may yield an advantage in classification. This concept has not been tested yet but represents a potential opportunity for obtaining better results, at coarser resolutions, than those we have reported.

One of the common types of metrics used in the classifier is a ratio of the peak heights of two spectral features. It is well established that absorbance is measured accurately, provided that the full width at half height (FWHH) of the peak is larger than the resolution, $\Delta\tilde{v}$. The Ramsay resolution parameter, $\rho$, is a useful measure that was originally developed for grating spectrometers but has been shown to be applicable to FT-IR spectrometers as well.[77] Since most bands are broad with peak absorbance values lower than $\sim0.7$, their peak absorbance values are not expected to be adversely impacted from the measurement process. With decreasing resolution (increasing $\Delta\tilde{v}$), however, broadening within complex peak shapes may lead to observed changes in the apparent absorption at a specific wavenumber. The change itself may not have a significant influence on the classifier performance as it depends on several such metrics. A second type of metric calculates the area under the curve of the peak. The area is not expected to be impacted significantly for most peaks. The third type of metric that we have used is the center of gravity of a spectral region that encompasses one band. While spectral analyses ordinarily attempt to locate the peak position and use it as a metric, we chose the center of gravity of an individual band or unresolved multiplet for its sensitivity to both position and asymmetrical shape changes in the complex spectral envelopes observed in biological samples. Since the classifier is based on center of gravity of a feature and not on the wavenumber of the peak maximum, it is a very robust measure that is relatively unaffected by spectral resolution. Normally, the measurements at low resolution would become susceptible to noise. Here, however, we demonstrate further that the SNR needed is sufficiently high for the degradation to not be significant in classification.

### 6.2.1 Generalization of developed algorithms to instruments and practical approaches

The characterization of classification with regard to spectrometer performance (SNR) and spectral resolution provides information to optimize parameters on one spectrometer. It is unclear, however, if the calibration would transfer to another spectrometer. We contend that the potential for a successful transfer is high as the classification process is relatively insensitive to resolution, implying that it would only be weakly sensitive to apodization or to small inaccuracies in wavelength scale. Similarly, if the SNR of acquired data is used as control, perturbations due to fixed pattern noise in focal plane array detectors, or in the different use of electronic filters by different manufacturers, are likely to be insignificant in classifying tissue samples correctly. Various instrument manufacturers also set the nominal optical resolution differently in their instruments. The issue of spatial resolution, of course, is more complex. Nevertheless, any resolution setting around the diffraction-limited case will likely provide consistent results. To our knowledge, there has been no comparison of classifier performance across mid-IR FT-IR imaging spectrometers as yet using algorithms developed on one specific instrument. The developed protocol provides for such a framework and detailed results are awaited from ongoing work.[78]

# 7 OUTLOOK AND PROSPECTS

An exciting period in imaging tissues spectroscopically, comparable with clinical practice using low power optical microscopy, is emerging. An ultimate goal of such studies is to provide a key technology for emerging molecular pathology but considerable work needs to be accomplished before spectroscopic imaging can become a clinical reality. The approach promises greatly reduced error rates, automation, and economic benefits over current pathology practice, leading ultimately to better patient care. Looking to the future, the preliminary studies and methodology development described here forms a base to develop technology for diagnosing cancers in premalignant stages prior to their apparent changes observable by conventional means, for predicting the prognosis of the lesion, and for intraoperative imaging in real time. Fundamental studies in drug discovery and mechanisms of molecular interactions are further examples that would be enabled by progress in this area. Since the biological questions and applications are so important, several exciting developments surely lie ahead with more practitioners. While progress is rapidly being made toward practical applications, the success of these translational efforts radically changes the practice of pathology and alters the clinical management of cancer patients.

# ACKNOWLEDGMENTS

The authors would like to acknowledge our close collaborator, Dr Stephen M. Hewitt, for numerous useful discussions and guidance. Dr Hewitt was the pathologist of record for most of our studies summarized in this chapter. Funding for this work was provided to RB in part by University of Illinois Research Board and by the Department of Defense Prostate Cancer Research Program. This work was also funded in part by the National Center for Supercomputing Applications and the University of Illinois, under the auspices of the NCSA/UIUC faculty fellows program. IWL would like to acknowledge the National Institutes of Health intramural program for support.

# END NOTES

[a.] There is no advantage of faster scanning once the modulation frequency has reached optimum level for MCT detectors (1 kHz). The reduced time to observe signal then decreases the SNR, because of the reduced duty-cycle efficiency.

[b.] It is noteworthy that we are examining trends in the absorbance spectra. Strictly, SNR should be measured in single beam spectra to relate rigorously to theory.

# ABBREVIATIONS AND ACRONYMS

| | |
|---|---|
| AU | Absorbance Units |
| AUC | Area Under the Receiver Operating Characteristic Curve |
| DRE | Digital Rectal Examination |
| FDR | False Discovery Rate |
| FPA | Focal plane array |
| FT-IR | Fourier Transform Infrared |
| FWHH | Full Width at Half Height |
| H&E | Hematoxylin and Eosin |
| IR | Infrared |
| MCT | Mercury Cadmium Telluride |
| mid-IR | Mid-Infrared |
| MNF | Minimum Noise Fraction |
| MRSI | Magnetic Resonance Spectroscopic Imaging |
| MSI | Mass Spectroscopic Imaging |
| PSA | Prostate-Specific Antigen |
| rms | Root-Mean-Square |
| ROC | Receiver Operating Characteristic |
| SFS | Sequential Forward Selection |
| SIMCA | Soft Independent Modeling of Class Analogies |
| sm | Smooth Muscle |
| SNR | Signal-to-Noise Ratio |
| TMA | Tissue Microarray |

# REFERENCES

1. S.H. Woolf, *N. Engl. J. Med.*, **333**, 1401–1405 (1995).

2. S.M. Gilbert, C.B. Cavallo, H. Kahane and F.C. Lowe, *Urology*, **65**, 549–553 (2005).

3. P.F. Pinsky, G.L. Andriole, B.S. Kramer, R.B. Hayes, P.C. Prorok and J.K. Gohagan, *J. Urol.*, **173**, 746–751 (2005).

4. P.A. Humphrey, 'Prostate Pathology', American Society for Clinical Pathology, Chicago (2003).

5. A.W. Partin, L.A. Mangold, D.M. Lamm, P.C. Walsh, J.I. Epstein and J.D. Pearson, *Urology*, **58**, 843–848 (2001).

6. S.J. Jacobsen, S.K. Katusic, E.J. Bergstralh, J.E. Oesterling, D. Ohrt, G.G. Klee, C.G. Chute and M.M. Lieber, *JAMA*, **274**, 1445–1449 (1995).

7. M.B. Amin, D. Grignon, P.A. Humphrey and J.R. Srigley, 'Gleason Grading of Prostate Cancer: A Contemporary Approach', Lippincott Williams & Wilkins, Philadelphia (2004).

8. P.L. Nguyen, D. Schultz, A.A. Renshaw, R.T. Vollmer, W.R. Welch, K. Cote and A.V. D'Amico, *Urol. Oncol. Sem. Orig. Invest.*, **22**, 295–299 (2004).

9. W.M. Murphy, I. Rivera-Ramirez, L.G. Luciani and Z. Wajsman, *J. Urol.*, **165**, 1957–1959 (2001).

10. J.I. Epstein, P.C. Walsh and F. Sanfilippo, *Am. J. Surg. Pathol.*, **20**, 851–857 (1996).

11. K.N. Syrigos, E. Karapanagiotou and K.J. Harrington, *Anticancer Res.*, **25**, 4527–4533 (2005).

12. J.I. Epstein, W.C. Allsbrook Jr, M.B. Amin and L.L. Egevad, *Adv. Anal. Pathol.*, **13**, 57–59 (2006).

13. R. Bhargava, *Anal. Bioanal. Chem.*, **389**, 1155–1169 (2006).

14. R.A. Weinberg, 'The Biology of Cancer', Garland Science (2006).

15. A. De La Taille, A. Viellefond, N. Berger, E. Boucher, M. De Fromont, A. Fondimare, V. Molinié, D. Piron, M. Sibony, F. Staroz, M. Triller, E. Peltier, N. Thiounn and M.A. Rubin, *Hum. Pathol.*, **34**, 444–449 (2003).

16. P.W. Hamilton, P.H. Bartels, R. Montironi, N.H. Anderson, D. Thompson, J. Diamond, S. Trewin and H. Bharucha, *Anal. Quant. Cytol. Histol.*, **20**, 443–460 (1998).

17. I.W. Levin and R. Bhargava, *Annu. Rev. Phys. Chem.*, **56**, 429–474 (2005).

18. M. Navratil, G.A. Mabbott and E.A. Arriaga, *Anal. Chem.*, **78**, 4005–4019 (2006).

19. R.M. Caprioli, T.B. Farmer and J. Gile, *Anal. Chem.*, **69**, 4751–4760 (1997).

20. P. Chaurand, M.E. Sanders, R.A. Jensen and R.M. Caprioli, *Am. J. Pathol.*, **165**, 1057–1068 (2004).

21. J. Kurhanewicz, D.B. Vigneron, H. Hricak, P. Narayan, P. Carroll and S. Nelson, *Radiology*, **198**, 795–805 (1996).

22. E.N. Lewis, P.J. Treado, R.C. Reeder, G.M. Story, A.E. Dowrey, C. Marcott and I.W. Levin, *Anal. Chem.*, **67**, 3377–3381 (1995).

23. M. Diem, M. Romeo, S. Boydston-White, M. Miljkovic and C. Matthaus, *Analyst*, **129**, 880 (2004).

24. R. Mendelsohn, E.P. Paschalis and A.L. Boskey, *J. Biomed. Opt.*, **4**, 14–21 (1999).

25. L.H. Kidder, V.F. Kalasinsky, J.L. Luke, I.W. Levin and E.N. Lewis, *Nat. Med.*, **3**, 235–237 (1997).

26. D.I. Ellis and R. Goodacre, *Analyst*, **131**, 875–885 (2006).

27. W. Petrich, *Appl. Spectrosc. Rev.*, **36**, 181–237 (2001).

28. P.G. Andrus, *Technol. Cancer Res. Treat.*, **5**, 157–167 (2006).

29. C. Krafft and V. Sergo, *Spectroscopy*, **20**, 195–218 (2006).

30. C. Petibois and G. Deleris, *Trends Biotechnol.*, **24**, 455–462 (2006).

31. M.J. Walsh, M.J. German, M. Singh, H.M. Pollock, A. Hammiche, M. Kyrgiou, H.F. Stringfellow, E. Paraskevaidis, P.L. Martin-Hirsh and F.L. Martin, *Cancer Lett.*, **246**, 1–11 (2007).

32. G. Srinivasan and R. Bhargava, *Spectroscopy*, **22**, 30–43 (2007).

33. P.R. Griffiths and J.A. de Haseth, 'Fourier Transform Infrared Spectrometry', 2nd edition, John Wiley & Sons, New York (2007).

34. F.N. Keith and R. Bhargava, *Technol. Cancer Res. Treat.*, Submitted.

35. E. Gazi, J. Dwyer, P. Gardner, A. Ghanbari-Siakhali, A.P. Wade, J. Myan, N.P. Lockyer, J.C. Vickerman, N.W. Clarke, J.H. Shanks, C. Hart and M. Brown, *J. Pathol.*, **201**, 99–108 (2003).

36. E. Gazi, M. Baker, J. Dwyer, N.P. Lockyer, P. Gardner, J.H. Shanks, R.S. Reeve, C. Hart,

N.W. Clarke and M. Brown, *Eur. Urol.*, **50**, 750–761 (2006).

37. T.J. Harvey, A. Henderson, E. Gazi, N.W. Clarke, M. Brown, E.C. Faria, R.D. Snook and P. Gardner, *Analyst*, **132**, 292–295 (2007).

38. C. Paluszkiewicz, W.M. Kwiatek, A. Banas, A. Kisiel, A. Marcelli and A. Piccinini, *Vib. Spectrosc.*, **43**, 237–242 (2007).

39. M.J. German, A. Hammiche, N. Ragavan, M.J. Tobin, L.J. Cooper, S.S. Matanhelia, A.C. Hindley, C.M. Nicholson, N.J. Fullwood, H.M. Pollock and F.L. Martin, *Biophys. J.*, **90**, 3783–3795 (2006).

40. E. Gazi, J. Dwyer, N.P. Lockyer, J. Miyan, P. Gardner, C.A. Hart, M.D. Brown and N.W. Clarke, *Vib. Spectrosc.*, **38**, 193–201 (2005).

41. R. Bhargava, S.M. Hewitt and I.W. Levin, *Nat. Biotechnol.*, **25**, 31–33 (2007).

42. R. Bhargava, D.C. Fernandez, S.M. Hewitt and I.W. Levin, *Biochim. Biophys. Acta – Biomembr.*, **1758**, 830–845 (2006).

43. D.C. Fernandez, R. Bhargava, S.M. Hewitt and I.W. Levin, *Nat. Biotechnol.*, **23**, 469–474 (2005).

44. P. Crow, N. Stone, C.A. Kendall, J.S. Uff, J.A.M. Farmer, H. Barr and M.P.J. Wright, *Br. J. Cancer*, **89**, 106–108 (2003).

45. R. Bhargava, S.Q. Wang and J.L. Koenig, *Appl. Spectrosc.*, **52**, 323–328 (1998).

46. J.A. Swets, *Science*, **240** (4857), 1285–1293 (1988).

47. P. Lasch and D. Naumann, *Biochim. Biophys. Acta*, **1758**, 814–829 (2006).

48. M. Jackson, L.P. Choo, P.H. Watson, W.C. Halliday and H.H. Mantsch, *Biochim. Biophys. Acta*, **1270**, 1–6 (1995).

49. A.J. Sommer and J.E. Katon, *Appl. Spectrosc.*, **45**, 1633–1640 (1991).

50. G.L. Carr, *Rev. Sci. Instrum.*, **72**, 1613–1619 (2001).

51. B.O. Budevska, *Vib. Spectrosc.*, **24**, 37–45 (2000).

52. J. Lee, E. Gaz, J. Dwyer, M.D. Brown, N.W. Clarke, J.M. Nicholson and P. Gardner, *Analyst*, **132**, 750–755 (2007).

53. M. Jackson, *Faraday Discuss.*, **126**, 1–18 (2004).

54. A. Cricenti, R. Generosi, M. Luce, P. Perfetti, G. Margaritondo, D. Talley, J.S. Sanghera, I.D. Aggarwal, N.H. Tolk, A. Congiu-Castellano, M.A. Rizzo and D.W. Piston, *Biophys. J.*, **85**, 2705–2710 (2003).

55. D. Vobornik, G. Margaritondo, J.S. Sanghera, P. Thielen, I.D. Aggarwal, B. Ivanov, J.K. Miller, R. Haglund, N.H. Tolk, A. Congiu-Castellano, M.A. Rizzo, D.W. Piston, F. Somma, G. Baldacchini, F. Bonfigli, T. Marolo, F. Flora, R.M. Montereali, A. Faenov, T. Pikuz, G. Longo, V. Mussi, R. Generosi, M. Luce, P. Perfetti and A. Cricenti, *Infrared Phys. Technol.*, **45**, 409–416 (2004).

56. T. Hirschfeld, *Appl. Spectrosc.*, **33**, 525–527 (1979).

57. D.L. Wetzel, *Vib. Spectrosc.*, **29**, 183–189 (2002).

58. P. Colarusso, L.H. Kidder, I.W. Levin, J.C. Fraser and J.F. Arens, *Appl. Spectrosc.*, **52**, 106A–120A (1998).

59. C.M. Snively and J.L. Koenig, *Appl. Spectrosc.*, **53**, 170–177 (1999).

60. R. Bhargava and I.W. Levin, *Anal. Chem.*, **73**, 5157–5167 (2001).

61. D.F. Ransohoff, *Nat. Rev. Cancer*, **4**, 309–314 (2004).

62. R. Bhargava and I.W. Levin (eds), 'Spectrochemical Analysis Using Infrared Multichannel Detectors', Blackwell Publishing, Oxford, 56–84 (2005).

63. See special issue of *Biochim. Biophys. Acta – Biomembr.*, **1758**, (2006).

64. B.R. Wood, L. Chiriboga, H. Yee, M.A. Quinn, D. McNaughton and M. Diem, *Gynecol. Oncol.*, **93**, 59–68 (2004).

65. S. Boydston-White, T. Gopen, S. Houser, J. Bargonetti and M. Diem, *Biospectroscopy*, **5**, 219–227 (1999).

66. R.A. Shaw, F.B. Guijon, V. Paraskevas, S.L. Ying and H.H. Mantsch, *Anal. Quant. Cytol.*, **21**, 292–302 (1999).

67. J.R. Mansfield, L.M. McIntosh, A.N. Crowson, H.H. Mantsch and M. Jackson, *Appl. Spectrosc.*, **53**, 1323–1333 (1999).

68. L.M. McIntosh, M. Jackson, H.H. Mantsch, M.F. Stranc, D. Pilavdzic and A.N. Crowson, *J. Invest. Dermatol.*, **112**, 951–956 (1999).

69. J. Einenkel, W. Steller, U. Braumann, L. Horn and C. Krafft, *Nat. Biotechnol.*, **25**, 29–31 (2007).

70. J. Kononen, L. Bubendorf, A. Kallioniemi, M. Barlund, P. Schraml, S. Leighton, J. Torhorst, M.J. Mihatsch, G. Sauter and O.P. Kallioniemi, *Nat. Med.*, **4**, 844–847 (1998).

71. R.L. Camp, L.A. Charette and D.L. Rimm, *Lab. Invest.*, **80**, 1943–1949 (2000).

72. Y. Benjamini and Y. Hochberg, *J. R. Stat. Soc. Ser. B*, **57**, 289–300 (1995).

73. Y. Pawitan, S. Michiels, S. Koschielny, A. Gusnanto and A. Ploner, *Bioinformatics*, **21**, 3017–3024 (2005).

74. N. Stone, C. Kendall, J. Smith, P. Crow and H. Barr, *Faraday Discuss.*, **126**, 141–157 (2004).

75. R. Bhargava, S.Q. Wang and J.L. Koenig, *Appl. Spectrosc.*, **54**, 486–495 (2000).

76. R. Bhargava, S.Q. Wang and J.L. Koenig, *Appl. Spectrosc.*, **54**, 1690–1706 (2000).

77. R.J. Anderson and P.R. Griffiths, *Anal. Chem.*, **47**, 2339–2339 (1975).

78. X. Llora, R.K. Reddy and R. Bhargava, unpublished work, manuscript in preparation.

# Spectral Histopathology of the Human Cervix

## Don McNaughton, Keith Bambery and Bayden R. Wood

*Monash University, Melbourne, Victoria, Australia*

## 1 BACKGROUND

Until the early 1990s, before breast cancer became the predominant cancer site, cervical cancer was the most frequent neoplastic disease among women in developing countries.[1] Worldwide there are more than 273 000 deaths from cervical cancer each year, and it accounts for 9% of female cancer deaths.[1] In the developed world, screening tests are significantly reducing the incidence of cervical cancer. In the United States, some 60 million screening tests were performed in 2005 and it is estimated that screening programs and follow-up intervention have reduced the incidence by ca. 80%.[2]

Cervical cancer develops usually over an extended period of time in the lining of the lower part of the uterus, the cervix. In this process, normal cervical cells gradually become abnormal or precancerous and then progress to cancerous. The process in its early stages is termed *dysplasia* and the term used to describe these abnormalities is *cervical intraepithelial neoplasia* (CIN). CIN is classified in a three-tier system according to the degree of abnormality as progressing through CIN 1, 2, and 3. These conditions precede carcinoma in

situ (CIS), a cancer that remains in the epithelial or outer membrane, and may progress to invasive carcinoma, where the cancer spreads to healthy tissue. These invasive carcinomas may develop in the squamous cells that line the cervix, termed *squamous cell carcinoma*, or in the surface cells of the glands, termed *adenocarcinoma*. Up to 90% of carcinomas are squamous cell carcinomas.

A more recent grading system is a two-tier Bethesda system that uses low-grade squamous intraepithelial lesion (LSIL), which encompasses both human papilloma virus (HPV) and CIN 1 and high-grade squamous intraepithelial lesion (HSIL), which encompasses CIN 2 and 3.

The currently accepted technique for diagnosing exfoliated cells is the Papanicolou (Pap) smear test where cells are collected with a Cytobrush™ and/or Ayre spatula from the cervical transformation zone, fixed in ethanol, and stained with the Papanicolou stain. Squamous intraepithelial abnormalities are classified using either the two-tier Bethesda system for exfoliated cells or the three-tier carcinoma intraepithelial neoplasia system for tissue biopsies.[a] Those graded LSIL are generally low risk in terms of proceeding to invasive carcinoma.[3] HSILs correspond to CIN II

*Vibrational Spectroscopy for Medical Diagnosis*. Edited by Max Diem, Peter R. Griffiths and John M. Chalmers.
© 2008 John Wiley & Sons, Ltd. ISBN 978-0-470-01214-7.

and CIN III and have a higher risk of proceeding to invasive disease. Persistent HPV infection is necessary but not sufficient for the later development of cervical carcinogenesis,[4] and the western world is now introducing HPV vaccination for young females as the most cost-effective approach to cervical cancer prevention.[2] Despite this, there will be a long-term need for screening for older generations.

Pap smear tests have reduced mortality by up to 70–80%[2,5,6] and tend to have a high specificity that ranges from 86–100% for CIN I and higher. However, despite its success, cytological screening has limitations, with many studies reporting low sensitivities (37–54%) for CIN I and higher specificities that vary significantly from country to country.[7] The misdiagnoses are often attributed to poor sample collection and slide preparation and errors in interpretation.[8] The solution to minimize errors in cytology screening is to improve the quality of sample acquisition, slide preparation, and overall diagnostic performance.[8] False-negative results have important medical, financial, and legal implications. In some countries, false-negative cytology specimens are a leading reason for medical malpractice litigation.[8] Consequently, alternative techniques aimed at reducing false-negative results and eliminating subjective diagnosis by cytological screeners have been pursued. Thinprep™ (Cytyc Corporation, Marlboro, MA), a procedure that removes much of the obscuring cellular material and deposits the sample as a monolayer, has improved diagnosis but not to the extent desired and at an increased cost.

As a consequence of the significant failure rate, a technique that is rapid and does not rely on human decision making would be a welcome asset to any clinical laboratory. PAPNET®, which relies on an artificial neural network (ANN) analysis of visual images to prescreen Pap smears, is one such alternative, although because of the marked increased cost for small gain it has not been adapted in the state of Victoria, Australia. Other alternatives under investigation include laser-induced fluorescence spectroscopy,[9] and a technique known as the *polarprobe*, which uses a combination of "electrical impulses and four different wavelengths of light". Both of these techniques rely heavily on a multivariate statistical approach in their data analysis as does infrared spectroscopy, but, unlike infrared spectroscopy, these do not provide direct molecular information. While the polarprobe technique has gained a good deal of commercial interest and is available in Australia as an adjunct screening tool,[10] many medical scientists remain skeptical of its blackbox nature.

Fourier transform infrared (FT-IR) spectroscopy first came to attention as a possible means of detecting cervical cancer in the early 1990s when Wong et al.[11,12] investigated the pressure-tuned infrared spectra of exfoliated cervical cells and reported spectral differences between samples from patients diagnosed normal and samples from patients diagnosed with dysplasia or cancer using cytology. Subsequent work using infrared microspectroscopy by the Diem group[13] and independently by McNaughton and coworkers[14] indicated that the observed spectral changes were not necessarily indicative of the different biomolecular composition of dysplastic cells per se but arose from other factors such as inflammation, the number of dividing versus nondividing cells,[15] and the overall divisional activity of the cells.[16] Regions of the spectrum originally thought useful for diagnosis were also obscured by biological components such as mucin, erythrocytes, and leukocytes.[13,14] It was concluded that individual spectral differences were not useful diagnostics and so multivariate statistical techniques and ANNs were applied to the analysis of the spectra of exfoliated cells in an attempt to circumvent these confounding variables.[17–21] These techniques are powerful in that they provide information on the important variables that distinguish normal from diseased samples, but they are limited in that they do not provide information on the cervical cell types and their stage of differentiation and maturation within the cervix. The stage of the menstrual cycle at cell collection was also shown to be a factor that could affect analysis.[22] Over a number of years, these studies demonstrated that a detailed understanding of the spectral features of the cell types and spectral variations resulting from differentiation, maturation, and cell cycle stages is a prerequisite for the interpretation of the spectral differences between normal and dysplastic cytological diagnosed samples. Much of the problem lies in the nature of the sample containing exfoliated cells, especially prior to any cleanup stage and so much of the recent work has concentrated on examining tissue specimens to characterize cell types, cell stages, and disease types. For this purpose, infrared point-to-point mapping and imaging of excised tissue sections have become the techniques of choice.

## 2 SAMPLE PREPARATION

Cervical tissue samples in all studies to date were obtained by cone biopsy sections from patients diagnosed by cytology with high-grade cervical dysplasia or carcinoma, and mounted in paraffin blocks for sectioning. A rigorous fixation and sectioning protocol designed to maintain cell morphology and minimize chemicals that may obscure diagnostically important bands is critical in obtaining high quality spectral maps of tissue sections. Our work has primarily utilized formalin-fixed, paraffin-embedded tissue because of its availability and relevance in the pathology laboratory. The importance of a rapid fixation approach cannot be overemphasized for the preservation of cell morphology and for the removal of water that would otherwise obscure the conformationally sensitive amide I mode. The various steps in tissue preparation for FT-IR spectroscopy are divided into the four stages of fixation, paraffin processing, sectioning, and dewaxing. The first three are common to both normal staining techniques and IR microscopy, while the last step, dewaxing, is critical for infrared work for the removal of paraffin, which would otherwise obscure the $C-H$ stretching $(3000-2800\,\mathrm{cm}^{-1})$ and deformation $(\sim 1450\,\mathrm{cm}^{-1})$ regions. Dewaxing is achieved by washing the tissue section three times in clean xylene. It is imperative that the xylene is clean

for each wash otherwise the paraffin permeates back into the tissue.

The choice of substrates for IR imaging of tissue is between expensive water-resistant crystal plates for transmission, such as $BaF_2$, $CaF_2$, ZnSe, which must be cleaned for reuse, and reflective surfaces. Transmission, where sections as thick as $10\,\mu m$ can be used without spectral saturation, has been used by Steller *et al.*[23] in their work, while in our work[24–27] we have used $Ag/SnO_2$-coated slides developed by Kevley Technologies™ (usually referred to as *low-e glass slides*) for measurement of transflection spectra. Visible light is transmitted by these slides while IR light is reflected, allowing the tissue to be visually examined or even stained for examination after spectroscopy. These slides in addition offer the advantage of affordability ($\sim$US \$2 per slide), and can be stored for further examination and for possible legal reasons. For transflection spectroscopy, sections must be cut with a thickness of $4–5\,\mu m$ to avoid spectral saturation. Although plain reflective metal surfaces can be used for spectroscopy, when comparison with tissue pathology is required they are not of much practical use.

The choice of spatial resolution, substrate, and mode of sample preparation is also affected by physical effects. Both Mie scattering and dispersion artifacts can cause considerable distortion in the resultant spectra. Mie scattering is prevalent when single cells are examined and the effects depend on the physical size of the cells and the cellular nucleus.[28] The effects in tissue samples are less noticeable but still need to be taken into account. Mie scattering in tissue causes broad baseline variability, which, as described below, can usually be eliminated or at least reduced by data pretreatment. Dispersion artifacts appear when the sample is particularly thin and/or when the refractive index of the sample is close to that of the substrate. It is a significant problem in tissue work in transflection spectroscopy, where the spectral distortion for thin regions of sample is large. There are methods for reducing these artifacts,[29] which are sometimes useful and they are not present if the embedding matrix, usually paraffin, is left in the sample. Either way, an awareness of these artifacts is essential in the decision tree of data analysis.

# 3 INSTRUMENTAL CONSIDERATIONS

Early work by the authors in the area of cervical tissue was carried out by point-to-point raster mapping using apertures of $20\,\mu$m $\times$ $20\,\mu$m with a step size of $10\,\mu$m. This aperture is essentially the spatial limit for single-point spectroscopy on a bench-top instrument before the effects of diffraction and low sensitivity preclude useful spectroscopy. Some 750 000 spectra of 10 samples were recorded in this study[27] requiring ca. 6 months, 7 days a week, and 12 h per day for collection. The spatial resolution limit, which precludes the investigation of small cells or cellular organelles, can be approached by using a synchrotron source,[30] which allows for microscopy in the $2000-1000\,\text{cm}^{-1}$ region at ca. $5\,\mu$m. This, however, requires up to 16 times longer acquisition times for imaging the same sample area.

Full rapid-scan imaging instruments, based on focal plane array (FPA) or linear array detectors, overcome the time limitation and to a minor extent the spatial resolution limitation since each pixel serves as its own limiting aperture. Liquid nitrogen cooled FPAs with up to $128 \times 128$ detector pixels allow for the collection of many thousands of $4-8\,\text{cm}^{-1}$ resolution spectra in a few minutes. The trade-off to achieve this is a reduced signal-to-noise ratio (SNR) and a reduced spectral range with a low wavenumber cutoff of ca. $900\,\text{cm}^{-1}$. The nominal spatial sampling varies with instrument design from ca. $3.9\,\mu$m pixel$^{-1}$ (Bruker Hyperion) to $5.5\,\mu$m pixel$^{-1}$ (Varian Stingray), although in practice this is not usually achieved and the practical spatial resolution, as pointed out by Wood *et al.*,[24] is not as good as that of synchrotron-based single-point instruments where confocal microscope systems theoretically allow for a 30% increase in spatial resolution over single-aperture microscopes.[31] For experiments

where single cells within tissue are under examination, this is an important consideration. Linear array systems based on 16 pixel arrays of mercury cadmium telluride (MCT) detectors offer SNR commensurate with single-point detectors and an increased spectral bandwidth to $750\,\text{cm}^{-1}$, albeit with nominal spatial resolutions of $6.5-12.5\,\mu$m pixel$^{-1}$ (Jasco IMV-4000) to $6.5-25\,\mu$m pixel$^{-1}$ (PerkinElmer Spotlight 300) and a need to collect the data by raster mapping. For biological samples like cells and tissue, the increased bandwidth is not usually of concern because most of the spectral information in such samples lies between $3500-2800\,\text{cm}^{-1}$ and $1800-900\,\text{cm}^{-1}$.

# 4 DATA PRETREATMENT AND ANALYSIS

Once hyperspectral data blocks have been collected, there are a number of tools available to process the data, including the software packages available with the various instruments described above. Although these packages are adequate for preliminary investigation of tissue and cell samples, most groups use more specialized software either written by themselves using such packages as MATLAB® or purpose-developed imaging programs such as CytoSpec.[32] The first essential task in preprocessing is to remove the effects of Mie scattering, dispersion artifacts, sloping baselines, or simply poor quality spectra and this can be divided into a number of phases. First, poor quality spectra, resulting from sample areas that are too thin or thick, noisy pixels, or those with atmospheric water vapor contamination need to be removed. Second, a preprocessing routine to remove baseline effects must be applied. There are three major alternatives for this. A simple baseline correction using the rubber band technique of Bruker Optic GmbH[33] or point-to-point linear or polynomial fitting prior to normalization can be applied, but for the generally large baseline variation across the many individual pixels of the image a common baseline correction is usually inadequate, while treating each spectrum individually is too tedious and time consuming. An alternative is spectral derivatization and normalization to eliminate baseline and thickness variation effects. When using *low-e* glass slides or ZnSe transmission plates at diffraction-limited resolution, dispersion and scattering artifacts also become important and these must also be eliminated or minimized in the preprocessing phase. The best alternative to correct for baseline variations, scattering, and thickness effects together is extended multiplicative scattering (EMSC), which has recently been applied to mid-infrared spectroscopy by Thennadil *et al.*[34] and applied to tissue imaging by Kohler *et al.*[35] We found this to be as good as or better than the other methods in most of the cases.

The final phase is image construction. FT-IR image data can be processed in a univariate mode where chemical maps (also called *functional group maps*) based on peak intensity, peak area, or peak ratios can be routinely generated with the software supplied with these type of instruments or using proprietary software such as CytoSpec. While these methods can provide information on the distribution and relative concentration of a particular functional group and hence a specific major biomolecule, they are not very useful in terms of classifying anatomical and histopathological features within the tissue matrix and multivariate image reconstruction is required. Typical methodology includes unsupervised hierarchical cluster analysis (UHCA), K-means clustering, principal components analysis (PCA), linear discriminant analysis, ANNs, and fuzzy C-means clustering. These methods are aimed at classifying spectra based on similarity and thus are used to discern anatomical and histopathological features based on underlying differences in the macromolecular chemistry of the different cell and tissue types that constitute the sample. We and others[23,26,27] have found UHCA to be the most useful for direct pathological comparison, although this severely restricts the size of images due to the computing overhead involved in calculating the distance matrices. Using a desktop PC with a Windows™ (Microsoft Corporation) operating system, the largest images that can be processed practically with UHCA

are $128 \times 128$; hence, there is a need for pixel aggregation when larger images are collected. Pixel aggregation, of course, leads to reduced spatial resolution, but if the object is an overview of the sample for comparison with histology or pathology a reduction in spatial resolution may be acceptable. Steller *et al.*[23] have managed to handle large image sizes by combining fuzzy C-means clustering with UHCA, where the fuzzy clustering is used to reduce the number of clusters in the UHCA. Our approach to this problem has been to train an ANN on the results of UHCA and build images using the ANN on new hyperspectral image data. This methodology is described briefly later in this chapter.

# 5 NORMAL CERVICAL PATHOLOGY

The first step in the development of IR imaging as a tool in cervical pathology is to generate spectra of all the cell types found in normal cervical tissue and build a reliable correlation between cell type and spectral type. This has been carried out by both mapping[27] and imaging.[25,26] In the latter studies, FT-IR hyperspectral datasets from FPA detector images were subjected to UHCA (*D*-values, Ward's algorithm) to investigate the tissue architecture of the cervical epithelium. The samples were obtained from cone biopsies, where the resultant tissue was determined to be normal by hospital pathology. After removal of poor quality spectra and conversion to second derivatives, UHCA was performed on a number of spectral regions in order to determine an appropriate reduced spectral range to minimize the analysis time while retaining cell type discrimination. Using just the amide I region ($1700-1570 \, \text{cm}^{-1}$) gave results very similar to those found for the full fingerprint range, although care needed to be taken to account for dispersion artifacts, whereas using just the $\nu_{\text{asym}}PO_2^-$ region ($1300-1200 \, \text{cm}^{-1}$) gave results that were not totally reliable. Further work has continued using the full spectral region to ensure that a full characterization of the spectrum is achieved. In the final step, all spectra in a cluster are assigned the same color, a pseudo color image is built, and the mean spectrum of each cluster is extracted. Each mean spectrum represents all the spectra in a cluster and is used to interpret the biochemical differences between the clusters, although individual spectra within the cluster are added to the database to ensure that a range of spectra attributed to each cell type is included for subsequent ANN training and analysis. The number of clusters to be used is chosen by the user who determines the optimum number for each tissue sample to ensure that every cell type and disease state is recognized.

The hemotoxylin and eosin (H&E) stained tissue section in Figure 1 shows the normal structure of ectocervical tissue with the seven defined regions of connective tissue: basement membrane, basal cells, parabasal cells, intermediate cells, superficial cells, and exfoliating cells. The nuclear-to-cytoplasm ratio of basal and parabasal cells is high and they are low in glycogen content, while both intermediate and superficial layers are known to be high in glycogen in the cytoplasm and intercellular bridge regions and have a low nuclear-to-cytoplasm ratio.[36] Figure 2 shows two pseudo color UHCA maps of an ectocervical section diagnosed as

normal by pathology, together with a photomicrograph of the adjacent H&E stained section and a univariate map derived from the ratio of the baseline-corrected integrated intensities of the glycogen ($1173–964$ cm$^{-1}$) and amide II (protein) ($1588–1481$ cm$^{-1}$) bands. Univariate maps are useful in showing regions of high macromolecular concentration and the areas of high glycogen content highlighted by the red regions in the color maps in Figure 2(b). These regions indeed are the superficial and intermediate layers, with the intermediate layer extremely high in glycogen and the superficial layer less intense. The glycogen concentration rises again on the periphery where the superficial layer becomes a flattened layer of cells for exfoliation. Figure 2(b) highlights how infrared imaging can replace stains such as the periodic acid Schiff (PAS) stain normally used to stain for glycogen. Given the carcinogenic and corrosive nature of Schiff's reagent due to pararosaniline and HCl, IR imaging has distinct advantages over PAS and many other standard chemical stains. The added advantage of infrared imaging methodology is that it can replace a number of stains using just one section of unstained tissue. IR spectroscopy is also extremely useful in identifying protein secondary

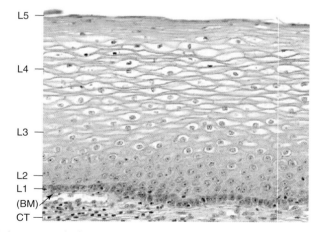

**Figure 1.** Structure of the ectocervical squamous tissue: CT = connective tissue, BM = basement membrane, L1 = basal cells (one layer), L2 = parabasal cells (two layers), L3 = intermediate cells (around eight layers), L4 = superficial cells (five or six layers), and L5 = exfoliating cells. [Reproduced by permission of the International Agency for Research on Cancer, Lyon, France.]

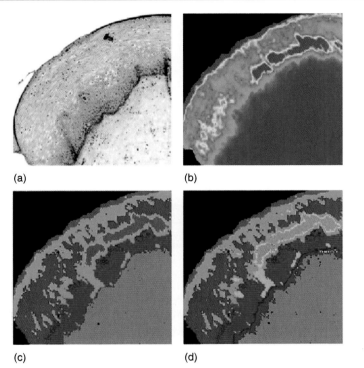

**Figure 2.** Cervical epithelium section (a) H&E stained section showing the five major cellular tissue types; (b) map based on ratio of glycogen (1173–964 cm$^{-1}$) to protein (1588–1481 cm$^{-1}$); (c) UHCA image using five clusters; and (d) UHCA image using eight clusters.

structure and indeed infrared univariate mapping has also been shown to be an excellent way of determining changes in protein structure in prion diseases and has been presented as a replacement for multiple staining in this context.[37]

Given that this section of normal ectocervical tissue contains essentially seven anatomical layers with the basement membrane too thin for the spatial resolution of IR imaging systems to be observed and the exfoliating cells and basal layer also very thin, one would expect at least five to seven clusters to be necessary and more if there are any remaining scattering, baseline, or thickness variation effects in the spectra. Using five clusters (Figure 2c), most of the cell types apparent in the stained cervical epithelium tissue section, i.e., the layer close to exfoliation, superficial, intermediate, parabasal, and connective tissue correlate well with the color coded clusters,

although the narrow basal cell region is not well defined. The thin exfoliating layer with its high glycogen content stands out as a distinct cluster and there are further correlations with the glycogen map in the intermediate and superficial layers. In Figure 2(d), where eight clusters are chosen, the basal region is also apparent as a separate cluster, while in both images the intermediate layer shows two distinct cluster types with the major spectral difference between the two being the amount of glycogen. There is also penetration of the superficial spectral type into the intermediate and even the parabasal layer, although at eight clusters this penetration area is shown to be spectrally different. In order to correlate pathology with infrared spectra, it is often necessary to go to higher cluster numbers to extract a spectral type. For sections displaying normal pathology, there is also further differentiation of the connective

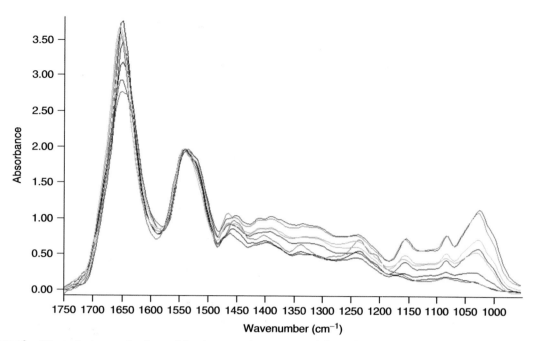

**Figure 3.** Mean cluster spectra from eight cluster analyses extracted from the UHCA image of Figure 2(d). Spectra are color coded to match the parent cluster regions and normalized to the intensity maximum of the amide II band.

tissue apparent at high cluster numbers, although the inclusion of further clusters in the analysis can lead to differentiation based on baseline variation, the absorption lines of atmospheric water vapor, and artifacts introduced by the mosaic nature of the images, and the decision maker must be aware of all these variables.

Figure 3 shows the mean spectra from the eight-cluster image, where the intensity of the glycogen bands ($1145-980 \, cm^{-1}$) correlates well with the glycogen levels discussed above. The close similarity of the basal cell mean spectrum to that of connective tissue is also apparent, leading to the necessity of including a larger number of clusters. Other spectral differences observed in the mean cluster spectra are in the bands due to protein, where small spectral shifts, intensity changes, and broadening are apparent in amide I, and in bands due to nucleic acids, where the $\nu_{asym}PO_2^-$ intensity shows great variation. These changes are as expected and the mean spectra

from the analysis using only the protein region are very similar, indicating that using a reduced spectral region leads to reduced computing time and gives almost identical discrimination for most applications. For our purpose of building a spectral databank of cell and disease types, as pointed out above, the full spectral range is optimum.

# 6 ABNORMAL SPECIMENS

The same type of analysis, as discussed above, can be carried out on microtomed sections from samples diagnosed with differing degrees of dysplasia and those with carcinoma; we have successfully done this on tissue from patients diagnosed with high-grade dysplasia showing a potential metastatic inclusion within the glandular endometrium.[24,27] The correlation of the cell types and dysplastic regions within the stained section and the cluster maps was easily apparent in this study and the mean extracted spectra for each of the clusters displayed a broad range of differences across the spectrum, indicating significant molecular differences between the different regions within the tissue. Typical differences in diseased tissue or tissue diagnosed as dysplastic were bands indicative of collagen at 1229, 1239, and 1250 cm$^{-1}$, while other spectra were relatively devoid of collagen but showed an increase in the symmetric and antisymmetric phosphate modes attributable to nucleic acids. In particular,

cells associated with carcinoma of the cervix have quite pronounced bands due to $\nu_{sym}(PO_2^-)$ and $\nu_{asym}(PO_2^-)$ at 1081 and 1240 cm$^{-1}$, respectively. For the samples studied, regions of leukocytes, erythrocytes, and the underlying connective tissue appeared as distinct clusters in the image, showing that the confounding variables that precluded useful diagnosis from attempts to diagnose by IR microscopy could easily be accounted for.

Figure 4 contains UHCA images and a micrograph from a sample diagnosed as CIN III

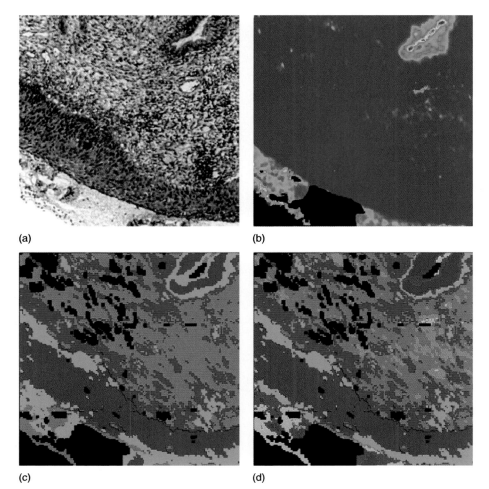

(a)

(b)

(c)

(d)

**Figure 4.** Cervical tissue from the region of the transformation zone showing squamous nuclear atypia through the whole depth of the sample diagnosed as CIN III. (a) H&E stained section; (b) map based on ratio of glycogen (1173–964 cm$^{-1}$) to protein (1588–1481 cm$^{-1}$); (c) UHCA cluster map using five clusters; and (d) UHCA map using eight clusters.

(severe dysplasia). The report of the optical microscopy study for the sample stated: "The cervical tissue from the region of the transformation zone shows full thickness squamous nuclear atypia. Atypical squamous epithelium extends into the underlying glands, which are otherwise normal." Both five- and eight-cluster images show the squamous epithelial as one cluster (red), with differentiation of a gland (dark blue) in the top right of both images c and d. The UHCA clusters again correlate well with the stained section. As can be seen from Figure 5, the mean cluster spectra show more similarity than those from a normal section with low glycogen throughout the tissue types with the exception of the spectra obtained from the gland (dark blue and purple). The UHCA discrimination now is based principally on amide I and amide II

band changes arising from protein compositional variations. A band at $1455 \, \text{cm}^{-1}$ due to incomplete removal of the paraffin-sectioning medium is present in the spectra from this particular section. While it is presently considered preferable that xylene washing is continued for sufficient time to ensure complete removal of all paraffin residue from the sectioned samples before FT-IR imaging is performed, this cannot always be guaranteed. We find that if UHCA is performed with the region $1390–1520 \, \text{cm}^{-1}$ excluded from the analysis that essentially identical UHCA tissue discrimination is achieved. In fact, if the samples are not deparaffinized at all, then very good discrimination can still be achieved and, in addition, the presence of the paraffin results in a better homogeneity of the sample's point-to-point optical thickness and consequently a beneficial

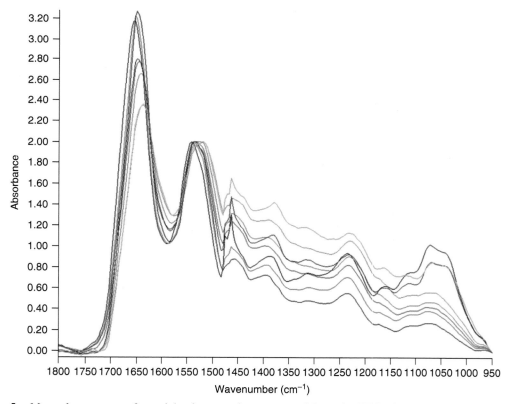

**Figure 5.** Mean cluster spectra from eight-cluster analyses extracted from the UHCA image of Figure 4(d). Spectra are color coded to match the parent cluster regions and normalized to the intensity of the amide II band.

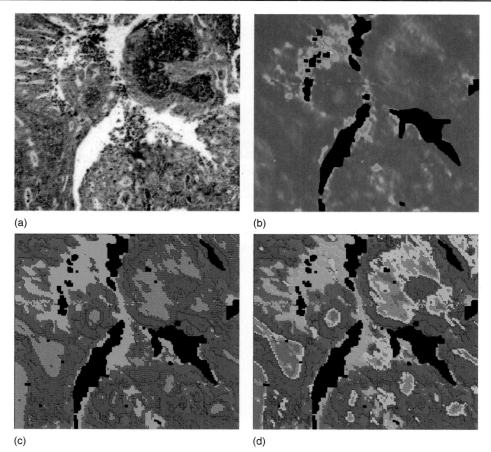

**Figure 6.** Differentiated endocervical adenocarcinoma: (a) H&E stained section; (b) map based on ratio of glycogen $(1173-964\,cm^{-1})$ to protein $(1588-1481\,cm^{-1})$; (c) UHCA image using eight clusters; and (d) UHCA image using eight clusters.

reduction in the presence of spectral dispersion artifacts.

Finally, Figure 6 shows images from an adenocarcinoma and even at five clusters there is differentiation of the features that are apparent in the stained section. Again, the mean cluster spectra show more similarity when compared with those from the normal sample in Figure 2.

It appears that the spectra of abnormal tissue cell types in general show less spectral variation when compared with those from normal tissue and that abnormal and normal cell types can be thought of as two specific groups from a spectral viewpoint.

# 7 ANN ANALYSIS

Although the use of a FT-IR imaging system reduces the time taken for obtaining images to a matter of minutes, the UHCA takes considerable processing time, as mentioned above, and becomes the limiting step in analysis. To overcome this, we have explored ANN imaging as a means of reducing this time. We have built a database of spectra from each cell type, which is used to train an ANN. ANNs are computational models inspired by biological decision-making structures such as the brain. An ANN is a nonlinear adaptive-learning information

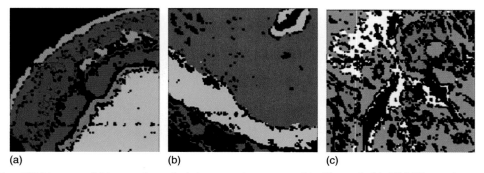

**Figure 7.** ANN images of (a) normal cervical tissue section presented in Figure 2, (b) CIN III sample as presented in Figure 4, and (c) adenocarcinoma as presented in Figure 6.

processing system comprised of an interconnected set of functional links called *neurons*. We classify our spectra using the Neurodeveloper™ (Synthon GmbH) software[38] to perform supervised training of neural networks based on the resilient backpropagation (Rprop) algorithm. Figure 7(a) shows an ANN image of the normal sample of Figure 2 constructed by training the ANN on data libraries of spectra extracted from both the normal and abnormal specimens (Figures 2, 4, and 6) for each of 13 UHCA clusters we have identified with distinct histopathological or anatomical features. The correlation between Figures 2(d) and 7(a) is excellent with the major difference being that some pixels have not been assigned to a spectral type (black pixels). Figure 7(b) and (c) shows ANN images of the CIN III sample and adenocarcinoma sample presented in Figures 4 and 6, respectively. Again, there is excellent agreement between H&E, UHCA, and ANN images. The use of ANN imaging, which classifies in a fraction of a minute, has greatly decreased the processing time and produces results to date that are as good as the UHCA images presented above.

# 8 3D IMAGING

In all the images presented above, the tissue was sectioned to ca. 4 µm thickness in order to measure spectra with the optimal SNR for analysis. Although extremely useful, this approach has a number of drawbacks for eventual diagnosis.

First, the section thickness is smaller than the diameter of many of the cells in the sample so each section consists of many fractions of cells. Second, the orientation of the tissue in the block can lead to orientation effects of the type described by Wood and McNaughton.[39] For this reason, and to allow for observation of the extent of abnormality in a full tissue sample, 3D imaging is a useful development that we have recently described.[25] In this study, a sample of cervical tissue sample exhibiting villoglandular adeno-carcinoma was sliced by microtome into 4 µm sections. One group of four sections was mounted on glass slides and stained with the routine histopathology (H&E) stain for light microscope examination, while an adjacent group of four sections was mounted on the Kevley™ "low-e" IR reflective microscope slides and imaged with a Varian Stingray™ FT-IR microscope system. For each of the four sections, a 16-tile (4 × 4) FT-IR image mosaic was constructed from FPA recordings collected as 16-pixel aggregates. Thus, the final spatial resolution in each image was approximately 22 µm, which was used to provide FT-IR images that covered an area of tissue large enough to encompass several examples of anatomically different tissue types. Construction of univariate and multivariate 3D images is described in Reference 25 and we restrict the major discussion here to multivariate UHCA-based images.

The four FT-IR images were stitched together side by side to give a single large 2D image frame using a MATLAB® routine specifically written for the task. UHCA was used to generate four clusters from second derivative spectra over the $1272-950\,cm^{-1}$ spectral window and the resultant cluster map reorganized back into the individual 2D cluster maps of the original four adjacent sections. The four sections together with an H&E section of a fifth adjacent section are shown in Figure 8 where the correlation of all four sections with each other and the stained section is apparent. The SCIRun software suite[40] was then used to render 3D cluster maps. Two views of the final 3D image are shown in Figure 9. MPEG movies (see http://www.mpeg.org/) of the full rotating images, which allow the full power of the images to be recognized, can be viewed on

the web at http://www.biomedcentral.com/1471-2342/6/12.

The 3D image shows excellent correlation with the anatomical and histopathological features indicated in Figure 8. The 3D UHCA map enables one to visualize the extent of penetration of the anatomical features and the degree of variation from section to section and enables visualization of tissue sections that cannot normally be analyzed using conventional mid-IR spectroscopic techniques due to the limited depth penetration of IR radiation. Such maps also enable the analysis of whole cells, rather than sections of cells, and minimize the effects of orientation artifacts that can arise during tissue sectioning. By rendering

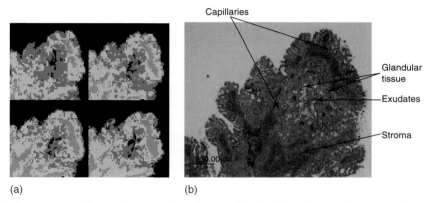

(a)                                    (b)

**Figure 8.** 2D cluster maps of four adjacent sections from a villoglandular adenocarcinoma sample, together with an adjacent H&E stained section from Reference 25.

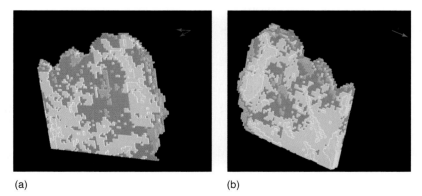

(a)                                    (b)

**Figure 9.** 3D cluster map from Reference 25: two views of 3D cluster maps constructed from the sections in Figure 8. The view is looking toward the section 1 side of the sampled volume in (a) and toward the section 4 side in (b).

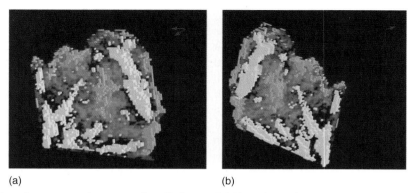

(a)                                           (b)

**Figure 10.**   3D semitransparent cluster map from Reference 25. Two views of 3D cluster maps identical to the maps in Figure 9 but with the stroma cluster removed from the plot and with glandular tissue now depicted in semitransparent blue. The view is looking toward the section 1 side of the sampled volume in (a) and toward the section 4 side in (b).

the image in semitransparent mode, individual clusters can also be studied as shown in Figure 10, where the stroma cluster has been removed from the plot and the glandular tissue is depicted in semitransparent blue. With this semitransparent capability, clusters in the center of the 3D FT-IR image can be visualized and used to observe the shape and penetration of important anatomical and histopathological features. These images can also be rotated and the images recorded in a movie to allow for easy visualization of the features in the data cube. Sample movies are available in Reference 25.

The steps and time involved in spectral acquisition and treatment to generate 3D FT-IR univariate images is as follows:

- Approximately 10 min to acquire each 2D FT-IR image from each of four individual tissue sections.
- Two minutes to run the MATLAB® stitching program.
- The registration of the images in the SCIRun program was quite time consuming in this study because it was carried out by hand, whereas it is envisioned that the development of suitable software for automation of the process could reduce the required time to a few minutes.
- Approximately 1 min is necessary for SCIRun to stack, interpolate, and render a single 3D image frame.

The timescale required for eventual 3D image construction is envisaged as less than 1 h from commencement of sample scanning on the FT-IR microscope stage with an additional hour for SCIRun to generate 3D movies composed of hundreds of individual 3D image frames. For 3D UHCA cluster map images, significantly more time is required because of the computationally intense nature of UHCA. Consequently, construction of UHCA 2D maps from large collections of tissue sections would be prohibitively slow to have immediate value as diagnostic technique. However, with the ANN alternative approach to UHCA described above, there is potential for rapid provision of 3D clusters images with the retention of the high correlation to morphological features as seen in Figure 7.

## 9   CONCLUSIONS

With the rapid data collection possible with linear or square array detector infrared imaging spectrometers, the next decade should see mid-IR imaging come to the fore in the field of pathology. We have shown the power and reliability of current FPA systems when combined with present software for cervical pathology; with the further development of data and image handling programs and the expected increase in computing power and speed over the next decade, there should be no technical or instrumental barrier to the introduction of these techniques to the pathology laboratory.

## ACKNOWLEDGMENTS

We thank the National Health and Medical Research Council (NHMRC) of Australia for financial support and our many collaborators who made this work possible and, in particular, Michael Quinn, Corey Evans, Max Diem. Mr Finlay Shanks is thanked for instrumental support and Mr Clyde Riley and Dr Virginia Billson (Royal Women's Hospital, Melbourne) for sectioning and histopathology advice respectively. Dr Wood is funded by an Australian Synchrotron Research Program Fellowship Grant and a Monash Synchrotron Research Fellowship.

## END NOTE

[a.] The standard cytological and pathological grading schemes are summarized in the Glossary.

## ABBREVIATIONS AND ACRONYMS

| | |
|---|---|
| ANN | Artificial Neural Network |
| CIN | Cervical Intraepithelial Neoplasia |
| CIS | Carcinoma in Situ |
| EMSC | Extended Multiplicative Scattering |
| FPA | Focal Plane Array |
| FT-IR | Fourier Transform Infrared |
| HPV | Human Papilloma Virus |
| HSIL | High-Grade Squamous Intraepithelial Lesion |
| H&E | Hemotoxylin and Eosin |
| IR | Infrared |
| LSIL | Low-Grade Squamous Intraepithelial Lesion |
| MCT | Mercury Cadmium Telluride |
| Pap | Papanicolou |
| PAS | Periodic Acid Schiff |
| PCA | Principal Components Analysis |
| SNR | Signal-to-Noise Ratio |
| UHCA | Unsupervised Hierarchical Cluster Analysis |

## REFERENCES

1. J. Ferlay, F. Bray, P. Pisani and D.M. Parkin, 'Cancer Incidence, Mortality and Prevalence Worldwide', IARC Press, Lyon (2004).

2. R. Roden and T.C. Wu, *Nat. Rev. Cancer*, **6**, 753–763 (2006).

3. T.C. Wright and R.J. Kurman, in 'Papillomavirus Reviews: Current Research on Pappilomaviruses', ed C. Lacey, Leeds University Press, Leeds, 215–225 (1996).

4. F.X. Bosch, A. Lorincz, N. Munoz, C.J. Meijer and K.V. Shah, *J. Clin. Pathol.*, **55**, 244–265 (2002).

5. G.H. Williams, P. Romanowski, L. Morris, M. Madine, A.D. Mills, K. Stoeber, J. Marr, R.A. Laskey and N. Coleman, *Proc. Natl. Acad. Sci. U.S.A.*, **95** (25), 14932–14937 (1998).

6. N.S. Larson, *J. Natl. Cancer Inst.*, **86**, 6–7 (1994).

7. D.C. McCrory, D.B. Matchar, L. Bastian, S. Datta, V. Hasselblad, J. Hickey, E. Myers and K. Nanda, 'Evaluation of Cervical Cytology. Evidence Report/Technology Assessment', Report No.: 5, Agency for Health Care Policy and Research, Rockville (1999).

8. E.L. Franco, E. Duarte-Franco and A. Ferenczy, *Can. Med. Assoc. J.*, **164** (7), 1017–1025 (2001).

9. K. Tumer, N. Ramanujam, J. Ghosh and R. Richards-Kortum, *IEEE Trans. Biomed. Eng.*, **45**, 953–961 (1998).

10. E.S. Ooi and S. Chapman, *Med. J. Aust.*, **179** (11/12), 639–643 (2003).

11. P.T.T. Wong, R.K. Wong, T.A. Caputo, T.A. Godwin and B. Rigas, *Proc. Natl. Acad. Sci. U.S.A.*, **88**, 10988–10992 (1991).

12. P.T.T. Wong, R.K. Wong and M.F.K. Fung, *Appl. Spectrosc.*, **47** (7), 1058–1063 (1993).

13. L. Chiriboga, P. Xie, V. Vigorita, D. Zarou, D. Zakim and M. Diem, *Biospectroscopy*, **4**, 55–59 (1997).

14. B.R. Wood, M.Q. Quinn, B. Tait, M. Ashdown, T. Hislop M. Romeo and D. McNaughton, *Biospectroscopy*, **4**, 75–91 (1998).

15. M. Diem, S. Boydston-White and L. Chiriboga, *Appl. Spectrosc.*, **53** (4), 148A–161A (1999).

16. S. Boydston-White, T. Gopen, S. Houser, J. Bargonetti and M. Diem, *Biospectroscopy*, **5**, 219–227 (1998).

17. B.R. Wood, M.A. Quinn, F.R. Burden and D. McNaughton, *Biospectroscopy*, **2** (3), 143–153 (1996).

18. M. Romeo, F.R. Burden, B.R. Wood, M.A. Quinn, B. Tait and D. McNaughton, *Cell. Mol. Biol.*, **44**, 179–187 (1998).

19. M. Romeo, B.R. Wood, M.A. Quinn and D. McNaughton, *Vib. Spectrosc.*, **72**, 69–76 (2003).

20. M.A. Cohenford, T.A. Godwin, F. Cahn, P. Bhandare, T.A. Caputo and B. Rigas, *Gynecol. Oncol.*, **66**, 59–65 (1997).

21. M.A. Cohenford and B. Rigas, *Proc. Natl. Acad. Sci. U.S.A.*, **95**, 15327–15332 (1998).

22. M. Romeo, B.R. Wood and D. McNaughton, *Vib. Spectrosc.*, **28**, 167–175 (2002).

23. W. Steller, J. Einenkel, L.C. Horn, U.-D. Braumann, H. Binder, R. Salzer and C. Krafft, *Anal. Bioanal. Chem.*, **384**, 145–154 (2006).

24. B.R. Wood, K.R. Bambery, L.M. Miller, M. Quinn, L. Chiriboga, M. Diem and D. McNaughton, *Proc. SPIE*, **5651**, 78–84 (2005).

25. B.R. Wood, K.R. Bambery, C.J. Evans, M. Quinn and D. McNaughton, *BMC Med. Imaging*, **6** (12), (2006).

26. K.R. Bambery, B.R. Wood, M.A. Quinn and D. McNaughton, *Aust. J. Chem.*, **57** (12), 1139–1143 (2004).

27. B.R. Wood, L. Chiriboga, H. Yee, M.A. Quinn, D. McNaughton and M. Diem, *Gynecol. Oncol.*, **93** (1), 59–68 (2004).

28. B. Mohlenhoff, M. Romeo, M. Diem and B.R. Wood, *Biophys. J.*, **88** (5), 3635–3640 (2005).

29. M. Romeo and M. Diem, *Vib. Spectrosc.*, **38** (1–2), 129–132 (2005).

30. M.J. Tobin, M.A. Chesters, J.M. Chalmers, F.J.M. Rutten, S.E. Fisher, I.M. Symonds, A. Hitchcock, R. Allibone and S. Dias-Gunasekara, *Faraday Discuss.*, **126**, 27–39 (2003).

31. G.L. Carr, *Rev. Sci. Instrum.*, **72** (3), 1613–1619 (2001).

32. P. Lasch, A Matlab based application for infrared imaging, http://www.cytospec.com (accessed Nov 2007).

33. M. Pirzer and J. Sawatzki, inventors; Bruker Optik GmbH, assignee. Method and Device for Correcting a Spectrum. USA (September 21, 2006).

34. S.N. Thennadil, H. Martens and A. Kohler, *Appl. Spectrosc.*, **60**, 315–321 (2006).

35. A. Kohler, C. Kirschner, A. Oust and H. Martens, *Appl. Spectrosc.*, **59**, 707–716 (2005).

36. L. Frappart, B. Fontanière, E. Lucas and R. Sankaranarayanan, 'Histopathology and Cytopathology of the Uterine Cervix—Digital Atlas', International Agency for Research on Cancer, Lyon (2004).

37. A. Kretlow, O. Wang, J. Kneipp, P. Lasch, M. Beekes, L. Miller and D. Naumann, *Biochim. Biophys. Acta, Biomembr.*, **1758**, 948–959 (2006).

38. T. Udelhoven, M. Novozhilov and J. Schmitt, *Chemom. Intell. Lab. Syst.*, **66**, 219–226 (2003).

39. B.R. Wood and D. McNaughton, 'FPA Imaging and Spectroscopy for Monitoring Chemical Changes in Tissue', in "Spectrochemical Analysis Using Multichannel Infrared Detectors", eds R. Bhargave and I. Levin, Ames, Iowa Blackwell, Oxford (2005).

40. Scientific Computing and Imaging Institute, The SCIRun Problem Solving Environment, (2004) http://www.software.sci.utah.edu/scirun.html.

# Raman Spectroscopy as a Potential Tool for Early Diagnosis of Malignancies in Esophageal and Bladder Tissues

## Nicholas Stone, Catherine Kendall and Hugh Barr

*Gloucestershire Hospitals NHS Foundation Trust, Gloucester, UK*

## 1 INTRODUCTION

Over the past few years, a number of groups have been working toward real-time, noninvasive techniques that utilize light to study abnormalities in tissue.[1] Recent technological developments have made it possible to obtain significant amounts of biochemical or architectural data from extremely complex biological tissue in very short timescales (milliseconds to seconds). Optical diagnosis relies upon measurement of the interaction of light photons with the constituents of biological tissue. The resultant data can provide an evaluation of histochemistry or morphology. This information can aid in the deduction of the pathological state of the tissue, and hence lead to a diagnosis.

Light can interact with tissue in a number of ways, including elastic and inelastic scattering; reflection of boundary layers; and absorption, leading to fluorescence and phosphorescence. All of these can be utilized in some way to measure abnormal changes in tissue. Many authors have used the term 'optical biopsy' when describing these techniques. Optical biopsy is a misnomer because no tissue is removed in the analysis; however, it does help to convey to the lay person the general principle of using light to detect cancerous transformations in tissue.

Initial optical biopsy systems, utilizing tissue fluorescence, have been used as an adjunct to current investigative techniques, mainly to improve targeting of blind biopsy.[2-5] Future prospects utilizing molecular-specific techniques may enable complete replacement of biopsy, with objective optical detection providing a real-time, highly sensitive and specific measurement of histological state of the tissue. However, until its efficacy is proven it is most likely that optical detection will be used as a complementary technique to improve targeting of biopsy selection.

The clinical requirements for an objective, noninvasive, real-time probe for the accurate and repeatable measurement of pathological state of the tissue are overwhelming. There is a clinical

*Vibrational Spectroscopy for Medical Diagnosis.* Edited by Max Diem, Peter R. Griffiths and John M. Chalmers.
© 2008 John Wiley & Sons, Ltd. ISBN 978-0-470-01214-7.

need for optical diagnosis in a number of important areas:

1. Situations where sampling errors severely restrict the effectiveness of excisional biopsy, such as the high failure rates associated with blind biopsies, whereby the clinician has to randomly select sites for sample collection. This method is used to screen for premalignant conditions such as ulcerative colitis and Barrett's esophagus.
2. Situations where conventional excisional biopsy is potentially hazardous; examples of vulnerable regions include the central nervous system, vascular system, and articular cartilage.
3. Situations where there is a need for immediate diagnosis during an investigative procedure; the use of spectroscopic methods of diagnosis could eliminate the need for many secondary procedures by enabling treatment to take place directly following diagnosis. This is especially useful with the development of treatments utilizing light energy, such as photodynamic therapy and laser ablation. This is likely to improve patient outcomes and decrease waiting times by reducing the number of costly procedures required.
4. Identification of tumor margins during surgical resection, thus enabling a more accurately targeted resection to be performed.
5. Times when a surgeon with any doubt over a diagnosis could validate a previous diagnosis prior to excision of an organ or lesion using a noninvasive optical probe.

Recently, it has been suggested that Raman spectroscopy has the potential to identify markers associated with malignant change.[6-8] This may lead to its use as a diagnostic tool for the study of the evolution of precancerous and cancerous lesions in vivo. The direct analysis of biochemical tissue constituents using vibrational spectroscopy may provide a new methodology for noninvasive detection of disease, together with new information for classifying, grading, and evaluating the progression of malignant neoplasms.

## 2   RAMAN SPECTROSCOPY

Raman spectroscopy is a very powerful, noninvasive analytical technique, which has only relatively recently been applied to biomedical samples.[9-11] An important characteristic of Raman spectra is that the intensity of an individual peak in the Raman spectrum is linearly proportional to the concentration of the molecular constituent(s) giving rise to that peak. Thus, a precise and quantitative molecular fingerprint is obtained.

Raman spectroscopy is the analysis of inelastic scattered photons excited by monochromatic laser radiation. The Raman spectrum is a plot of inelastically scattered intensity, as a function of energy difference between the incident and scattered photons. The loss (or gain) in photon energies corresponds to the difference in the final and initial vibrational energy levels of molecules participating in the interaction. The observed bands in a Raman spectrum are relatively narrow, easy to resolve, and sensitive to molecular structure, conformation, and environment. Figure 1 shows a set of typical Raman spectra recorded from biological molecules using an 830-nm wavelength laser for excitation. Raman spectroscopy provides information about both the chemical and morphological structure of the tissue in near real time[12] and can be used as a noninvasive optical method of tissue characterization. Most biological molecules are Raman active with their own characteristic spectral fingerprint. Proteins, nucleic acids, cell membranes, single cells, and tissues can all be studied.

In general, biological tissues are inhomogeneous in composition and highly scattering; the full analysis of Raman signals thus requires an understanding of tissue optical parameters and photon propagation in turbid media. Raman signals are inherently weak and, in addition, early diagnosis of disease requires detection of tissue molecular constituents present in low concentrations. This is accentuated by the fact that lasers with high intensity cannot be used to observe weak signals from tissues because of the potential for sample damage. It should also be noted

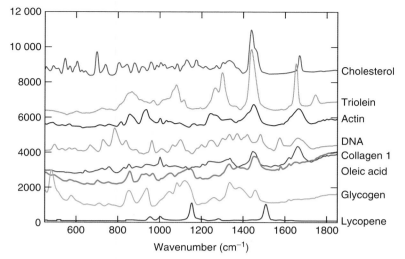

**Figure 1.** A selection of Raman spectra from biochemical tissue constituents (spectra have been offset for clarity).

that the complex nature of tissue composition results in absorption of light throughout the entire UV–visible region, and subsequent intense fluorescence emission strongly interferes with weak Raman signals.

Fluorescence is one of the primary obstacles encountered in the Raman spectroscopy of biological tissue. This can be ameliorated by the use of near-infrared (NIR) laser excitation and dispersive spectrometers equipped with array detectors that allow the rapid detection of spectra with low background fluorescence. NIR radiation generally does not induce electronic absorption in tissue chromophores; consequently, fluorescence is weak relative to the Raman signal. Holographic notch filters, or more recently metal oxide edge filters, are used to filter the elastic (Rayleigh) scattered light from the signal.[13] In combination with a single monochromator, the optical efficiency of the system is increased compared with that of the double and triple monochromators commonly used previously. The development of charge coupled device (CCD) detectors and solid-state semiconductor lasers has enabled the development of portable NIR Raman systems that produce spectra with high signal-to-noise ratio (SNR) in short integration times. Compatibility with fiber optics and the development of new probes is making in vivo diagnostics with Raman spectroscopy a real possibility.

By the start of the 1990s, various groups were using Raman spectroscopy to distinguish between normal and neoplastic tissues. The first studies looked at differentiating between normal tissue and advanced cancers in the breast[3] and gynecological organs.[7,8] As techniques were refined, interest moved toward diagnosing neoplastic change at progressively earlier stages. To date, in vitro studies have also been undertaken to differentiate between different pathologies in a number of other tissues including colon,[14,15] esophagus,[16–18] brain (*see also* **Neuro-oncological Applications of Infrared and Raman Spectroscopy**),[19] skin,[20,21] and larynx.[11]

Previous work performed by our group has involved the optimization of a commercially available Raman microspectrometer to enable accurate, repeatable, Raman spectra from tissue to be acquired in short timescales.[17] Our group has also demonstrated the use of Raman spectroscopy to provide a molecular fingerprint of a number of epithelial tissues, both in normal and premalignant and malignant states. This involved the study of tissue samples to identify the differences in the Raman spectra measured from homogeneous samples covering the full spectrum of disease in a number of organs. These include esophagus,[16,22] colon,[23] larynx,[11] bladder,[24] prostate,[25] and breast lymph nodes.[26] Use of multivariate statistical analysis of the spectra has demonstrated that very subtle changes from one pathology group to the next can be observed and employed to distinguish between the groups. As many as nine different pathology groups have been discriminated from both snap-frozen (fresh) and formalin-fixed tissue.[17]

The objective of this chapter is to provide an overview of this work and to identify the possibilities of targeting (pre)cancerous lesions in two examples of epithelial tissues. Furthermore, a discussion is reported on the use of spectral analysis to provide detailed, disease-specific biochemical information.

# 3 THE CLINICAL NEED

One in three people in the UK will be struck down by cancer in their lifetime and three-quarters of them will succumb to the disease.[27] As the mean age of the population increases—due to improvements in diet and medical treatments for the traditional killers such as cardiovascular disease—the number of people affected by cancer will increase.

The majority of cancers develop in epithelial tissues; these are the tissues that line hollow organs, and are thus exposed over a lifetime to numerous assaults from chemical and pathogenic carcinogens. Epithelial tissue can be specialized for a number of functions and depending on the organ will contain different cell architectures. Although the pathogenesis of most cancers is not fully understood, some are known to develop through a premalignant state. Figure 2 shows the possible carcinogenesis pathway for infiltrating ductal carcinoma of the breast.

Current methods of detecting malignancies rely upon surveillance of at-risk populations or diagnostic investigations following presentation with suspicious symptoms. By the time these symptoms are present it is often too late to facilitate a full cure. However, most cancers need not be killers if they can be exposed at an early stage to

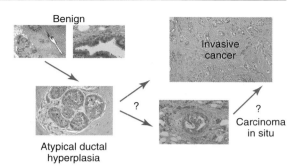

**Figure 2.** Schematic diagram of possible paths for carcinogenesis in the breast. This example shows H&E stained tissue sections from normal tissue to benign conditions such as atypical ductal hyperplasia and then demonstrates that the route to invasive ductal carcinoma may or may not pass through a carcinoma in situ phase.

the wide range of improving treatments. Therefore, the primary requirement for a successful treatment of any malignancy is early detection.

## 3.1 Current methods for detection and diagnosis

Over the last century there have been myriad of technological breakthroughs leading to improvements in medical diagnosis and therapy. Most in vivo diagnostic techniques measure the changes in architecture and morphology of tissue that follow initiation of the disease process, and others are able to provide some information on functionality. However, the spatial resolution of the techniques is usually of the order of millimeters. Therefore, it is apparent that using current medical technology, the detection of malignant change at a cellular level (where a successful treatment is most likely) will rely upon the removal of tissue samples to enable microscopic, histopathological analysis.

### 3.1.1 Histopathology

The "gold standard" for detection of malignancies and premalignancies is excisional biopsy followed by histopathological analysis. This technique relies upon sectioning the tissue less than one cell thick ($<10\,\mu$m) and staining with hematoxylin and eosin (H&E). This provides some functional information in addition to the morphology of the tissue that can be viewed under a conventional light microscope. The principle of histology has changed little since its conception in the nineteenth century.[28]

A further problem associated with the gold standard involves the targeting of biopsy samples from microscopic abnormal lesions. It is often the case with many precancerous lesions that they will be too small to be detected by endoscopy alone. Therefore, a random tissue selection procedure is usually employed. This can lead to a high probability of missing abnormal tissue by random biopsy. Van Sandick *et al.*[29] demonstrated this in 1998 during surveillance of patients with Barrett's esophagus, a condition whereby the risk of developing esophageal adenocarcinoma is increased manyfold.

Routine techniques for the detection of cancer measure the changes in structure and morphology in tissue that follow initiation of the disease process. However, it is likely that these architectural changes are driven by early biochemical changes within tissue. The qualitative analysis of such variations provides important clues in the search for a specific diagnosis, and the quantitative analysis of biochemical abnormalities is important in measuring the extent of the disease process, in designing therapy, and in evaluating the efficacy of treatment. Furthermore, histopathological analysis inherently involves a time delay between the investigation, diagnosis, and commencement of treatment, leading to further medical procedures with inherent risks, patient anxiety, and costs involved.

Histopathologists are able to provide classification into large subgroups of cancer and noncancer. However, recent inter- and intraobserver studies have demonstrated that pathologists can demonstrate acceptable levels of agreement for the two major comparative groups, positive and negative for cancer, with a $\kappa$ value of 0.8, where a $\kappa$ value of 1 represents full agreement and 0 represents chance agreement. The division into the four groups, namely, normal, mild precancerous change, severe precancerous change, and cancer, has revealed that there are poorer levels of agreement (intraobserver $\kappa$ values of 0.64; interobserver $\kappa$ values of 0.43).[30] The clinical decision of whether to perform treatments involving radical surgical excision, radiotherapy, or chemotherapy, which carry risks of mortality and substantial complications, rests on this pathological diagnosis.

# 4 ESOPHAGUS

In 2002 in the UK 3% of all cancers diagnosed were found in the esophagus (7080) and 5% of all deaths were from esophageal cancer (7008).[27] This incidence is rising rapidly in the developed world.

Long periods of exposure to gastroesophageal reflux can lead to a change in the cells lining the esophagus, whereby the squamous epithelium lining the distal esophagus is replaced by columnar epithelium.[31] This is called Barrett's esophagus and is a metaplastic condition, in which the cells are essentially normal but they do not belong in that location. Prospective studies have shown that the prevalence of columnar cell-lined esophagus in patients exhibiting reflux was between 11 and 12%.[32,33] Morphologically, the columnar epithelium of Barrett's esophagus can resemble the mucosa of the gastric fundus, cardia, antrum, or intestine. The columnar-lined mucosa is more likely to be resistant to the effects of reflux than squamous-lined mucosa.

Figure 3 shows some typical H&E stained esophageal sections used to discriminate between normal squamous tissue, metaplastic tissue (Barrett's phagus), precancerous changes (dysplasias), and adenocarcinoma. Note that the lesions develop only on the surface of the epithelium, which is around $500\,\mu$m thick. It is only when cancer has taken hold that invasion into deeper tissues can take place.

Barrett's esophagus is associated with an increased risk of esophageal adenocarcinoma.[34–38]

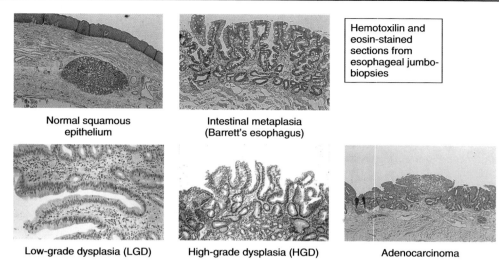

**Figure 3.** Photomicrograph of stained histology of the esophagus to demonstrate depth of lesion development (various magnifications).

Adenocarcinomas occurring in Barrett's esophagus make up between 54 and 68% of all esophageal cancers[39,40] and the rate of incidence is rising more rapidly than for any other cancer.[41,42] These carcinomas typically occur in older (mean age 57 years),[43] male (male : female ratio = 5.5 : 1),[43] white patients.[36,44]

Dysplasia is the most frequently used marker of increased cancer risk in Barrett's esophagus. It is an abnormal change in the maturation and development that can occur in all three types of Barrett's epithelia. These are named after the metaplastic cells they exhibit, namely cardiac, fundic, or intestinal metaplasia (IM) (see Table1 for a full list of the major esophageal pathology groups). However, it is more commonly seen in areas of intestinal metaplasia (IM) and it is unlikely that esophageal adenocarcinoma occurs except in patients with IM.[45–47] Follow-up studies in patients with Barrett's esophagus have shown that only patients with dysplasia progress to adenocarcinoma.[48] A number of studies have shown that over 50% of patients operated on for high-grade dysplasia (HGD) will be found to have invasive carcinoma.[47,49,50] Therefore, the potential benefits of removing an esophagus exhibiting dysplasia must be weighed against the relatively

high mortality associated with esophagectomy (5–15%)[37] and the poor outcome in patients who present with invasive adenocarcinoma of the esophagus (23% survival at 1 year and 7% survival at 5 years in the England and Wales between 1986 and 1990).[51]

Histopathology is currently the gold standard for diagnosis of dysplastic changes in the esophagus. However, it can be limited by the subjective nature of biopsy interpretation and targeting,[52] especially in the grading of dysplasia, which has no set limits to progression (i.e., follows a continuum). Most patients with Barrett's esophagus will not develop dysplasia or adenocarcinoma.[48,53] Reliable methods for identifying those patients with Barrett's esophagus who will progress to carcinoma do not exist at present. Therefore, techniques that enable a greater understanding of the biological processes leading to carcinogenesis will be extremely valuable.

## 5 BLADDER

The incidence of bladder cancer in the UK is 8% in men and 3% in women.[54] It is the fourth most common cancer in men and the eighth in

**Table 1.** List of the pathology groups studied in the esophagus and bladder. The names in bold represent the pre-cancerous/cancerous groups.

| Organ/group | Esophagus | Bladder |
| --- | --- | --- |
| 1 | Normal | Normal |
| 2 | Cardiac metaplasia | Cystitis |
| 3 | Fundic metaplasia | **Carcinoma in situ** |
| 4 | Intestinal metaplasia | **Transition cell carcinoma G1** |
| 5 | **Low-grade dysplasia** | **Transition cell carcinoma G2** |
| 6 | **High-grade dysplasia** | **Transition cell carcinoma G3** |
| 7 | **Adenocarcinoma** | **Squamous carcinoma** |
| 8 | **Squamous cell dysplasia** | **Adenocarcinoma** |
| 9 | **Squamous cell carcinoma** | |

women. Each year there are 9000 cases diagnosed in men and 3600 diagnosed in women in the UK; even though the figures are high, there has been a reduction in the age-standardized incidence since the 1980s. It is thought that the reason for this is the reduction in smoking and the banning of aromatic amines in the 1980s. Both of these are known risk factors for the development of bladder cancer.[16,55]

Ninety percent of bladder cancers are transitional cell carcinomas of the bladder, and the majority of these are superficial. They tend to be found in people over the age of 65. The standard method for the diagnosis of bladder cancer is by means of a cystoscopy and biopsy, which ordinarily involves a general anesthetic. Once diagnosed, patients are staged by means of the tumor-nodes-metastases (TNM) classification and entered into a surveillance scheme, if no further treatment is required. The surveillance program involves 6 monthly or yearly cystoscopies for at least 10 years.

The gold standard noninvasive test is urine cytology. Cells isolated from urine are stained with a Papanicolaou stain. Urine cytology has 95% specificity, but has only 40–60% sensitivity (the sensitivity increases with the grade of the tumor). This is because of the significant inter- and intraobserver variability, as well as the fact that urine cytology is altered by urinary tract infection (UTI), bladder instrumentation, indwelling catheters, radiotherapy, and intravesical chemotherapy.[56] The *sensitivity* of any test is the capability of that test to detect individuals with

a disease. High sensitivity is required when the disease to be diagnosed is serious, and should not be missed, and false-positive results do not have adverse consequences for the patient. *Specificity* is a measure of how well the test correctly identifies the negative cases. High specificity is required when the disease to be diagnosed is serious, and false-positive results may have serious adverse consequences for the patient.

Unfortunately, when viewed with a cystoscope some of the bladder tumors, especially carcinoma in situ and flat superficial tumors, can look just like cystitis or normal bladder. In view of this, a test that has a greater sensitivity and specificity than urine cytology is urgently needed, to minimize any unnecessary anesthetic procedures.

# 6 EXPERIMENTAL

## 6.1 Sample handling

Numerous epithelial tissues have been studied by our group and others. Esophageal and bladder tissues have been selected for this work as examples of organs exhibiting slowly developing cancers, with endoscopic access and of significance in the mortality and incidence statistics.[27]

Harvesting of specimens for our laboratory studies involved the collection of additional biopsy samples, during routine investigations, following informed written consent. Ethical approval has been provided by the Gloucestershire Local Research Ethics Committee for this work.

The specimens were mounted on acetate paper, placed in a 2 ml cryovial (BDH Corning Ltd.), and snap frozen in liquid nitrogen. Histological sections were cut from each of the samples using a freezing-microtome and stained with H&E. The remaining biopsy blocks were stored at $-80\,^{\circ}$C until spectroscopic studies could be carried out. Prior to carrying out spectral measurements, the specimens were passively warmed to room temperature. No other sample pretreatment was performed. By mounting the specimens immediately, it was possible to locate the mucosal surface of the tissue following sectioning and storage; furthermore, contamination from fixing and mounting agents has been minimized.

The H&E stained sections were sent for histopathological analysis by two to three expert pathologists, who were blinded to patient history and each other's results. Samples were histopathologically graded following the procedures defined by the panel of pathologists in advance, following current international guidelines for premalignant/malignant diagnosis in these tissues.[57] Those samples demonstrating mixed pathologies (nonhomogeneous) or any histopathological disagreement were rejected. Table 1 lists the pathology groups included in the study from the two tissues under investigation.

The importance of gaining a consensus histopathological opinion on all specimens used cannot be overstated. It has been shown by Kendall *et al.*[16] that using specimens with only a majority of opinion from two out of the three expert pathologists seriously downgrades the performance of the model. This is especially the case for pathology groups where there is greater interobserver variability.

## 6.2 Spectroscopy procedures

Biopsy blocks were removed from the $-80\,^{\circ}$C freezer and warmed passively to room temperature. Tissue samples were oriented in the in vivo geometry (tissue surface facing the laser beam) and placed on a UV-grade calcium fluoride slide (Crystran, UK). In vitro Raman scattering measurements were obtained with a customized Renishaw System 1000 Raman microspectrometer (Renishaw plc, Wotton-under-Edge, UK). An (Mid-infrared) MIR-Plan Olympus (80×, 0.75 NA for esophagus or 20×, 0.4 NA for bladder biopsies) ultra-long-working-distance (ULWD) lens was used to focus the laser beam (power at sample of $31 \pm 0.8$ mW for the 80× and around 110 mW for the 20×) to a spot on the tissue surface and to collect the backscattered photons in nonconfocal mode. The laser spot was elliptical and had dimensions of approximately $3 \times 15\,\mu$m for the 80× objective and $12 \times 60\,\mu$m for the 20× objective. The entrance slit to the monochromator was set to 50 $\mu$m. A minimum of 10 spectra were acquired from the epithelium surface of each sample, using a random sampling procedure. This provides data from discretely different clusters of cells from the same patient and pathology grouping (assuming that the sample is pathologically homogeneous). The scattered Raman signal was integrated between 10 s (for 80× objective) and 30 s (for 20× objective) and measured over a spectral range of $200-2100\,\text{cm}^{-1}$ with respect to the laser excitation frequency.

A Renishaw semiconductor laser provided laser illumination at 830 nm, with up to 350 mW output. Customized metal oxide edge filters were used for optimum rejection of the elastic scattered light. A single 300 lines per mm dispersion grating (Kaiser Optical Systems, Inc., Ann Arbor, USA) was required to separate the collected light into its spectral components. A deep depletion CCD detector (Renishaw) was utilized for sensitive detection at NIR wavelengths. This combination facilitated the acquisition of spectra with high SNR in a matter of seconds without the requirement of moving the grating, thus improving wavelength repeatability and speed.

## 6.3 Raman mapping

Mapping of tissue sections has also been performed by taking consecutive sections (10 $\mu$m) from the biopsy blocks, with the first section being used for the stained histology and the second mounted on $CaF_2$ for mapping studies. An $80\times$ ULWD objective was used to map the specimens in steps of $100\,\mu m$ by moving the automated xyz stage (Prior Scientific, Inc.). Each point spectrum was acquired in $30\,s$. An autofocus routine was used to prevent out-of-focus spectra from being obtained when mapping large areas. This involved a stepwise approach either side of the last $z$ value, to maximize the signal at $1450\,cm^{-1}$ (a strong peak in all Raman spectra of tissue samples).

## 6.4 Calibration and spectral processing

Wavenumber calibration of the Renishaw Raman system 1000 microspectrometer was achieved using a neon–argon standard lamp source. A number of atomic emission lines were chosen to calibrate the abscissa of the spectrometer. They were selected to cover the spectral range required and to be sufficiently resolvable for the system to measure them accurately without any confusion from nearby lines. Calibration was carried out over the absolute wavenumber range of $10\,000–12\,100\,cm^{-1}$ (corresponding to Raman shifts of 2050 to $50\,cm^{-1}$).

The most significant change encountered was that the instrument sensitivity varies slowly with wavelength. This is mainly caused by the gradual change in quantum efficiency of the detector across the spectral range. A smaller contribution to the response function originates from etaloning in the CCD, modal interference in optical filters, and variation in the CCD pixel sensitivity. This can cause a rapid variation in the sensitivity of a Raman instrument with changing wavelength.

The energy transfer function of the system was evaluated by measuring the slowly varying spectrum of a tungsten-filament lamp. The lamp source had been calibrated against a secondary standard by the National Physical Laboratory (NPL, UK) and the user was provided with a spectrum of absolute intensities versus wavelength at a fixed distance from the source. It was possible to calculate the energy transfer function of the system by dividing the absolute spectrum of the lamp by the spectrum measured using the spectrometer (see Figure 4 (middle row)). This transfer function can then be used to correct the measured tissue spectra for the energy-dependent sensitivity of the system by multiplying the raw tissue spectrum by the transfer function. Figure 4 (bottom) shows the effect of correction of the energy sensitivity of spectrometer on a typical esophageal Raman spectrum.

More recently a number of groups have utilized fluorescent glass standards. These provide long-term stability and have the advantage that their spectra are measured from the same sampling geometry as tissue spectra, whereas lamp emission standards can introduce subtle artifacts based on off-axis rays reaching the spectrometer.

## 6.5 Raman spectroscopy for grading of esophageal disease

Work performed by our group in Gloucester aimed to overcome any problems with the gold standard histopathology by bringing together three expert pathologists for each tissue type studied. Their expertise has been useful in providing blind pathological opinion on all samples used for our studies. The results provided were combined to achieve a consensus pathological opinion.

Large sample numbers were collected, but it can be seen from Figure 5 that many were discarded prior to analysis. Figure 5 is a schematic diagram of the process of histopathological analysis of 150 esophageal biopsy samples. First, each pathologist histopathologically analyzed all the samples in turn, and then either a majority or consensus of opinion of the histopathological state in each specimen was taken. Any sample without this agreement was discarded from the model. Gray circles represent unknown or mixed pathology samples; green circles are normal tissue, blue circles are fundic Barrett's (FB), cyan circles are cardiac Barrett's (CB), purple circles are IM, pink circles are low-grade dysplasia (LGD), red circles are HGD, black circles are adenocarcinoma, yellow circles are squamous dysplasia, and orange circles are squamous cell carcinoma (SCC).

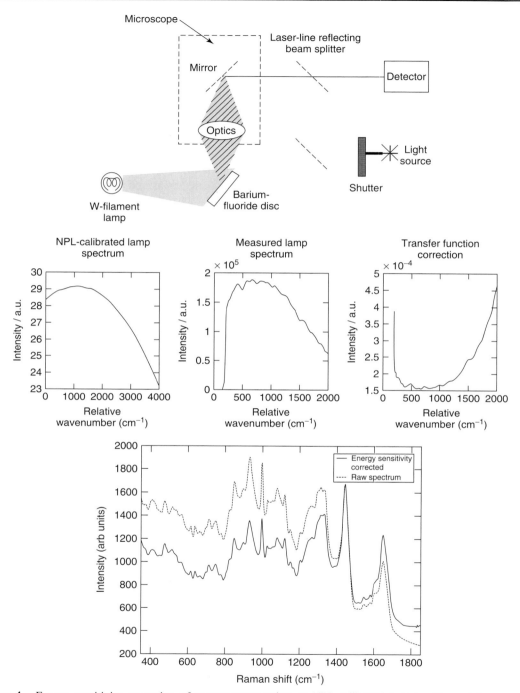

**Figure 4.** Energy sensitivity correction of spectrometers using an NPL-calibrated tungsten-filament lamp source: (top) apparatus arrangement; (middle) absolute lamp spectrum, measured spectrum, and calculated correction function; (bottom) the effect of the correction on a typical Raman tissue spectrum.

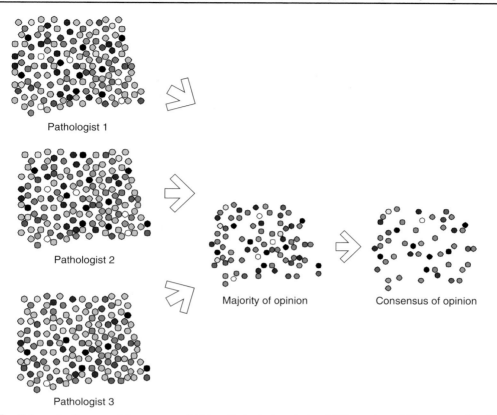

Pathologist 1

Pathologist 2

Pathologist 3

Majority of opinion

Consensus of opinion

**Figure 5.** Schematic diagram of the process of histopathological analysis of 150 esophageal biopsy samples. See text for color definitions.

From 150 biopsy samples, between 80 and 90 samples were homogeneous (depending on pathologist); therefore, following majority selection there were 80 samples remaining and following consensus selection there were only 50 samples remaining. By looking at the color-coded biopsy symbols in Figure 5, it can be seen that there is little disagreement in the normal samples (green), but significant disagreement in premalignant (or dysplastic) specimens.

### 6.5.1 Spectral analysis and testing

A mean Raman spectrum calculated from hundreds of esophageal tissue spectra from all pathology groups is shown in Figure 6. The peak positions are labeled and a list of tentative peak assignments is made in Table 2. Mean spectra

measured from each esophageal pathology group were calculated and are shown in Figure 7. Tentative peak assignments have been further outlined elsewhere.[8,11,58] Note the similarities in the mean spectra acquired from esophageal biopsy tissue. Normal spectra (green) are quite distinct from cancerous tissue spectra (black). However, the stages in between exhibit more subtle spectral changes.

For applications of spectroscopy in medicine, a number of authors have used empirical methods such as peak heights, peak areas, and peak ratios to glean diagnostic information from highly complex spectra. In a small number of cases where there is a single molecular species that directly accompanies any abnormality, this procedure will be sufficient. However, in the majority

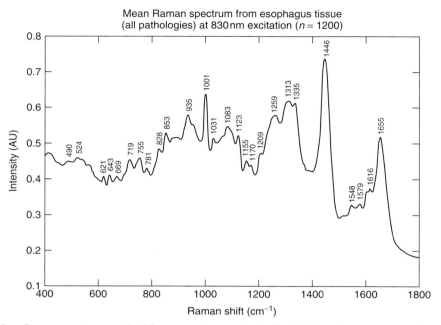

**Figure 6.** Mean Raman spectrum acquired from normal, metaplastic, dysplastic, and cancerous esophageal tissue with an 830-nm excitation source. Time of acquisition = 30 s.

of cases biochemical changes preceding and accompanying morphological changes are extremely complex. Therefore, methods of extracting information from Raman spectra should use as much information as possible to facilitate the most accurate prediction of histopathology. Multivariate techniques have been explored to enable utilization of the maximum available information for prediction of diagnosis.

The calculation of principal components to describe the greatest variance in the spectra from the mean of the data set makes it possible to substantially reduce the number of data points in the analysis, thus enabling prediction algorithms to effectively include all the biochemical data that varies from one sample to the next. Principal components analysis (PCA) finds combinations of variables, which describe the major trends (sources of independent variation) in the data.[59]

PCA is an unsupervised data transformation procedure that creates new variables in the directions of maximal variance, but not necessarily in the directions that are most useful for diagnosis.

Group membership for a particular spectrum can be predicted by comparing the spectrum to a number of reference spectra, by using some spectral distance measure, and classifying it to the most similar ones from a particular group. In many cases, the principal component space is not optimal for the separation of the desired groupings. Further multivariate techniques can be applied that incorporate information about the origin of the samples in the model (supervised classification). This information can be used to find the directions in the spectral space that provide maximal distances between groups. Linear discriminant analysis (LDA) or some derived form of this method is most commonly used. The linear discriminant function was originally derived by Fisher[60] as a technique for classifying an object or individual into one of two possible groups on the basis of a set of observations or measurements. LDA finds the direction in spectral space that is optimal for separating two groups. It selects the linear combination of vectors in spectral space that give the maximum

**Table 2.** Major vibrational modes identified in esophageal samples.

| Peak position (cm$^{-1}$) | Major assignments |
|---|---|
| 490 | Glycogen |
| 524 | S—S disulfide stretch in proteins |
| 621 | C—C twisting mode of phenylalanine |
| 643 | C—C twisting mode of tyrosine |
| 669 | C—S stretching mode of cystine |
| 719 | C—N (membrane phospholipid head)/nucleotide peak |
| 755 | Symmetric breathing of tryptophan |
| 781 | Cytosine/uracil ring breathing (nucleotide) |
| 828 | Out of plane ring breathing tyrosine/O—P—O stretch DNA |
| 853 | Ring breathing mode of tyrosine and C—C stretch of proline ring |
| 920 | C—C stretch of proline ring/glucose/lactic acid |
| 935 | C—C stretching mode of proline and valine and protein backbone ($\alpha$-helix conformation)/glycogen |
| 1001 | Symmetric ring mode of phenylalanine |
| 1031 | C—H in-plane bending mode of phenylalanine |
| 1083 | C—N stretching mode of proteins (and lipid mode to lesser degree) |
| 1123 | C—C stretching mode of lipids/protein C—N stretch |
| 1155 | C—C (& C—N) stretching of proteins (also carotenoids) |
| 1170 | C—H in-plane bending mode of tyrosine |
| 1209 | Tryptophan and phenylalanine $\nu$(C—C$_6$H$_5$) mode |
| 1240–1265 | Amide III (C—N stretching mode of proteins, indicating mainly $\alpha$-helix conformation) |
| 1313 | CH$_3$CH$_2$ twisting mode of collagen/lipids |
| 1335 | CH$_3$CH$_2$ wagging mode of collagen and polynucleotide chain (DNA-purine bases) |
| 1446 | CH$_2$ bending mode of proteins and lipids |
| 1548 | Tryptophan |
| 1579 | Pyrimidine ring (nucleic acids) and heme protein |
| 1603 | C=C in-plane bending mode of phenylalanine and tyrosine |
| 1616 | C=C stretching mode of tyrosine and tryptophan |
| 1655 | Amide I (C=O stretching mode of proteins, $\alpha$-helix conformation)/C=C lipid stretch |

value for the ratio between intergroup variance and intragroup variance.

The linear discriminant function was calculated for maximal group separation and each individual spectral measurement was projected onto the model (using leave-one-out cross validation) to obtain a score. The scores for each individual spectrum projected onto the model and color coded for consensus pathology are shown in Figure 8.

Esophageal models have been constructed to demonstrate the possibility of biopsy targeting with Raman spectroscopy[22] and full pathological discrimination in the esophagus with nine pathology groups.[16] Each of the models has included all the data measured to cover all

pathology groups found in the organ. For the three-group model shown in Figure 8, the sensitivities and specificities range from 84 to 97% and 93 to 99%, respectively, with an overall correct figure of 93%. For the all-group consensus model[16] in the esophagus, this reduces to 89% correct overall, with a sensitivity of 73–100% and specificity of 92–100%. This demonstrates some misclassifications likely to be owing to the biochemical continuum between the pathology groups and the difficulty of obtaining sufficient specimens with a consensus in groups such as LGD.

Furthermore, a comparison has been made between the three-pathologist consensus on all samples with an independent pathologist's

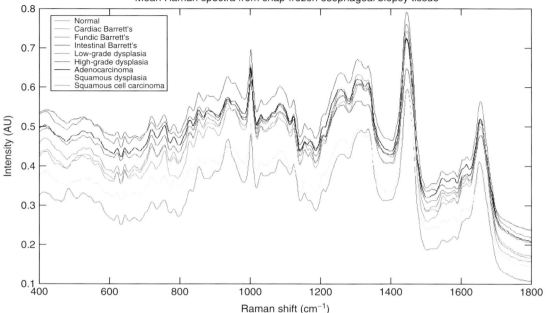

**Figure 7.** Mean Raman spectra from fresh-frozen esophageal biopsy specimens. The spectra have been corrected for spectrometer energy sensitivity.

opinion and the Raman prediction on all samples.[16] This work demonstrated the possibility that Raman spectroscopy could perform discrimination of tissue specimens to a similar level to a well-trained pathologist, both around a $\kappa$ value of 0.8, which is very good. However, the implementation of a device making a decision based on molecular spectra could lead to significant improvements in intercenter reproducibility of results.

One further approach to analyzing clinical spectra for discrimination of pathology has been outlined by Stone *et al.*[58] The linear discriminant functions can use misclassification costs to take into account any relevant clinical requirements. For example, if a diagnosis of cancer based upon a Raman spectral measurement would lead to organ removal, one would want to be certain that there were no false positives. Therefore, one would make high misclassification costs for the noncancer group. However, if Raman spectroscopy was used in its biopsy-targeting

mode, then the operator would not wish to miss any potentially cancerous tissue. Therefore, false negatives would be less acceptable. In this case the misclassification costs for the cancerous group would be raised. This methodology has the effect of moving the decision line from the geometric center or Mahalanobis distance between the group centroids towards one of the centroids, thus making the predictive decision for one group more likely. An example is shown in Figure 9.[60]

## 6.6 Raman spectroscopy for grading of bladder disease

Similar results to those obtained with the esophagus have been achieved with the bladder specimens collected at cystoscopy.[24,58,61] Biopsy-targeting models and full pathological discrimination models have been constructed and tested with leave-one-out cross validation. The three-group biopsy-targeting model comparing normal with

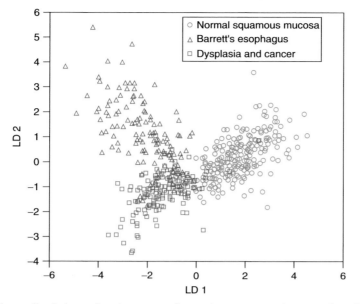

**Figure 8.** Plot of linear discriminant function scores for each spectrum, when tested against the model using leave-one-out cross validation. Normal squamous (310), Barrett's (145), and dysplasia (162) and cancer spectra were included.

cystitis and all cancers and precancers produced a model with sensitivities and specificities in the range 90–95% and 95–98%, respectively. When attempting to distinguish all the pathologies in the bladder (see Table 1), sensitivities of 71–100% and specificities of 94–100% were achieved.

## 6.7 Staging of bladder disease

One interesting further result from the bladder studies has demonstrated some possibility of using the Raman signal measured from the urothelium (or inner surface) of diseased bladder to provide information on how advanced the disease has become, i.e., the stage or level of invasion.[61] Figure 10 shows the cancer groups included in this evaluation. This was only performed as a small pilot study, but the results look very promising with 93% of spectra correctly predicted for bladder cancer stage. This leads to the possibility that, when a lesion is found during investigative cystoscopy, first the presence of benign or cancerous disease can be identified (pathology

grade) and second the depth of invasion (stage) could be determined. It is the stage that the clinical treatment decision will be based upon.

## 6.8 Probe-based tissue measurements

The next step in taking forward this methodology to in vivo clinical work involved the testing of clinically applicable robust in vivo probes. In this case we utilized the Enviva Raman probe (Visionex, Inc.).[62] This study investigated the use of the probes on fresh-frozen excised bladder tissue to evaluate whether the substantial background signal from the probes (even with all the filters in place)[15] would prevent spectral discrimination of disease.

The major problem associated with Raman-fiber probes is that Raman signals are generated by the fibers themselves. The fiber-signal is proportional to the length of the fiber and to the excitation light intensity. It can have magnitudes equal to and often greater than that of the sample under study. Fiber signal is generated in the

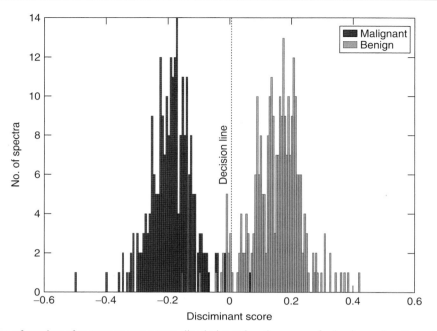

**Figure 9.** Plot of number of measurements versus discriminant function scores for benign and malignant breast tissue (as an example). The decision line was chosen without the use of misclassification costs, i.e., they are equal for both groups.

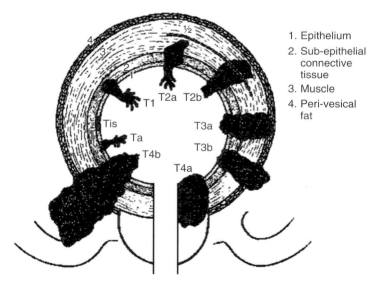

**Figure 10.** A schematic diagram outlining the different stages of invasive cancer in the bladder. Groups pTa, pT1, and pT2 were investigated in this study. The most significant clinical decision would be made between pT1 and pT2. [Reproduced from TNM classification of malignant tumors: Urinary bladder, 6th ed., with permission from John Wiley & Sons, © 2002.]

delivery fiber by the excitation light. In addition, a background signal is also generated in the collection fibers by the elastically scattered excitation light returning along the collection fiber. A practical probe design must prevent unwanted signal generated in the delivery fiber from illuminating the sample as well as preventing light elastically scattered by the tissue from entering the collection fibers and generating unwanted signal. In practice, holographic notch filters are used to reject the Rayleigh scattered light from the tissue and a narrow band-pass filter is used to exclude the silica Raman before the excitation light illuminates the tissue.

The system utilized for this study, performed at the Mayo clinic, Rochester, MN, consisted of a tunable laser diode (Model 8630, SDL, Inc.) emitting at 785 nm, a high-throughput holographic spectrograph (Holospec f/1.8, Kaiser Optical Systems), and a liquid-nitrogen-cooled CCD detector array (Series 2000, Photometrics, Inc.). The 785-nm wavelength was used because it provided a compromise among the Raman signal strength, detector sensitivity, and

reducing fluorescence intensity. The 2-mm diameter, gaser level 10, Raman fiber-optic probes (Enviva Raman probes, Visionex, Inc.) consist of a central delivery fiber (400-$\mu$m diameter), surrounded by seven collection fibers (300-$\mu$m diameter). The probe fibers have a numerical aperture of 0.22, and a tissue sample volume of 1 mm$^3$ is illuminated to a depth of 500 $\mu$m. These reusable probes are suitable for use in open, laparoscopic, and endoscopic procedures.

A total of 220 Raman spectra were recorded from 29 snap-frozen bladder samples collected with cystoscopic procedures. The mean spectra from benign and malignant groups following background subtraction with a fifth-order polynomial are shown in Figure 11. Note the difficulty in observing the fine spectral detail obtained compared with the microscope systems discussed previously. The spectra were correlated with the histological features and used to construct a PCA-fed-LDA diagnostic algorithm. This algorithm was tested for its capability to determine the pathologic finding of a sample from its Raman spectrum.

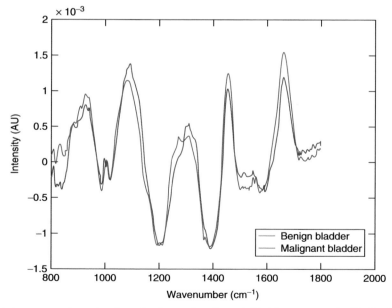

**Figure 11.** Mean preprocessed spectra from benign and malignant bladder, measured with the Emvision Visionex probe with 785-nm excitation.

The diagnostic algorithms for bladder tissues differentiated benign samples (normal and cystitis) from malignant samples (transition cell carcinoma (TCC)) with an overall accuracy of 84%. In this study, the clinical Raman system diagnosed bladder cancer in vitro with good diagnostic accuracy. As the Raman probe is suitable for use during endoscopic, laparoscopic, or open procedures, this study paves the way for further in vivo studies.

## 6.9   Biochemical interpretation of data

Recently several groups, including our own, have attempted to exploit the molecular basis of the tissue Raman spectra.[63-66] The main objective is to attempt to provide a greater understanding of the source of the pathological variation found within the spectral signals.

A number of tissue spectral data sets have been studied by fitting of basis spectra from pure substances. The mean spectra for the different pathologies and the constituents were ascertained allowing for further analysis to occur. The solid constituents were placed on a calcium fluoride slide and the liquid constituents were placed in a well plate. A 20× MIR-Plan ULWD Olympus microscope objective was then used to focus in on the substance and to measure the spectra. The laser was always at full power (ca. 100 mW) with the monochromator slit set to 50 μm (corresponding to a resolution of 10–12 cm$^{-1}$) for all spectra obtained. Different acquisition times were used for each constituent. This was optimized to produce spectra with a high SNR. The basis spectra were then time normalized to replicate the tissue spectra and their signal corrected for the concentration of the standard up to a signal representing a pure substance (i.e., a 95% pure standard would have the spectrum divided by 0.95).

Preprocessing of all the spectra was performed to correct for the energy sensitivity of the spectrometer. This was done using a NPL-calibrated tungsten-filament lamp source, as described above. The first-derivative spectra were then used for all the least squares fit calculations.

The fitting of the mean spectra of the constituents to the mean spectra of the different pathologies in the bladder[66] and the esophagus was performed using "ordinary least squares (OLS)" analysis, explained as follows:

$$X = CS + E \qquad (1)$$

Here, $S$ is the matrix of spectral components, $C$ is the matrix of concentrations to be predicted, $X$ is the measured spectrum, and $E$ is a residual. This can be used to provide a "best fit" of the spectral components or basis spectra found within the measured spectrum. The assumption is made that the residual is minimized and that the spectral components selected are the main components of the spectra.

$$C = \frac{(X - E)}{S} \qquad (2)$$

The disadvantage with this technique is that any colinearity in the components selected will skew the fit. A calculation of the orthogonality of the basis spectral set was made to enable evaluation of the importance of this issue. Observation of the residual, $E$, enabled the quality of the fit to be observed and any remaining features of the spectra to be included in the next iteration of the model. Those constituents with minimal contributions were left out of the next iteration.

An approximation of the error in the concentration prediction (due to noise in the tissue spectra) was made, with the assumption that the basis spectra had minimal noise content as they were measured using optimum conditions. This error was calculated for each constituent prediction in each tissue spectrum.[66] Figure 1 shows basis spectra of some of the components measured for use in the biochemical fitting of the tissue Raman spectra.

Figure 12 shows the calculated relative "concentrations" of biochemical constituents found in the esophagus tissue spectra. Note the specific variations found with pathology. Glycogen levels are high in normal tissue and much reduced as cell division used up the energy store of the cells. The DNA-to-cytoplasm (actin) ratio increases as the

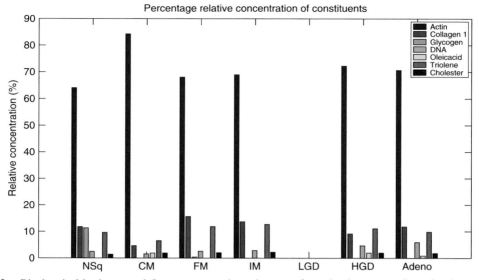

**Figure 12.** Biochemical basis spectral fit to mean esophageal spectra from the data set used previously to construct the LDA models (same pathology groups: normal squamous (NSq), cardiac metaplasia (CM), fundic metaplasia (FM), intestinal metaplasia (IM), high-grade dysplasia (HGD), and adenocarcinoma (Adeno)). Note there are no low-grade dysplasia (LGD) spectra in this data set. The relative concentration levels were normalized to a total of 100%.

grade of pathology increases to malignancy. Both of these biochemical changes are those that would be expected, and the DNA-to-cytoplasmic ratio is one of the markers used by histopathologists to identify rapidly dividing malignant cells.

Using a similar method, the gross biochemical changes with pathology in the bladder were obtained.[66] The results have shown that for pathologies within the bladder, the DNA content increases as the tissue progresses from normal to malignant and the collagen content decreases. This is as expected, as the nuclear-to-actin ratio increases from normal to malignant tissue and DNA is abundant in the nucleus. Also, the cells become more abundant; therefore, the extracellular matrix, which is abundant in collagen, is reduced. However, when the tumor becomes large one expects more collagen. We found that using collagen types I and III gave a better spectral fit than by including collagen type IV, suggesting that types I and III are more abundant in the bladder tissue. Cholesterol was used in the model as a potential component of necrosis. As a tumor becomes bigger, there tend to be

necrotic areas within the tumor where it is too big for its blood supply. Interestingly, there is an increase in cholesterol level not only on moving from benign to malignant disease but also with increased severity of disease. The actin level remains reasonably constant throughout the pathology groups. However, the actin represents the cytoplasm and as the nuclear-to-cytoplasm ratio increases with malignancy, one would expect the actin level to decrease slightly. Other more subtle changes were also observed but not commented upon here.

Note that the resulting concentrations have been calculated with OLS, which can be subject to some errors, if significantly collinear constituent spectra are used or if the data are of poor quality (noisy). Great care was taken to minimize these possible issues by acquiring good quality basis spectra with SNRs greater than 20 and attempting to include only relevant basis spectral components. Furthermore, it is possible with OLS to achieve the result of negative concentrations. In this case, these were minimized by optimizing the output concentrations and the residual plot.

An alternative approach investigated elsewhere is to use non-negative least squares and semi-parametric approaches to calculate the predicted constituent concentrations.[65] These techniques were not deemed necessary for this study. The basis spectra fit well to the tissue spectra with very little in the way of negative concentration prediction. However, ongoing work is being performed to evaluate the effect of the omitted variable bias in overfitting the residual with the components selected.

## 6.10   Mapping of specimens

Mapping of numerous tissue sections has been performed and the spectral data were studied to enable an understanding of the spatial distribution of the constituents of interest.[67] This allows for the possibility of spatially correlating histopathology with tissue constituents and attempting to understand how spatially distinct the biochemical changes are from the histologically defined boundaries. This could significantly enhance the understanding of the carcinogenesis process. Figure 13 shows the spatial distribution of eight tissue constituents across an esophageal tissue section. The section exhibits both normal squamous areas and HGD (effectively carcinoma in situ). Note the raised levels of glycogen in the normal tissue and the higher choline[68] and DNA in the dysplastic tissue. This correlates with the expected pathogenesis and demonstrates a first step in understanding the biochemical distribution of disease measured with Raman spectroscopy.

## 6.11   System intercomparison

A key aspect of Raman spectroscopy for diagnosis will be the capability to transfer models

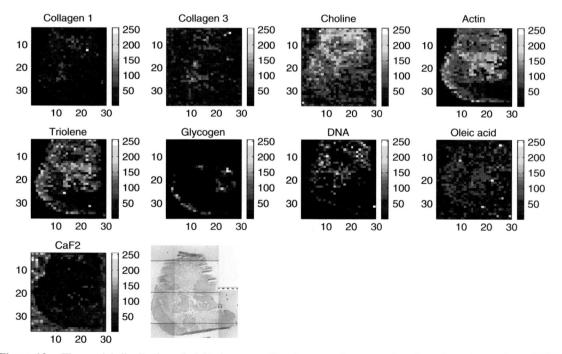

**Figure 13.**   The spatial distribution of eight tissue constituents across the mapped section of esophagus tissue (H&E stained section also shown). The substrate, $CaF_2$, is included as it contributes somewhat to the tissue spectrum measured from 10-μm-thick sections.

trained on one spectrometer to those on another system. In this example, two Renishaw system 1000 microspectrometers were used, both with similar optics and 830-nm laser sources. A set of esophageal tissue biopsy specimens was measured in the following way. The spectra were measured from the same samples with two spectrometer systems using three objectives. An initial comparison of the classification performance of models developed for each configuration was made.[69] Six individual three-group models were developed, with normal squamous, Barrett's, and neoplasia pathology groups. The spectra were calibrated for energy sensitivity according to the system and objective with which they were measured.

The average spectra measured on both systems for each of the objectives are shown in Figure 14. The average energy sensitivity corrected spectra measured with all the objectives are presented for both systems in Figure 15. Encouragingly, there are very few differences in the mean esophageal spectra between one system and the next. This is, however, not the case for different objectives. Each objective obviously samples different volumes of tissue and hence, as the epithelial

tissue is not homogeneous with depth, its spectra vary with sampling depth.

Figures 16 and 17 show the three-group model constructed from all spectra measured with each objective and system. The shapes used in the plots help visualization of the distribution in the models of each configuration. The objectives have subclusters in the discriminant scores plots, whereas the systems have an even distribution in each cloud. This leads to the conclusion that the intersystem transferability can be a reality, provided the same objective (sampling volume) is used and the spectra are corrected for system transfer function.

These results are encouraging for the future development of Raman spectroscopy as a diagnostic tool. In this case, Raman spectroscopy will need to be implemented using multiple systems in multiple centers using a standard training model for the prediction of new data. It has been demonstrated that spectra measured using different systems and measurement parameters can be combined into a successful classification model with the minimum of spectral correction. It is thought that an additional system and objective

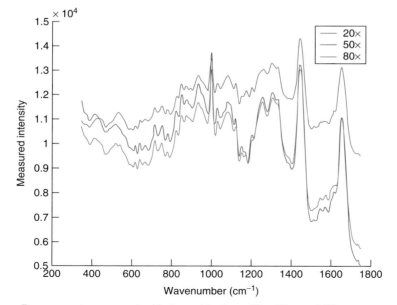

**Figure 14.** Average Raman spectra measured with three objectives, 20×, 50×, and 80×, on two systems.

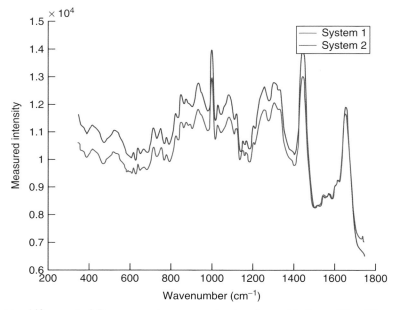

**Figure 15.** Average shift-corrected Raman spectra measured using three objectives, 20×, 50×, and 80×, on two Raman systems.

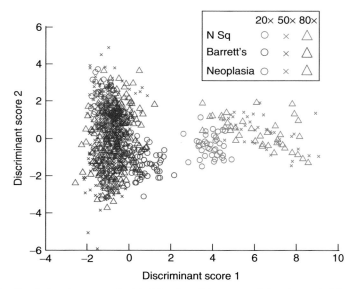

**Figure 16.** Plot of the discriminant scores calculated for the all in one model: color coded for pathology information (green for normal squamous, blue for Barrett's, and red for neoplasia) and shape coded according to the objective used for the measurement on both systems (circles for 20×, crosses for 50×, and triangles for 80× objective).

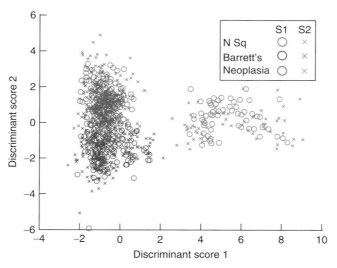

**Figure 17.** Plot of the discriminant scores calculated for the all in one model: color coded for pathology information (green for normal squamous, blue for Barrett's, and red for neoplasia) and shape coded according to the system used for the measurement with all three objectives (circles for system 1 and crosses for system 2).

correction factor could be applied to improve the prediction results.

# 7   DISCUSSIONS

The Raman spectra obtained from epithelial tissues are highly complex in nature, consisting of a superposition of Raman scattering peaks from a myriad of tissue constituents. Furthermore, the spectra from each tissue are inherently similar. Therefore, the utilization of any differences between the spectra from one pathological group to the next for classification of disease requires complex analytical procedures to exploit the subtle variations.

There are many discrepancies in the literature regarding peak assignments. It is likely that most bands will be made up of contributions from many biological molecules co-added; thus, peak shapes and positions may vary. Moreover, a system resolution of around $10-12 \, cm^{-1}$ (for the 300 lines per mm grating and $50 \, \mu m$ slit configuration) will introduce some discrepancies in comparison to spectra measured at higher resolution. Utilization of tentative peak assignments has enabled some

sense to be made of the biochemical changes that accompany the progress toward cancer in the epithelial tissues studied. These include increases in the concentration of nucleic acids, expression of proteins in different conformations, and reductions in the concentration of glycogen and carotenoids. More advanced techniques utilizing basis spectral fits will enable a more accurate measure of gross biochemical variation with tissue pathology. Significant further work is required to validate these studies.

The multivariate spectral predictive models used in this study have been able to discriminate between benign and neoplastic/malignant tissue to a high level of accuracy. It has been possible with the use of misclassification costs to optimally adjust the prediction of the model to provide higher levels of sensitivity to the diseased state. This approach could be very useful in the clinical application of Raman spectroscopy for biopsy targeting. However, the clinical risks of this approach will need to be evaluated. For example, more false positives occurring in biopsy targeting will produce no risk, whereas false negatives will mean abnormal samples will be missed. On the other hand, if Raman spectroscopy were being used for in vivo diagnosis, so that a surgical resection of the esophagus would result if a positive result was obtained, then the risk of false positives would be significant. Therefore, the algorithm architecture will depend on the application. Furthermore, there are alternative multivariate diagnostic algorithms available that utilize all of the information in the spectra. These include cluster analysis and artificial neural networks that may be worth further investigation.

The histopathological grading systems used by pathologists have been developed historically utilizing architectural and morphological differences to recognize by experience when a tissue has undergone a change that is likely to be clinically significant. The boundaries between groups have often been arbitrarily set and therefore in the utilization of a technique, such as Raman spectroscopy, that has the ability to quantify the molecular constituents present, there is likely to be some discrepancies between the techniques. This is because cells are undergoing a continuum of biochemical change from normal to premalignancies to invasive malignancies.

Published studies have indicated high levels of discrepancy between even highly experienced histopathologists in the classification of precancerous lesions. Spectral models for this study were trained with consensus pathology data, removing any samples with disagreement or lack of homogeneity in pathology. These heterogeneous samples were discarded to remove the possibility of introducing errors from accidentally sampling different areas of the specimens.

There is a significant difficulty in using an imperfect "gold standard" for calibration of novel diagnostic techniques. Classification algorithms will only be as good as the training they receive. In addition, considering that histopathology is regarded as giving the correct diagnosis, it will be very difficult to demonstrate that a new technique provides a more accurate representation of the true diagnosis. Improvements in detection and diagnosis of cancers and precancers using new technologies will only be possible if histopathological results are carefully interpreted, with the proviso that the "gold standard" diagnosis

is not necessarily correct. Furthermore, development of biochemical measurements of tissue state such as immunohistochemical markers, including p53 and telomerase, is likely to provide greater evidence for provision of an objective, repeatable diagnosis.

The greatest obstacle to automated medical diagnosis will most likely be the transferability and repeatability of Raman spectra acquired with different systems. If automated diagnostic algorithms are to be used in clinical practice, it will be necessary to provide an off-the-shelf spectral library and an optimized diagnostic algorithm to provide a figure for probability of the sample being associated with a particular disease or tissue type. This implies that spectra measured on all other systems will be identical. This will never be the case. However, variations can be minimized and spectra can be corrected to achieve the ideal that is used for calibration of these classification models. One method of correction for spectrometer variations has been described by Mann and Vickers.[70] Preprocessing of spectra has been shown to reduce the effects caused by spectrometer variations. These include shift correction, energy sensitivity correction, wavenumber calibration, and smoothing filters.

There are some additional technical problems that need to be solved to enable routine use of Raman spectroscopy as an in vivo clinical tool. The use of fiber-optic systems based upon NIR diode laser illumination, filtered probes, and a single spectrometer/CCD detector arrangement appears to be the most promising, although radical new designs may be necessary to eradicate the fiber-optic silica Raman signal. With quick integration times, due to high sensitivity and low tissue fluorescence background signal, the NIR system appears to offer the greatest benefits for use in a clinical environment. Current in vivo probe designs tend to sample from the first few hundred micrometers. However, since NIR photons are highly scattered in tissue, imaging of subsurface lesions may become possible with improved understanding of photon migration in tissue.

## 7.1 Future prospects

Raman spectroscopy can be applicable to many different cancers in the human body, including those found in solid organs. However, each of these will require individual investigation following rigorous protocols to build spectroscopic libraries and predictive algorithms for diagnosis. The greatest impact of Raman spectroscopy will be in the assistance of biopsy targeting during surveillance in organs that develop cancer via precancerous lesions, such as ulcerative colitis in the colon, Barrett's metaplasia in the esophagus, and infection with human papilloma virus in the cervix. In many cancers, the early cellular changes will be invisible and detection will rely upon a random biopsy-targeting procedure. Evidently, there will be a fair chance of missing abnormal lesions and large numbers of biopsy samples will be generated for histopathology departments to process. The use of a noninvasive spectroscopic targeting method to preselect samples will be likely to significantly improve detection rates and reduce the collection of normal tissue samples.

A complementary approach may involve the use of Raman spectroscopy coupled to a confocal microscope for use as a histopathology tool. This may eventually lead to automated sample sorting, which could be achieved with minimal changes to the current sample preparation procedures. Samples could potentially be analyzed prior to sectioning and staining for histopathological analysis to remove the huge number of normal samples produced in these procedures, leaving the histopathologists to deal with the samples of medical importance. Both of these prospects will take time for acceptance by the medical community and large multicenter trials will be required to demonstrate efficacy. Moreover, the most significant problem for histopathology is the high inter- and intraobserver discrepancy between classification of the samples. Raman pathology tool would be likely to assist pathologists in providing a repeatable diagnosis from one site to the next and thus increase the chances of the adoption of screening and surveillance tests, provided intersystem repeatability and extensive work on diagnostic algorithm training have been performed.

In principle, Raman spectroscopic classification of diseases that exhibit a significant biochemical modification should be possible in any tissue from which a spectrum can be measured. The exploitation of this technique will require multidisciplinary teams in many centers to investigate the differences exhibited in diseased and normal samples of all relevant tissues. In practice, it may prove more valuable to use an in vivo Raman probe in combination with a complementary imaging technique, such as ultrasound, endoscopy, magnetic resonance imaging, fluorescence imaging, or optical coherence tomography. This would facilitate the survey of larger areas with less specific techniques and then allow application of a more targeted approach in abnormal areas.

## ACKNOWLEDGMENTS

We thank Paul Crow, Consuelo Hart Prieto, and Geeta Shetty (Surgical Research Fellows) for their contribution to this chapter and Joanne Hutchings and Martin Isabelle (PhD students) for useful discussions in advancing this work. Significant contributions have been made to this work by the following senior pathologists: Neil Shepherd, Karel Goboes, Bryan Warren, Jeremy Uff, Keith McCarthy and Jay Christie-Brown, who have been involved in the classification of samples used and in discussing and defining the protocols for this work. The histopathology technicians have worked extremely hard on additional sample preparation.

## ABBREVIATIONS AND ACRONYMS

| | |
|---|---|
| CB | Cardiac Barrett's |
| CCD | Charge Coupled Device |
| CM | Cardiac Metaplasia |
| FB | Fundic Barrett's |
| FM | Fundic Metaplasia |

| | |
|---|---|
| H&E | Hematoxylin and Eosin |
| HGD | High-grade Dysplasia |
| IM | Intestinal Metaplasia |
| LDA | Linear Discriminant Analysis |
| LGD | Low-grade Dysplasia |
| MIR | Mid-Infrared |
| NIR | Near-Infrared |
| NPL | National Physical Laboratory |
| NSq | Normal Squamous |
| OLS | Ordinary Least Squares |
| PCA | Principal Components Analysis |
| SCC | Squamous Cell Carcinoma |
| SNR | Signal-to-Noise Ratio |
| TCC | Transition Cell Carcinoma |
| TNM | Tumor-Nodes-Metastases |
| ULWD | Ultra-Long-Working-Distance |
| UTI | Urinary Tract Infection |

# REFERENCES

1. A.G. Bohorfoush, *Endoscopy*, **28**, 372–380 (1996).

2. M. Andrea, O. Dias and A. Santos, *Ann. Oto. Rhinol. Laryn.*, **104** (5), 333–339 (1995).

3. R.R. Alfano, G.C. Tang, A. Pradhan, W. Lam, D.S.J. Choy and E. Opher, *IEEE J. Quantum Electron.*, **23**, 1806–1811 (1987).

4. S. Andersson-Engels, J. Johansson, K. Svanberg and S. Svanberg, *Photochem. Photobiol.*, **53**, 807–814 (1991).

5. M. Panjehpour, R. Overholt, T. Vo-Dinh, R.C. Haggit, D.H. Edwards and P.F. Buckley, *Gastroenterology*, **111**, 93–101 (1996).

6. D.C.B. Redd, Z.C. Feng, K.T. Yue and T.S. Gansler, *Appl. Spectrosc.*, **47** (6), 787–791 (1993).

7. C.H. Liu, B.B. Das, W.L.S. Glassman, G.C. Tang, K.M. Yoo, H.R. Zhu, D.L. Akins, S.S. Lubicz, J. Cleary, R. Prudente, E. Celmere, A. Caron and R.R. Alfano, *J. Photochem. Photobiol. B, Biol.*, **16** (2), 187–209 (1992).

8. A. Mahadevan-Jansen and R. Richards-Kortum, *J. Biomed. Opt.*, **1**, 31–70 (1996).

9. A. Mahadevan-Jansen, M.F. Mitchell, N. Ramanujam, A. Malpica, S. Thomsen, U. Utzinger and R. Richards-Kortum, *Photochem. Photobiol.*, **68**, 123–132 (1998).

10. T.C. Bakker Schut, H. Van Dekken, H.W. Tilanus, H.A. Bruining and G.J. Puppels, in 'Spectroscopy of Biological Molecules: Modern Trends', eds P. Carmona, R. Navarro and A. Hernanz, Kluwer Academic Publishers, 455–456 (1997).

11. N. Stone, P. Stavroulaki, C. Kendall, M. Birchall and H. Barr, *Laryngoscope*, **110**, 1756–1763 (2000).

12. H. Barr, T. Dix and N. Stone, *Laser Med. Sci.*, **13**, 3–13 (1998).

13. K.P.J. Williams, G.D. Pitt, D.N. Batchelder and B.J. Kip, *Appl. Spectrosc.*, **48** (2), 232–235 (1994).

14. M.S. Feld, R. Manoharan, J. Salenius, J. Orenstein-Carndona, T.J. Romer, J.F. Brennan, R. Dasari and Y. Wang, *Adv. Fluoresc. Sensing Technol. SPIE*, **2388**, 99–104 (1995).

15. M. Shim, L.M. Song, N.E. Marcon and B.C. Wilson, *J. Photochem. Photobiol., A*, **72**, 146–150 (2000).

16. C. Kendall, N. Stone, N. Shepherd, K. Geboes, B. Warren, R. Bennett and H. Barr, *J. Pathol.*, **200** (5), 602–609 (2003).

17. N. Stone, 'Raman Spectroscopy of Biological Tissue for Application in Optical Diagnosis of Malignancy', PhD thesis, Cranfield University, (2001).

18. T.C. Bakker Schut, M.J.H. Witjes, H.J.C.M. Sterenborg, O.C. Speelman, J.L.N. Roodenburg, E.T. Marple, H.A. Bruining and G.J. Puppels, *Anal Chem.*, **72**, 6010–6018 (2000).

19. B. Schrader, S. Keller, T. Lochte, S. Fendel, D.S. Moore, A. Simon and I. Sawatzki, *J. Mol. Struct.*, **348**, 293–296 (1995).

20. H. Edwards, A. Williams and B. Barry, *J. Mol. Struct.*, **347**, 379–387 (1995).

21. G.W. Lucassen, R. Wolthuis, H.A. Bruining and G.J. Puppels, *Biospectroscopy*, **4**, S31–S39 (1998).

22. N. Stone, C. Kendall, N. Shepherd, P. Crow and H. Barr, *J. Raman Spectrosc.*, **33** (7), 564–573 (2002).

23. N. Stone, C. Kendall, N. Chandratreya, N. Shepherd and H. Barr, in 'Biomedical Spectroscopy: Vibrational Spectroscopy and Other Novel Techniques II', Vol. **SPIE 4614–21**, (2002).

24. P. Crow, C. Kendall, M. Wright, R. Persad, A. Ritchie and N. Stone, *Br. J. Urol.*, **90** (1), 71 (2002).

25. P. Crow, N. Stone, C. Kendall, J. Uff, J.A.M. Farmer, H. Barr and M.J.P. Wright, *Br. J. Cancer*, **89** (1), 106–109 (2003).

26. J. Smith, C. Kendall, A. Sammon, J. Christie-Brown and N. Stone, *Technol. Cancer Res. Treat.*, **2**, 4 (2003).

27. Cancer Research UK Incidence and Mortality Statistics, (2002) http://www.cancerresearchuk.org/aboutcancer/statistics/.

28. R. Virchow, 'Cellular Pathology as Based on Histology', (1850).

29. J.W. Van Sandick, J.J. van Lanshot, B.W. Kuiken, G.N. Tytgat, G.J. Offenhaus and H. Obertop, *Gut*, **43**, 216–222 (1998).

30. E. Montgomery, M.P. Bronner, J.R. Goldblum, J.K. Greenson, M.M. Haber, J. Hart, L.W. Lamps, G.Y. Lauwers, A.J. Lazenby, D.N. Lewin, M.E. Robert, A.Y. Toledanao and K. Washington, *Hum. Pathol.*, **32**, 268–378 (2001).

31. J.A. Jankowski, R.F. Harrison, I. Perry, F. Balkwill and C. Tselepis, *Lancet*, **356**, 2079–2085 (2000).

32. C. Winters, Jr, T.J. Spurling, S.J. Chobanian, D.J. Curtis, R.L. Eposito, J.F. Hacker III, D.A. Johnson, D.F. Cruess, J.D. Coteligam and M.S. Gurney, *Gastroenterology*, **92** (1), 118–124 (1987).

33. N.S. Mann, M.F. Tsai and P.K. Nair, *Am. J. Gastroenterol.*, **84**, 1494–1496 (1989).

34. A.P. Naef, M. Savary and L. Ozzelo, *J. Thorac. Cardiovasc. Surg.*, **70**, 826–835 (1975).

35. M.M. Berenson, R.H. Riddell, D.B. Skinner and J.W. Freston, *Cancer*, **41**, 554–561 (1978).

36. S.J. Spechler and R.K. Goyal, *N. Engl. J. Med.*, **315**, 362 (1986).

37. R.H. Riddell, 'Dysplasia and Regression in Barrett's Epithelium', in "Barrett's Esophagus: Pathophysiology, Diagnosis, and Management", eds S.J. Spechler and R.K. Goyal, Elsevier Science, New York, 188–197 (1985).

38. A.J. Cameron, *Gastroenterol. Clin. North Am.*, **26** (3), 487–494 (1997).

39. G.E.J. Gelfand, R.J. Finley, B. Nelems, R. Inculet, K.G. Evans and G. Fradet, *Arch. Surg.*, **127**, 1164–1167 (1992).

40. J. Ramamim and C.W. Cham, *Br. J. Surg.*, **80**, 1305 (1993).

41. P.J. Hesketh, R.W. Clapp, W.G. Doos and S.J. Spechler, *Cancer*, **64**, 526 (1989).

42. W.J. Blot, S.S. Devesa and J.F. Fraumeni, *JAMA*, **270**, 1320 (1993).

43. R.W. Sjögren and L.F. Johnson, *Am. J. Med.*, **74**, 3131 (1983).

44. R.C. Haggitt and P.J. Dean, 'Adenocarcinoma in Barrett's Epithelium', in "Barrett's Esophagus: Pathophysiology, Diagnosis, and Management", eds S.J. Spechler and R.K. Goyal, Elsevier Science, New York, 153–166 (1985).

45. H. Sanfey, S.R. Hamilton, R.R.L. Smith and J.L. Cameron, *Surg. Gynecol. Obstet.*, **161**, 570 (1985).

46. B.J. Reid and C.E. Rubin, *Gastroenterology*, **88**, 1152 (1985).

47. R.R.L. Smith, J.K. Boitnott, S.R. Hamilton and E.L. Rogers, *Am. J. Surg. Pathol.*, **8**, 563–573 (1984).

48. M. Miros, P. Kerlin and N. Walker, *Gut*, **32**, 1441–1446 (1991).

49. T.R. DeMeester, S.E.A. Atwood, T.C. Smyrk, D.H. Therkildsen and R.A. Hinder, *Ann. Surg.*, **212**, 528–542 (1990).

50. N.K. Altorki, M. Sunagawa, A.G. Little and D.B. Skinner, *Am. J. Surg.*, **161**, 97–100 (1991).

51. Office for National Statistics, 'Cancer Survival 1991–1998', Office for National Statistics, (31st March 2000) web page: http://www.ons.gov.uk.

52. B.J. Reid, R.C. Haggitt, C.E. Rubin, G. Roth, C.M. Surawicz, G. Van Belle, K. Lewin, W.M. Weinstein, D.A. Antonioli, H. Goldman, W. MacDonald and D. Owen, *Hum. Pathol.*, **19** (2), 166–178 (1988).

53. B.J. Reid, P.L. Blount, C.E. Rubin, D.S. Levine, R.C. Haggitt and P.S. Rabinovich, *Gastroenterology*, **102**, 1212–1219 (1992).

54. Cancer Research UK, 'Cancerstats Monograph', (2004).

55. J.P. Meyer and D. Gillat, *Trends Urol. Gynaecol. Sexual Health*, **8** (6), 25–29 (2003).

56. C.S. Stewart, 'Novel Detection Strategies for Transitional Cell Carcinoma', in "Renal, Bladder, Prostate and Testicular Cancer: An Update: The Proceedings of the VIth Congress and Controversies in Oncological Urology", eds K.H. Kirth, G.H. Mickisch and F.H. Schroder, eds. Parthenon Publishing Group, New York, (2001).

57. R.J. Schlemper, R.H. Riddell, Y. Kato, F. Borchard, H.S. Cooper, S.M. Dawsey, M.F. Dixon, C.M. Fenoglio Preiser, J.-F. Flejou, K. Geboes, T. Hattori, T. Hirota, M. Itabashi, M. Iwafuchi, A. Iwashita, Y.I. Kim, T. Kirchner, M. Klimpfinger, M. Koike, G.Y. Lauwers, K.J. Lewin,

G. Oberhuber, F. Offner, A.B. Price, C.A. Rubio, M. Shimizu, T. Shimoda, P. Sipponen, E. Solcia, M. Stolte, H. Watanabe and H. Yamabe, *Gut*, **47**, 251–255 (2000).

58. N. Stone, C. Kendall, J. Smith, P. Crow and H. Barr, *Faraday Discuss: Appl Spectrosc Biomed Probl*, **126**, 141–157 (2004).

59. S. Sharma, 'Applied Multivariate Techniques', John Wiley & Sons, (1996).

60. R.A. Fisher, *Ann. Eugen.*, **7**, 179–188 (1936).

61. P. Crow, J.S. Uff, J.A. Farmer, M.P. Wright and N. Stone, *BJU Int.*, **93**, 1232–1236 (2004).

62. P. Crow, A. Molckovsky, N. Stone, J. Uff, B. Wilson and L.-M. Wong Kee Song, *Urology*, **65** (6), 1126–1130 (2005).

63. K.W. Short, S. Carpenter, J.P. Freyer and J.R. Mourant, *Biophys. J.*, **88** (6), 4274–4288 (2005).

64. K.E. Shafer-Peltier, A.S. Haka, M. Fitzmaurice, J. Crowe, J. Myles, R.R. Dasari and M.S. Feld, *J. Raman Spectrosc.*, **33**, 552–563 (2002).

65. M.G. Sowa, M.S.D. Smith, C. Kendall, E.R. Bock, A.C.-T. Ko, L.-P. Choo-Smith and N. Stone, *Appl. Spectrosc.*, **60** (8), 877–883 (2006).

66. N. Stone, M.C. Hart Prieto, P. Crow, J. Uff and A.W. Ritchie, *Anal. Bioanal. Chem.*, **387** (5), 1657–1668 (2007).

67. G. Shetty, C. Kendall, N. Shepherd, N. Stone and H. Barr, *British Journal of Cancer*, **94**, 1460–1464 (2006).

68. J.C. Vilanova and J. Barcelo, *Abdom. Imaging*, **32** (2), 253–261 (2007).

69. C.A. Kendall, A study of Raman spectroscopy for the early detection and classification of malignancy in oesophageal tissue, PhD thesis, Cranfield University, (2002).

70. C.K. Mann and T.J. Vickers, *Appl. Spectrosc.*, **53** (7), 856–861 (1999).

# Neuro-oncological Applications of Infrared and Raman Spectroscopy

## Christoph Krafft and Reiner Salzer
*Dresden University of Technology, Dresden, Germany*

## 1 INTRODUCTION

There are at least a thousand diseases that can affect the nervous system, and approximately one in three people in the United States will be affected by one of them at some point in life. The seriousness of these diseases has led to a large emphasis on research into their causes, diagnoses, therapies, and prevention. Although it is likely that we will never completely understand the complexity of the brain, combining traditionally separate techniques will bring about completely new methods for studying and diagnosing problems and will allow researchers and clinicians to choose the best tools for the task at hand. Biophotonic technology has significantly contributed to the progress in brain research. Biophotonics—composed of the Greek phrases "bios" for life and "phos" for light—is the general term for the scientific discipline that applies light-based techniques to problems in medicine and life sciences. In our opinion, the role of vibrational spectroscopy has been underestimated so far in neuroscience. This chapter introduces the terminology relevant to neuro-oncology, compares biophotonic methods for neuro-oncological applications, and describes how Raman and infrared (IR) spectroscopy can contribute to a better diagnosis of the most frequent primary and secondary brain tumors. First results are presented for gliomas, brain metastases, and an animal model. However, the vibrational spectroscopic applications are not restricted to this field, but exemplify only their potential in neuroscience. They have already been successfully transferred to Alzheimer's,[1] Parkinson's,[2] and variants of Creutzfeld–Jacob's diseases,[3] which are not within the scope of this chapter.

### 1.1 Primary brain tumors

Tumors that begin in brain tissue are known as primary brain tumors. They are classified by the cell types from which they originate, which include glial cells, neurons, pinealocytes, the meninges, choroid plexus, pericytes of small blood vessels, and cells of the pituitary gland. Treatment of cancer in the brain is one of the most challenging problems in oncology as well as neurosurgery. Although primary brain tumors account for only 2% of all adult cancers,

*Vibrational Spectroscopy for Medical Diagnosis.* Edited by Max Diem, Peter R. Griffiths and John M. Chalmers.
© 2008 John Wiley & Sons, Ltd. ISBN 978-0-470-01214-7.

gliomas, which account for more than 50% of them, are the second leading cause of cancer-related death in the age range of 15–44 years.[4] Gliomas originate from glial cells or precursor cells and form a heterogeneous group encompassing many different histological types and malignancy grades.[5] So far, the histopathological diagnosis is the gold standard not only for subtyping gliomas into astrocytomas, oligodendrogliomas, mixed gliomas, and ependymomas but also for assessment of malignancy grades within the different glioma categories. For such grading, a combination of histological features like nuclear pleomorphism, mitotic activity, endothelial proliferation, and necrosis is used. Astrocytic gliomas, which are by far the most common in adults, are classified by increasing malignancy into grade II (diffuse astrocytoma), grade III (anaplastic astrocytoma), and grade IV (glioblastoma multiforme (GBM)). GBM either progresses from diffuse astrocytomas (secondary GBM) or arises de novo (primary GBM). The grade and type of the tumor are important to determine its effective management and provide an indication of the prognosis. In spite of all the advances in understanding the pathomechanism, diagnosis by tomographic imaging, and availability of new therapeutic tools, the medial survival rate of patients with GBM of 12 months has been prolonged only slightly and cure remains elusive. The typical median survival is more than 5 years for grade II gliomas and 2 to 5 years for grade III gliomas. None of the currently available surgical tools, including operative microscopy, lasers, and image-guided surgery, enables detection and removal of all tumor tissue. There are several success stories in treatment of other cancers, but similar success has not been achieved for gliomas. Characteristics of GBM that hinder efforts to cure it include the following: (i) access of chemotherapeutic agents to the tumor is more restricted than to other organs due to the blood–brain and blood–tumor barriers; (ii) aggressive local spread and diffuse infiltrating growth of the tumor; (iii) location of the tumor within the brain obviates radical extirpation; and (iv) the residual tumor grows rapidly and resists curative

treatment. Management of gliomas encompasses surgery to establish the histological diagnosis and to reduce the tumor burden and the mass effect, with the maintenance of the patient's neurological function in mind. Cure of GBM is unlikely unless the tumor is eradicated completely, which also involves other approaches such as chemotherapy, radiation therapy, or vector-mediated gene therapy as adjuncts to surgery.

Meningeomas are the most common tumors of the meninges. They are the second most frequent intracranial tumors after gliomas and account for approximately 20% of all primary brain tumors. There are many different subtypes of meningeomas, most of which are histologically benign: they exhibit slow growth pattern, and usually do not infiltrate brain tissue or cause metastases. Although predominantly benign, meningeomas can recur. Fifteen-year recurrence rates of over 30% have been reported after complete tumor resection.[6] Schwannomas are the third most frequent intracranial tumors that are derived from Schwann cells. Most of them are acoustic neurinomas with low malignancy of grade I.

## 1.2 Secondary brain tumors

Metastasis is the spread of cancer. Cancer that begins in other parts of the body may spread to the brain and cause secondary brain tumors, also called *brain metastases*. Brain metastases occur in about 15–40% of all cancer patients.[7] This proportion has increased in recent years partly due to the increased sensitivity of tomographic techniques to detect these tumors, and also to the longer survival of these patients. Secondary brain tumors are estimated to become symptomatic in about 80 000 patients per year in the United States compared to 18 000 with primary brain tumors. In almost 50% of patients undergoing surgery for brain metastases, the secondary brain tumor represents the first indication of cancer. The clinical presentation of brain symptoms may be due to the fact that brain metastases cause pronounced edema in tissue surrounding the metastasis. Thus, they may present symptoms at a much earlier stage than the primary or other secondary tumors. In lung cancer, for example, which is the most frequent primary site (51%) causing brain metastasis, other manifestations such as thoracic symptoms occur rather late.

Other frequent primary sites of brain metastases include breast cancer, colorectal cancer, renal cell carcinoma, and malignant melanoma. The diagnosis of brain metastasis in a patient with a known systemic cancer is a simple matter. However, in a significant number of brain metastases patients, the primary tumor is not known at the time of neurosurgery. Therefore, beside tumor resection, the scope of neurosurgical procedures is to collect tissue specimens for diagnosis. The majority of brain metastases are poorly differentiated adenocarcinomas of glandular epithelium, which show even less differentiation in brain metastases. Since epithelial tissue covers the surface of organs and is exposed to a broad range of aggressive chemical and physical conditions, these carcinomas are among the most common forms of cancer including breast, lung, colon, prostate, stomach, pancreas, and cervix. Unfortunately, metastatic adenocarcinomas from different primary sites have similar microscopic appearance, impeding histopathological diagnosis, and almost invariably demanding the application of additional methods. Immunohistochemistry, which detects cell and tissue specific antigens by antibody reactions, can complement histopathology. Although antibodies are available to classify the cell type and—in some cases—even the organ of the primary tumor, their use is limited. Consequently, in up to 15% of the patients with brain metastases, the primary tumor is not identified despite thorough investigation by standard screening techniques. Thus, brain metastases with unknown primary tumors are estimated to be almost equal in numbers to primary brain tumors.

## 2 BIOPHOTONIC METHODS IN NEURO-ONCOLOGY

Biophotonic methods can provide information regarding intrinsic tissue optical properties and the presence or absence of endogenous or exogenous chromophores. Optical detection offers several advantages such as diffraction-limited spatial resolution in the micrometer, and even submicrometer range, nondestructive sampling and fast data collection, and relatively inexpensive instrumentation. These advantages have been utilized in numerous neuro-oncological studies using absorption spectroscopy, fluorescence spectroscopy, reflection spectroscopy, and bioluminescence. Raman and mid-infrared (mid-IR) spectroscopy offer another advantage due to their capacity to gather detailed molecular information without the need for labeling. A general disadvantage of all optical techniques is their limited penetration depth in brain tissue, which is dependent on its absorption and scattering properties. High absorption exists in both the visible wavelength range (350–700 nm) mainly due to hemoglobin and at wavelengths greater than 900 nm due to lipids, proteins, and water. However, in the range 700–900 nm, the absorption of all biomolecules is weak; thus imaging using radiation in this interval maximizes tissue penetration while simultaneously minimizing autofluorescence from nontarget tissue. In addition, subcutaneous implanted brain tumors in small animals offer the possibility of noninvasive and in vivo optical imaging, because the required path length of light is much shorter. Besides this special case, clinical applications require the use of microscopes or hand-held probes during open surgery or coupling to fiber-optic probes for minimal-invasive procedures. Optical modalities are emerging as complements to noninvasive imaging methods such as magnetic resonance imaging (MRI), computed tomography (CT), or positron emission tomography (PET). First, optical methods offer prospects in intraoperative situations when tomographic methods cannot be used. Second, they enable noninvasive imaging of diseases and provide the opportunity for rapid and cost-effective activity studies to screen newly developed compounds, before the more costly radionuclide-based imaging studies.

## 2.1 Absorption spectroscopy

Light absorption can excite electronic transitions in molecules that yield quantitative chemical information about the chromophores and their concentration. Medical applications in neuroscience have been developed for the real-time determination of the blood volume, total hemoglobin content, and oxygenation state that are based on the absorption profile of oxygenated hemoglobin in the range 880–920 nm and of deoxygenated hemoglobin in the range 740–780 nm.[8] The sensitivity and specificity of near-infrared (NIR) spectroscopy for hemoglobin were used to image functional brain stimulation and activation through measurement of changes in hemoglobin concentrations. A study of 12 patients was reported[9] on how noninvasive NIR spectroscopy can complement functional MRI to monitor glioma patients during recovery after neurosurgery.

Growth and expansion of gliomas are highly dependent on vascular neogenesis. An association of microvascular density and tumor energy metabolism is assumed in most human gliomas. Intraoperative NIR spectroscopy was applied to 13 glioma patients in order to elucidate the relationship between microvascular blood volume, oxygen saturation, histology, and patient survival.[10] High intratumoral blood volume and high oxygen saturation in GBM correlated with a low median survival time of 10 months. Limitations of oxygen saturation data acquired using NIR technology include the following: (i) only tumors with appearance on the pial surface are accessible for noninvasive NIR spectroscopy probes and (ii) total blood volume of the whole tumor can usually not be assessed since intraoperative NIR measurements are performed on the tumor surface to a depth of only 4 mm.

## 2.2 Fluorescence spectroscopy

Excited electronic states in molecules can return to their ground state upon emission of red-shifted photons. Fluorescence spectroscopy of tissues and cells can be divided into autofluorescence,

which probes information from endogenous fluorophores, and induced fluorescence, where fluorescent probes are reacted with the moiety of interest. Fluorescence-based techniques are the dominant biophotonic methods in neuro-oncology.

### 2.2.1 Autofluorescence

Autofluorescence with 337 nm excitation can discriminate solid tumors from normal brain tissues based on their reduced fluorescence emission at 460 nm compared with normal tissue. False positive rates constitute a general problem because not all regions showing reduced fluorescence are tumors.[11] A hand-held optical spectroscopic probe was applied in clinical trials that combined autofluorescence with diffuse reflection spectroscopy.[12,13] Spectra of brain tumors and infiltrating tumor margins were separated from spectra of normal brain tissues in vivo using empirical algorithms with high sensitivity and specificity. Blood contamination was found to be a major obstacle.

Multiphoton-excited autofluorescence using NIR femtosecond laser pulses allows reconstruction of three-dimensional microanatomical images of native tissues at a subcellular level without the need for contrast enhancing markers or histological stains. Multiphoton autofluorescence microscopy could visualize solid tumor, the tumor-brain interface, and single invasive tumor cells in gliomas of mouse brains and of human biopsies. Furthermore, fluorescence lifetime imaging of endogenous fluorophores provides an additional parameter in so-called 4D microscopy, since the different decay times of the fluorescent signal could differentiate tumor and normal brain tissue.[14]

### 2.2.2 Induced fluorescence

In order to improve selectivity, fluorescent markers have been applied to malignant brain tissue. These markers include indocyanine green,[15] fluorescein,[16] and fluorescein-albumin.[17] Fluorescent porphyrins accumulate in malignant brain tissue after administration of the metabolic precursor 5-aminolevulinic acid (ALA).[18] These fluorophores are taken up by gliomas where breakdown of the blood–brain barrier has occurred, but not in normal brain tissue. The technique can be integrated in an operating microscope and the fluorescence images permitted intraoperative detection of the brain tumors and a safer tumor resection. Results from a human clinical trial showed that ALA-derived fluorescence with peaks at 635 and 704 nm could predict the presence of tumor tissue with a specificity of 100% and a sensitivity of 85% with respect to histopathology. A multicenter clinical phase III trial of 322 patients recently revealed that this approach led to improved progression-free survival of patients with malignant gliomas.[19] However, identification of low-grade tumors and tumor margins could be problematic if the fluorophor was not taken up by tumor cells in areas with an intact blood–brain barrier. Moreover, the effectiveness of induced fluorescence is limited by photobleaching of the fluorophor. These problems have diminished the capability of induced fluorescence for brain tumor demarcation so far.

The NIR fluorescent dye Cy5.5 has an absorption maximum at 675 nm and an emission maximum at 694 nm and can be detected in tissue at subnanomole quantities and at sufficient depth for optical imaging studies. The human GBM cell line U87MG expresses $\alpha_v\beta_3$ integrin, which seems to play a critical role in regulating tumor growth and metastasis as well as tumor angiogenesis. The specific interaction of the cyclic peptide arginine–glycine–aspartic acid (RGD) to $\alpha_v\beta_3$ integrin was combined with the near-infrared fluorescence (NIRF) imaging detection of Cy5.5 to image the integrin expression and monitor anti-integrin treatment efficacy in subcutaneous glioma xenografts.[20,21] Another study introduced a NIRF deoxyglucose analogue for optical imaging of brain tumors based on Cy5.5.[22] The related compound 2-deoxy-2-[18F]fluoro-D-glucose ([18F]FDG) has extensively been used as a tracer for cancer studies by PET.

Quantum dots (QDs) as fluorescent biolabeling reagents might represent a new generation of fluorophores. QDs have unique optical and electronic properties such as size- and composition-tunable fluorescence emission from visible to NIR wavelengths, a large absorption coefficient across a wide spectral range, and very high level of brightness and photostability.[23] RGD peptide–labeled QDs were recently used for in vivo targeting and imaging of human GBM in mice induced by the cell line U87MG.[24] Preliminary work in a rat glioma model showed that intravenously injected QDs colocalize with brain tumors and can be detected by fluorescence imaging.[25]

### 2.2.3 Multifunctional nanoparticles

Very intriguing for potential clinical applications is the design of multifunctional nanoparticles that can be detected both by MRI and fluorescence imaging, allowing for the noninvasive preoperative assessment of the tumor and for the intraoperative visualization of tumor margins by optical imaging.[26,27] One of the main challenges in targeting brain tumors is the blood–brain barrier formed by intercellular tight junctions of the endothelial cells of the brain capillaries, which limits the access of drugs or targeting therapeutics to the brain tissue. It has been demonstrated that the blood–brain barrier may be overcome by particles with a size dimension smaller than approximately 50 nm or by lipid-mediated transport or receptor-mediated and polyethylene glycol (PEG)–assisted processes. Thus, PEG-linked nanoprobes may promise the penetration of the nanoprobe across the blood–brain barrier.[28]

## 2.3 Bioluminescence

Bioluminescence imaging (BLI) has emerged as a sensitive imaging technique for small animals. BLI exploits the emission of photons based on energy-dependent reactions catalyzed by luciferases with a maximum depth of 2–3 cm. Luciferases comprise a family of photoproteins that emit photons in the presence of oxygen and adenosine triphosphate (ATP) during metabolism of substrates such as luciferin into oxyluciferin. Luciferins are injected immediately before data acquisition. The light from these enzyme reactions typically has very broad emission spectra that frequently extend beyond 600 nm, with the red components of the emission spectra being the most useful for imaging by virtue of easy transmission through tissues.[29] In the context of neuro-oncology, BLI was used to monitor the formation of grafted tumors in vivo, to measure cell number during tumor progression and response to therapy,[30,31] and to monitor the proliferative activity of glioma cells and cell cycle in a genetically engineered mouse model of glioma in vivo.[32] The ability to image two or more biological processes in a single animal can greatly increase the utility of luciferase imaging by offering the opportunity to distinguish the expression of two reporters biochemically. Dual BLI was used to monitor gene delivery via a therapeutic vector and to follow the effects of the therapeutic protein TRAIL (tumor necrosis factor-related apoptosis-inducing ligand) in gliomas.[33,34] TRAIL has been shown to induce apoptosis in neoplastic cells and may offer new prospects for tumor treatment.

## 2.4 Reflection spectroscopy

In reflection spectroscopy, tissue is typically illuminated with a broadband light source. Analyzed photons are directly reflected or elastically scattered many times before the light returns to the collection probe. These principles were utilized in approaches which are called *optical tomography* and which are subdivided into optical diffraction tomography, diffuse optical tomography, and optical coherence tomography (OCT). OCT is analogous to ultrasound imaging except that reflected NIR radiation, rather than sound, is detected. OCT synthesizes cross-sectional images from a series of laterally adjacent depth scans using radiation with low coherence interference. OCT enables noninvasive real-time in vivo imaging of brain tissues up to 2 mm penetration depth with a spatial resolution of 10–15 μm and in ultra-high resolution (UHR) mode even better (0.9-μm axial, 2-μm lateral). Both time

domain (TD) and spectral-domain (SD) OCT of experimental gliomas in mice and human brain tumor specimens delineated normal brain, infiltration zone, and solid tumor based on the tissue microstructure and signal characteristics.[35] UHR-OCT discriminated between healthy and pathological human brain tissue biopsies by visualizing and identifying microcalcifications, enlarged cell nuclei, small cysts, and blood vessels.[36] OCT-based noncontact imaging of brain tissue during neurosurgical procedures is challenging by the fact that after opening of the dura the target volume follows the respiratory and arterial cycle, resulting in movements of several millimeters in amplitude. Slow scan times result in distortion of the tissue surface contour or may result in the area of interest moving out of the measurement window. Therefore, short scan acquisition times are crucial. In particular, SD-OCT allowed rapid scanning times of three images per second, which would be sufficient to suppress motion artifacts of the relatively slow movements of the brain exposed during operations. Reflection spectroscopy has its strength in clarifying the architectural tissue morphology. However, for many diseases, including cancer, diagnosis based on such architectural features is not reliable since the most important diagnostic indicators of neoplastic changes are features like accelerated rate of growth, mass growth, local invasion, lack of differentiation, anaplasia, and metastasis, which mainly occur on the subcellular and molecular level. Spectroscopic OCT imaging might provide additional chemical information and would further enhance the OCT potential.[37]

## 2.5 Raman spectroscopy

More bands are observed in vibrational spectra of cells and tissues than in other optical spectra because numerous vibrations of biomolecules can be excited simultaneously without using labels, giving a fingerprint-like signature. Therefore, more potential information about the biochemistry and composition of the underlying sample may be collected. Molecular vibrations excited via inelastic scattering of light give rise to Raman spectra. Problems arise from the inherent weak Raman intensities of biomolecules and from tissue autofluorescence, which often masks the Raman signals. Therefore, Raman spectroscopy of tissue is usually performed with NIR laser excitation, which minimizes tissue autofluorescence and allows a better penetration for incident and scattered light. Further requirements include high-throughput optics and sensitivity-optimized detectors, which became available during the past decade and enabled various biomedical applications.[38,39] In the authors' current system, light from a NIR diode laser at 785-nm wavelength with intensity of approximately 100 mW is coupled to a Raman microscope (HoloLab Series 5000, Kaiser Optical Systems, USA) or to a fiber-optic probe (Inphotonics, USA). The Raman scattered light from the samples is guided to a spectrograph (Holospec f/1.8, Kaiser Optical Systems, USA) and detected by a charge coupled device (CCD) detector (Roper Scientific, USA). Raman images are acquired in the mapping mode using a motorized sample stage.

## 2.6 Mid-infrared spectroscopy

Molecular vibrations can also be excited by absorption of radiation in the mid-IR range ($4000-400\,cm^{-1}$). As a consequence of different physical mechanisms, mid-IR and Raman spectra complement each other. Owing to the high water content of tissues and cells and the strong absorption of mid-IR radiation by water, the penetration depth is limited to few micrometers. Therefore, tissue samples for most mid-IR spectroscopic studies are cut in 5- to 20-μm-thick sections and subsequently dried. For data acquisition in transmission mode, tissue sections are transferred onto mid-IR transparent substrates such as calcium fluoride or barium fluoride. Whereas most Raman spectrometers operate in a dispersive mode, mid-IR spectrometers use the interferometric Fourier transform (FT) principle, which has the multiplex, throughput, and wavenumber accuracy advantages. The basic setup consists of a broadband radiation source, an interferometer, a sample chamber, which can also be a microscope, and a fast detector. Images are obtained by rastering the sample on an automated translation stage (see **Introduction to Spectral Imaging, and Applications to Diagnosis of Lymph Nodes** by Romeo, Dukor and Diem). In the past decade, Fourier transform infrared (FT-IR) imaging spectrometers were developed that enabled rapid assessment of tissue sections for biomedical applications.[39] The basic approach in implementing an IR imaging interferometer to obtain two-dimensional, spatially resolved IR spectra requires the use of IR-sensitive, multichannel detectors, termed focal plane array (FPA) detectors, that enable parallel collection of spectra. In the authors' current system, an FT-IR spectrometer IFS66 (Bruker Optics, Germany) is coupled to an IR microscope Hyperion (Bruker Optics, Germany) with $15\times$ magnification or to the macro sample chamber IMAC (Bruker Optics, Germany). With a $64 \times 64$ mercury cadmium telluride–based FPA detector, the field of view in the microscope is $270\,\mu m \times 270\,\mu m$ per image and each pixel corresponds to an area of $4.2\,\mu m \times 4.2\,\mu m$. The actual optical resolution is primarily determined by the diffraction limit and is about twice the pixel resolution. In the IMAC sample chamber, a $4\,mm \times 4\,mm$ area is imaged and each pixel corresponds to an area of $62.5\,\mu m \times 62.5\,\mu m$.

## 3 NEURO-ONCOLOGICAL RESULTS USING INFRARED AND RAMAN SPECTROSCOPY

### 3.1 FT-IR imaging of primary brain tumors

On-site pathological assessment of cryosections during surgery is performed to confirm initial diagnoses and to determine tumor margins. The latter are usually determined by a combination of CT and MRI before surgery, but they may deviate from the preoperative locations due to intraoperative brain shift or edema. Distinction between tumor and normal brain tissue is particularly important in neurosurgery in order to maximize tumor removal with minimal neurological damage. Cryosections are prepared from fresh frozen specimens. After thawing, they are stained by hematoxylin and eosin (H&E) and evaluated by a neuropathologist. However, such an expert is not always immediately available. Further problems that might impair an accurate diagnosis result from the inadequate quality and quantity of tissue specimens, and from intraobserver variability since pathological diagnoses are somewhat subjective.

Unstained cryosections can be analyzed by FT-IR imaging if they are prepared on suitable substrates. Therefore, our group initiated a research program to apply IR spectroscopy as a diagnostic tool to assess tissue sections of primary brain tumors. Its advantages include that it can easily be integrated into the clinical routine, it is rapid and objective as spectral data can be

collected and interpreted within minutes by automated algorithms, and it is nondestructive so the sample can be subjected to further analyses such as immunohistochemistry.

### 3.1.1 Infrared spectra of normal brain tissue and gliomas

IR spectra of normal brain tissue and gliomas with malignancy grades II to IV, based on approximately 150 000 IR spectra from 71 samples,[40] are summarized in Figure 1. Spectra are displayed for white matter of normal brain tissues, hemorrhage and leptomeninges, and the gliomas astrocytoma grade II, astrocytoma grade III, and GBM. The IR spectrum of gray matter is presented in Section 3.3. Three spectra for each tissue type demonstrate the intraclass variability. Multipoint baselines were subtracted from the IR spectra with baseline points near 3600, 3050, 2750, 1800, 1480, and 950 cm$^{-1}$. The baseline segments were obtained by linear interpolation between

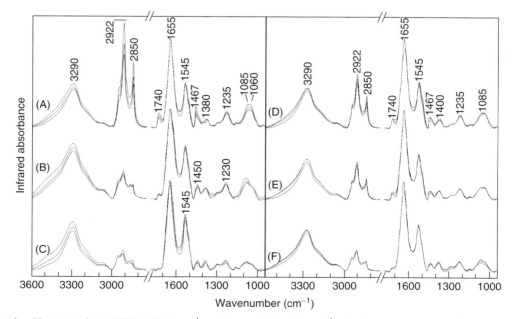

**Figure 1.** IR spectra from 3600 to 2750 cm$^{-1}$ and from 1800 to 950 cm$^{-1}$ of white matter in normal brain tissue (A), of leptomeninges (B), of hemorrhage (C), of astroyctoma grade II (D), of astrocytoma grade III (E), and of glioblastoma multiforme (F). For comparison, spectra are baseline corrected, normalized, and shifted to avoid overlap. Three spectra from each tissue type demonstrate the intraclass variability.

the selected points. Subsequently, IR spectra were normalized to equal intensities at $1655\,cm^{-1}$ to compensate for differences in sample thickness. The wavenumber range below $950\,cm^{-1}$ could not be observed because of the cutoff of the calcium fluoride substrate. The wavenumber ranges $2750-1800\,cm^{-1}$ and above $3600\,cm^{-1}$ are not displayed because of the absence of significant bands. Main spectral contributions are assigned to proteins and lipids. Protein bands at 1545, 1655, and $3290\,cm^{-1}$ originate from $C=O$ and $N-H$ vibrations of the peptide linkage and near $1400\,cm^{-1}$ from $C=O$ vibrations of $COO^-$ groups in amino acid side chains. Lipid bands at 1467, 2850, and $2922\,cm^{-1}$ originate from $C-H$ vibrations of fatty acids side chains, at $1740\,cm^{-1}$ from $\nu(C=O)$ vibrations of ester groups, and at 1085 and $1235\,cm^{-1}$ from $PO_2^-$ vibrations of phospholipids. Further spectral contributions are assigned to cholesterol near 1060 and $1380\,cm^{-1}$. White matter can easily be distinguished from all other tissues by the high intensity of the lipid and cholesterol bands which decrease in IR spectra of tumors in the order astrocytoma grade II > astrocytoma grade III > GBM. This observation is consistent with the decreased lipid content of gliomas that is also found by chromatography[41] and nuclear magnetic resonance (NMR) spectroscopy.[42] The protein bands of white matter and gliomas do not differ significantly, which makes the selection of the most intense protein band at $1655\,cm^{-1}$ for normalization plausible. Leptomeninges and hemorrhage show similar intensities of lipid bands to tumors. However, additional spectral contributions from collagen near 1230 and $1450\,cm^{-1}$ and from hemoglobin at $1545\,cm^{-1}$ distinguish their IR spectra from tumor spectra.

### 3.1.2 Classification of infrared spectra from gliomas by linear discriminant analysis

General problems of supervised classification models for spectroscopic data are the inherent heterogeneity of tissue and sample-to-sample or patient-to-patient variability, which requires a careful selection of training data. The intraclass variability is demonstrated by the three slightly different IR spectra per class in Figure 1. However, the spectral variances within tissue classes are smaller than those between different tissue classes. This aspect constitutes an important requisite for the development of any classification model. Brain tissues usually contain a variety of cell types even in normal samples. In tumor samples, various grades of neoplasia, tumor stroma, or necrotic tissue are found in addition. Classification models must consider these inherent biological variances. Since it is unrealistic that a classification model is trained to recognize all tissue classes, the model should contain a class called *not assigned* in order to reduce false positive or false negative assignments. A problem for the development of a classification model for gliomas is that the malignancy grades form a histological continuum; that is, there is a continuous transition between tumor grades. Therefore, it is extremely difficult to find an accurate criterion to separate tumor grades by spectroscopic features. Additionally, high-grade malignant gliomas may contain areas of lower cellularity, which are reminiscent of low-grade astrocytomas.

Two independent classification models were developed on the basis of the linear discriminant analysis (LDA) algorithm. The principle of this supervised algorithm is described in Section 3.5.3. The first model randomly selected a subset set of 400 IR spectra from FT-IR maps of 25 tissue sections representing the four tissue classes: normal, astrocytoma grade II, astrocytoma grade III, and GBM.[43] These spectra were then subjected to an LDA classification whose variables were optimized in an iterative process by a genetic optimal region selection routine. The number of variables ranged from 4 to 16. The four selected regions were located near 1060, 1080, $1400-1470$, and $1550\,cm^{-1}$. The main problem of this approach was the lack of neuropathological tissue evaluation to select training data and determine the classification accuracy of test data.

The main advantage of the second LDA classification model was that it was trained by

IR spectra from histopathologically confirmed tissue regions. Here, variables were manually selected on the basis of characteristic differences in the biochemical tissue composition.[44] Three variables were found to be sufficient to identify malignant gliomas of astrocytoma grade III and GBM. The first variable is the intensity ratio $2850\,cm^{-1}$ to $1655\,cm^{-1}$, which represents variations of the lipid-to-protein ratio between white matter and gliomas; the second variable is the intensity ratio $1545-1655\,cm^{-1}$, which is associated with hemoglobin in hemorrhage; and the third variable is the intensity ratio $(1230 + 1450)\,cm^{-1}-1655\,cm^{-1}$, which is associated with collagen in leptomeninges. These variables are called *molecular descriptors*.

After the LDA model was trained by the IR spectra in Figure 1 to classify six classes (normal tissue, astrocytoma grade II, astrocytoma grade III, GBM, hemorrhage, and leptomeninges), the classification procedure for FT-IR images involved the following steps:

1. a multipoint baseline was subtracted from each spectrum,
2. spectra with intensities below 0.1 absorbance units (AU) were removed from the images because they mainly originate from holes and fissures,
3. spectra with intensities above 1 AU were removed from the images because they might be affected by nonlinear detector response,
4. three molecular descriptors were calculated as described above, and
5. these variables were subjected to the LDA model.

During the course of the validation process, it turned out that the classes for normal tissue, astrocytoma grade II, astrocytoma grade III, GBM, and hemorrhage were well defined. However, IR spectra from impurities such as tissue freezing medium were also assigned to the leptomeninges class. Therefore, it seems that this class possesses the above-mentioned desired property for "not assigned" IR spectra and the class is called *other tissue* in the next section.

### 3.1.3 Application of infrared-based classification models to gliomas

The FT-IR images in our early papers were acquired by the mapping mode. Total acquisition times were dependent on the number of spectra and on the acquisition time per spectrum. Collecting five scans at a spectral resolution of $4\,cm^{-1}$ resulted in the measurement of approximately 600 IR spectra per hour.[44] The first LDA model which was optimized by the genetic algorithm gave classification accuracies of 85% for 5 normal specimens, 83% for 5 astrocytomas grade II, 88% for 5 astrocytomas grade III, and 47% for 10 GBM specimens using the above-mentioned four variables.[43] The distinction between normal tissue and GBM improved when more variables were considered.

The long data accumulation times for complete assessment of tissue sections in the order of hours are not compatible with the requirements of a rapid diagnostic tool. Instead of complete assessment, registration of few single IR spectra at selected positions constitutes a possibility to accelerate sample throughput. This approach was applied to cryosections from 51 independent patients.[44] The second LDA model with three variables, which was trained by histopathologically confirmed cryosections, recognized normal brain tissue with 100% accuracy and malignant gliomas with 93% accuracy. The remaining 7% spectra were assigned to hemorrhage and other tissue, but not to normal brain tissue.

Acquiring FT-IR images by the imaging mode using the instrumentation described in Section 2.6 constitutes a way to reduce accumulation times 100-fold down to a few minutes. Both classification models were transferred to FT-IR images of brain tissue and gliomas. One hundred and fifty-one FT-IR images from 57 patients were subjected to the genetic algorithm combined LDA approach. The overall success rate of the LDA classification using eight variables was 64%, with the highest success rate of 95% for normal tissue.[45] Although the positions and widths of the eight spectral regions varied, the previously described four variables near 1060, 1080,

1400–1470, and 1550 cm$^{-1}$ were again found to be the most frequently selected ones. It is important to note that the FT-IR images in this study were recorded using the microscope, which probed just small regions of 270 μm × 270 μm. Such small regions were rather homogeneous. Since they did not encompass the full heterogeneity of tissue sections, they required a preselection of regions in a manner similar to the single-spectra approach used previously.[44] FT-IR microspectroscopic imaging is discussed in more detail in Section 3.2.2.

The second LDA model using three manually selected variables from training spectra of specimens with confirmed diagnosis was validated by FT-IR images from a new set of specimens.[46] Multiple tissue specimens were obtained from the tumor center and from the tumor periphery of six glioma patients. The classification model was developed using data from mapping experiments.[44] The transfer of classification models between different instruments is a fundamental problem in analytical sciences that deserves special attention. Therefore, it was shown first that the spectra of a given region of tissue measured by single-point mapping and by hyperspectral imaging were essentially identical.

Two color-coded classification results of FT-IR images are compared with photomicrographs of unstained and H&E stained tissue sections in Figure 2. The size of the sample in panel (a) was 7.1 mm × 5.3 mm, which required two FT-IR images of 4 mm × 4 mm to be collected. The LDA model assigned more than 97% of all spectra correctly to malignant gliomas (top: 90% GBM, 9% astrocytoma grade III; bottom: 93% GBM, 4% astrocytoma grade III), whereas no spectra were assigned to normal tissue. Therefore, the assignment in panel (b) is consistent with a location of the tissue specimen near the tumor center of a GBM. The histomorphology of a consecutive 5-μm-thick H&E stained cryosection is included in panel (c). The overview indicates the coarse cellular, fibrous structure and high cell density of a malignant astroglial tumor, e.g., GBM. At high magnification, large polymorphic

tumor cells with atypical nuclei and some necrotic areas are visible, supporting this finding. The size of the sample in panel (d) was approximately 4 mm × 6 mm, which enabled almost complete coverage by two FT-IR images. The LDA model assigned the main part of the sample to astrocytoma grade II (top 65%, bottom 86%) and minor parts to astrocytoma grade III (top 26%, bottom 6%), to normal tissue (top 5%, bottom 6%), and to GBM (top 1.5%, bottom 0.1%). These assignments in panel (e) are consistent with a location at the periphery of the tumor. The transition to the tumor center seems to be near the top because of the increasing fraction of the malignant glioma astrocytoma grade III. The histomorphologic overview in panel (f) shows brain tissue with slightly increased cellularity. At high magnification, a few suspicious cells are evident, which may represent astrocytic tumor cells. It is important to note that, in the context of GBM, astrocytoma grades II and III are used as descriptive terms delineating the cellularity and morphological appearance of tumor areas in analogy to the morphological parameters used for grading by the World Health Organization (WHO). However, they are not identical with the respective WHO classification. In a tumor that contains areas of obvious malignancy (panel c) areas of low cellularity with only few tumor cells (panel f) do not represent a benign tumor astrocytoma grade II. Despite the morphologically benign appearance, these tumor cells may retain their malignant potential. A summary of the classification results of 54 tissue sections from altogether six patients in this validation set (one from astrocytoma grade III, two from oligoastrocytoma grade III, and three from GBM) will be presented in a forthcoming paper. So far, this classification model has been validated for malignant gliomas. Validation of low-grade astrocytomas failed because the number of astrocytoma grade II specimens is still low and the spectral properties of astrocytoma grade II, reactive gliosis, and gray matter significantly overlap. The classification model might be improved to recognize more tissue types or tissue types with

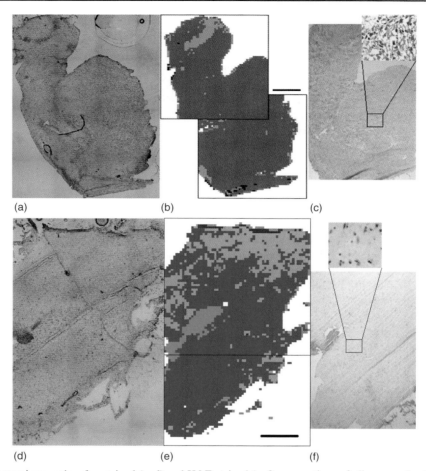

**Figure 2.** Photomicrographs of unstained (a, d) and H&E stained (c, f) cryosections of gliomas and color-coded LDA classification of FT-IR images (b, e). Color code: white matter of normal brain tissue (green), astrocytoma grade II (blue), astrocytoma grade III (orange), glioblastoma multiforme (red), hemorrhage (brown), and other tissue (gray). Bar = 1 mm.

smaller spectral variances by including more spectroscopic features as molecular descriptors. In the current model, three descriptors were sufficient because the spectral variances between the six tissue classes were quite pronounced. Eight descriptors are used in Section 3.3.2 in an LDA model for secondary brain tumors whose spectral variances are smaller. Another way of improvement would be to train alternative supervised classification algorithms such as artificial neural networks, support vector machines, or soft independent modeling of class analogies (SIMCA).

The last one is applied in Section 3.3.2 to classify FT-IR images of secondary brain tumors. Its principle is described in Section 3.5.4.

## 3.2   Raman imaging of primary brain tumors

Surgical removal of brain tumors is the most common initial treatment received by brain tumor patients because it relieves the mass effect of the tumor on neurological tissue and allows histopathological diagnosis of the tumor, which directly affects the direction of follow-up therapeutic strategy. It is generally accepted that aggressive surgical resection enhances the survival length and quality of life for brain tumor patients. To achieve this goal with minimal neurological damage, an accurate identification of brain tumor margins during surgery is required. Currently, neurosurgeons determine brain tumor margins intraoperatively by visual inspection and information provided by surgical navigation systems that are based on CT/MRI and/or intraoperative ultrasound (IOUS). However, limitations of these techniques include that

1. the true infiltrating margins may not be visible on CT/MRI images,
2. registration errors and intraoperative brain shift can degrade the spatial accuracy of the surgical navigation systems,
3. primary brain tumors, unlike most metastatic tumors, do not typically possess clear boundaries, i.e., the margins appear blurred in the IOUS images, and
4. it is often difficult even for experienced neurosurgeons to visually differentiate low-grade gliomas and associated tumor margins from normal brain tissue.

Because of their complexity, IOUS and intraoperative MRI cannot provide a continuous online imaging of the resection cavity, and integration of these technologies into microsurgical instruments or operating microscopes is limited. Applications of IOUS and MRI require a pause in the course of the brain tumor resection for intermittent analyses, which is realistic only for a limited number of investigations. Among the biophotonic techniques presented in Section 2, Raman spectroscopy is a candidate to analyze brain tissue nondestructively in real time and to provide intraoperative in vivo diagnosis. The potential advantages of Raman spectroscopy include that tissue can be characterized label-free at the microstructural and/or molecular level with spatial resolution in the single cell range, which permits accurate delineation of tumor margins, and sensitive and

specific identification of tumor remnants upon preservation of normal tissue.

### 3.2.1 Raman spectra of pristine brain tissue and gliomas

Raman images of pristine tissue specimens were collected ex vivo by a Raman spectrometer coupled to a microscope with a low magnification objective ($10\times$/NA 0.25).[47] $2\,mm \times 2\,mm$ regions of 2-mm-thick tissue sections, which were covered by a window to prevent drying during data acquisition, were scanned in two dimensions using a step size of $100\,\mu m$. The laser intensity of 60 mW was focused to a spot of approximately $60\,\mu m$ diameter. No tissue degradation was observed during the signal collection time of 20 s per spectrum. Figure 3 (top panel) shows the gray-scaled results of a $k$-means cluster analysis for three Raman images, which segmented the spectra according to their similarity in the spectral range $1800-1200\,cm^{-1}$ into two main groups. The cluster-averaged Raman spectra are displayed in the bottom panel of Figure 3. The principle of $k$-means cluster analysis is described in Section 3.5.2.

The two clusters of normal brain tissue in Figure 3(a) are assigned to white matter and gray matter. Since the lipid content of white matter is higher than that of gray matter, the Raman bands of lipids at $717\,cm^{-1}$ (choline group of phosphatidylcholine and sphingomyelin), at 1064, 1129, 1267, 1298, 1439, 1664, and $2883\,cm^{-1}$ (CC and $CH_2$ groups of fatty acids side chains) and of cholesterol at $700\,cm^{-1}$ dominate in the spectrum of white matter (Figure 3d, trace A). Raman spectra of brain lipids were presented in detail previously.[48] The band at $1005\,cm^{-1}$ is assigned to the aromatic amino acid phenylalanine of proteins. Other prominent protein bands near 1267 (amide III), 1660 (amide I), 1450, and $2900\,cm^{-1}$ ($CH_3$ bending and stretching bands of amino acids side chains, respectively) significantly overlap with lipid bands. In the Raman spectrum of gray matter (Figure 3d, trace B), the intensities of lipid and cholesterol bands decrease,

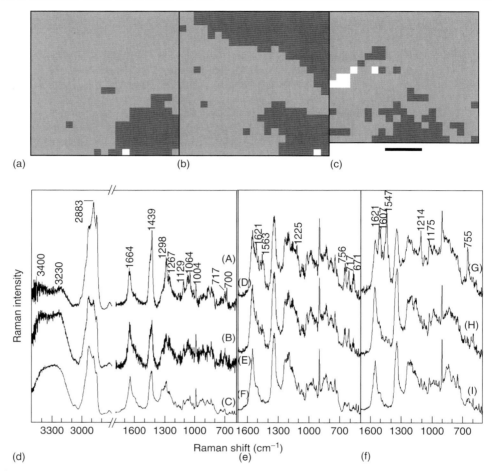

**Figure 3.** Raman images of pristine tissue (a) from normal brain, (b) from astrocytoma grade III and (c) from glioblastoma multiforme. Gray shades indicate two classes of a cluster analysis. Cluster-averaged Raman spectra (d) of normal brain (traces A, B) from 3500 to $2700\,\text{cm}^{-1}$ and from 1800 to $600\,\text{cm}^{-1}$. Cluster-averaged Raman spectra (e and f) of astrocytoma grade III (traces D, E) and of glioblastoma multiforme (traces G, H) from 1800 to $600\,\text{cm}^{-1}$. Fitted Raman spectra are included (traces C, F, I). Bar = $500\,\mu\text{m}$.

whereas the intensities of water bands at 3230 and $3400\,\text{cm}^{-1}$ increase.

The two clusters of astrocytoma grade III are assigned to tumor tissue with higher and lower hemoglobin concentrations (Figure 3b). The cluster-averaged Raman spectra are shown in Figure 3(e) as spectra D and E. The spectral contributions of hemoglobin at 671, 756, 1225, 1563, and $1621\,\text{cm}^{-1}$ are more intense compared to the Raman spectra of normal brain tissue and indicate hemorrhage or higher blood perfusion

that are typical for intracranial tumors. Further spectral differences are the decrease in lipid bands relative to protein bands and the increase in the choline band at $717\,\text{cm}^{-1}$ relative to the lipid bands. This observation is consistent with an accumulation of phosphatidylcholine in gliomas, which was also found by NMR spectroscopy[42] and chromatography.[41]

In contrast to Figure 3(a) and (b), the specimen of GBM, from which the Raman image in Figure 3(c) was obtained, was not snap frozen and thawed before Raman spectroscopy. Instead, the Raman image was recorded immediately after surgery. Analogous to Figure 3(b), the two clusters are assigned to tumor tissue with higher and lower hemoglobin content. The cluster-averaged Raman spectra are shown in Figure 3(f) as spectra G and H. The Raman spectrum of the lower hemoglobin-content GBM sample (trace H) is similar to the Raman spectrum of the lower hemoglobin-content astrocytoma grade III sample (trace E) except for a small decrease in the intensity of lipid bands relative to that of protein bands. However, the Raman spectrum of the higher hemoglobin-content GBM sample (trace G) contains intense additional bands at 755, 1175, 1214, 1547, 1607, and 1621 cm$^{-1}$, which are assigned to deoxygenated hemoglobin.[49] The snap-freezing and thawing procedures seem to affect hemoglobin more strongly than lipids and proteins.

Quantitative determination of lipid-to-protein ratios, based on single bands, which was performed for IR spectra, is less effective for Raman spectra because of the above-mentioned band overlap and because of the lower signal-to-noise ratio (SNR). An alternative method to determine the sample composition is to fit the entire spectral range to a set of reference spectra and to analyze the fit parameters instead of band intensities. Similar strategies were applied to determine the composition of coronary arteries,[50] human breast tissue,[51] and urological pathologies.[52] It turned out that four Raman spectra of proteins, lipids, cholesterol, and water were sufficient to fit the Raman spectra of brain tissue. The results of the fitting procedure for gray matter (trace B), for astrocytoma grade III (trace E), and for GBM (trace H) are presented in Figure 3 as fitted spectra (traces C, F, and I, respectively). The lipid-to-protein ratios are calculated as 0.44 for gray matter, 0.25 for astrocytoma grade III, and 0.22 for GBM. The observed decrease with increasing malignancy agrees with the IR data discussed in Section 3.1 and constitutes the basis for a supervised classification model after acquisition of more independent data sets. Deviations between the Raman spectra and the fitted spectra can be explained by differences in protein structures and composition, differences in lipid composition, and additional constituents such as hemoglobin. The fitting procedure can be improved by including basis spectra of reference material that is directly extracted from brain tissue and by including additional model compounds to the set of basis spectra.

### 3.2.2 Comparison of Raman and FT-IR microspectroscopic images of brain tissue

Lateral resolutions of 60 μm in vibrational spectroscopic images can be achieved without microscopes as described in Sections 2.6 and 3.4. For this reason, microspectroscopic images are in this context images acquired with a lateral resolution better than 10 μm, which is required to resolve small structures such as subcellular features in single cells or microcrystals. In order to achieve such a high resolution in Raman images, Raman spectrometers are coupled to microscopes with high magnification objectives (e.g., 100×/NA 0.9), which can focus the laser to a diffraction-limited spot size below 1 μm. Usually, the step size is set at the same size as the diameter of the laser spot. To reduce the number of spectra, Raman images can be collected at step sizes that are larger than the spot size. Although the applicability of wide-field Raman microscopic imaging with superior lateral resolution near 0.5 μm was demonstrated for brain tissue,[53] it has not been used for further biomedical studies so far. The lateral resolution in mid-IR spectroscopy (2.5–12.5 μm) is determined by the diffraction limit, and is approximately equal to the wavelength. In order to achieve such a high resolution in FT-IR mapping, mid-IR spectrometers are coupled to microscopes and used with apertures smaller than 10 μm × 10 μm, which results in a low SNR when a Globar®-type source is used. Far better results may be obtained when high-brightness mid-IR radiation from a synchrotron source is used; in this case, the position of the sample can be rastered with a much smaller step size. Another possibility to acquire FT-IR images with high lateral resolution is given by FT-IR spectrometers coupled to microscopes and linear

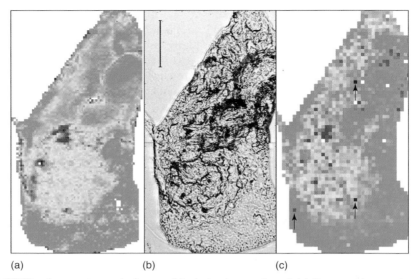

(a)                              (b)                              (c)

**Figure 4.** (a) FT-IR microspectroscopic image, (b) photomicrograph and (c) Raman microspectroscopic image of an unstained cryosection of an astrocytoma grade II brain tumor. Color codes represent the score plots of principal components analyses. Bar $= 100 \mu$m.

or two-dimensional array detectors, as discussed in Section 2.6.

Figure 4 compares a Raman image of an astrocytoma grade II tissue section, which was collected with laser focus of $1 \mu$m diameter and a step size of $7.5 \mu$m, and an FT-IR image, where each square corresponds to an area of $4.2 \mu$m $\times$ $4.2 \mu$m. Here, the different lipid-to-protein ratios are visualized by score plots of the second principal component (PC) after principal component analysis (PCA) (described in Section 3.5.1). High lipid concentrations in the central portion of the sample are indicated in both images by yellow to red colors, low lipid concentrations by green to blue colors. The main advantage of the FT-IR imaging approach is the shorter acquisition time. Acquisition of two sample FT-IR images and one background FT-IR image took approximately 30 min to image a region of $270 \mu$m $\times 540 \mu$m. Acquisition of $36 \times 72 = 2592$ Raman spectra with an exposure time of 1 min per spectrum took more than 2600 min corresponding to more than 43 h (additional time was needed to save the spectra and to move the sample to the next position). The spectral signatures of microscopic

features are more pronounced in the Raman image. The positions of microcrystals are indicated by arrows in Figure 4(c). They are identified as cholesterol by their Raman spectrum. A typical spectrum is shown in Figure 5 (trace A). In addition, Raman spectra of cholesterol ester (CE) microcrystals and of microcalcifications in Figure 5 were detected in other Raman images of tissue sections from GBM, meningeomas, and schwannomas.[53] None of these features have been found in FT-IR images so far which might be (i) a result of the diffraction-limited optical resolution being lower for mid-IR than Raman spectra and/or (ii) a consequence of the different collection geometries. Raman spectra are collected with $180°$ collection and the laser is focused on the top surface of the sample. Thus, the spectra of microcrystals can be readily observed if the laser is focused on them. IR spectra are usually collected in transmission mode and potential spectral contributions of microcrystals overlap with underlying spectral contributions of tissue. Also, if the microcrystals are smaller than the diffraction limit of the IR measurement, their contribution is merged with the surrounding material. Calcifications, such

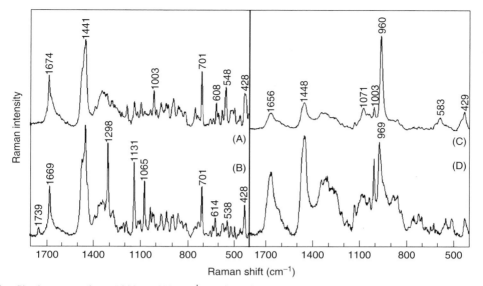

**Figure 5.** Single spectra from 1800 to 400 cm$^{-1}$ obtained from microcrystals in Raman microspectroscopic images of brain tumor cryosections: cholesterol (trace A), cholesterol ester (trace B), hydroxyapatite (trace C), and tricalcium phosphate (trace D).

as hydroxyapatite with a strong band at 960 cm$^{-1}$ (trace C) and tricalcium phosphate with a strong band at 969 cm$^{-1}$ (trace D), are detected in some brain tumors due to the un-differentiation of tumor cells. The detection of CE (trace B) can be considered as a marker for intracranial tumors. The $\nu$(C=O) band of esters at 1739 cm$^{-1}$, fatty acids bands at 1065, 1131, 1298, and 1442 cm$^{-1}$, and CE-specific bands at 428, 538, 614, and 1669 cm$^{-1}$ clearly distinguish this spectrum from the Raman spectrum of cholesterol (trace A). Whereas only trace amounts of CE are present in normal brain tissue, its concentration was reported to increase up to 100 times in tumors.[54] The detection of microcrystals is facilitated by the drying process, which induces crystallization of hydrophobic compounds such as cholesterol and CE. Spectral contributions of proteins at 1003, 1448, and 1656 cm$^{-1}$ are present at varying intensities.

Other Raman microspectroscopic imaging studies distinguished necrotic tissue from vital GBM tumor[55] and meningeoma from normal dura mater in dried sections of brain tissue.[56] Raman spectra of necroses contained more intense lipid bands with additional spectral features assigned to cholesterol microcrystals and hydroxyapatite deposits, as displayed in Figure 5. Raman spectra of vital tumor were dominated by protein bands, with some spectral contributions from glycogen and fatty acids. Raman spectra of dura mater were dominated by collagen bands. The intensities of collagen bands decrease in meningeomas, which enable identification of tumors in tissue sections[53,56] and pristine tissue.[47] Additional spectral contributions of CE and hydroxyapatite in tissue sections of meningeomas were reported in both independent studies.[53,56]

In summary, the applicability of Raman and IR microspectroscopic imaging—with an optical resolution better than 10 μm as defined in the first paragraph—as a screening tool for brain tumor sections seems to be more favorable for small samples or small sampling areas. Complete assessment of extended tissue samples by microspectroscopic imaging is extremely time consuming using the state-of-the-art instruments presented in Sections 2.5 and 2.6 and generates a huge amount of data that constitutes a challenge for analysis and classification. For example, the 4 mm × 6 mm tissue section in Figure 2(d) would require acquisition of approximately 330 FT-IR microspectroscopic images or 400 × 600 = 240 000 spectra with a step size of 10 μm. Selection of small sampling areas within an extended, inhomogeneous sample is problematic as there is a high risk to oversee important information such as the detection of tumor cells outside the probed region.

## 3.3 FT-IR imaging of secondary brain tumors

As described in Section 1.2, there is a strong need to develop new diagnostic methods to determine the origin of brain metastases with unknown primary tumors because standard screening techniques fail in up to 15% of cases. The extracerebral primary tumor is the limiting factor for the prognosis of ca. 80% of patients. Therefore, an early identification of the primary tumor is important in order to select an organ-specific therapy. Metastatic cells contain the molecular information of the primary cells. Since FT-IR spectroscopy probes chemical and structural properties of tissues at the molecular level, FT-IR imaging combined with multivariate classification constitutes a new approach to identify the primary tumor of brain metastases. Analogous to the FT-IR images of gliomas in Section 3.1.3, this study reports FT-IR images with a resolution of 62.5 μm × 62.5 μm per pixel. The advantage is that 4 mm × 4 mm sample areas can be probed in a few minutes. The lateral resolution is sufficient to identify the molecular information of the main tissue classes in heterogeneous tissues.

### 3.3.1 Infrared spectra of the most frequent brain metastases

Figure 6 compares three IR spectra each from white and gray matter of normal brain tissue, from brain metastases of renal cell carcinoma (RCC), lung cancer (LC), colorectal cancer (CC) and breast cancer (BC). Spectral changes between the three spectra demonstrate sample or patient-to-patient variations between specimens of the same tissue type. The IR spectra of white matter in Figure 6 (trace A) closely resemble the spectral features of white matter spectra in Figure 1, showing bands of phospholipids, glycolipids, and cholesterol at 1060, 1080, 1235, 1380, 1467, 1740, 2850, and 2922 cm$^{-1}$. (The spectra are not identical as they were extracted from different data sets.) The intensities of lipid bands decrease in the IR spectra of gray matter (trace B). However, they are still more intense than the corresponding bands in the IR spectra of brain metastases. Therefore, the lipid-to-protein ratio seems to distinguish normal brain tissue from brain metastases similar to the case for gliomas described in Section 3.1. Spectral contributions of glycogen increase the bands at 1026, 1080, 1153, and 3290 cm$^{-1}$ in IR spectra of RCC (trace C). As expression and accumulation of glycogen is a typical property of renal cells, the detection of these IR bands in brain metastases is indicative of RCC as the primary tumor. IR spectra of normal brain tissue, of LC, CC, and BC do not show the spectral fingerprint of glycogen, which can therefore be considered as a molecular marker for RCC. Other epithelial tissues also accumulate glycogen. As they are known to cause brain metastases less frequently, this fact reduces the specificity for diagnosing RCC only marginally. The molecular markers are less pronounced in LC, CC, and BC, and the IR spectra are similar at first sight. However, careful inspection of difference spectra reveals significant differences. Arrows in the difference spectrum (trace D−B) point to positive bands, which indicate that the amide II band is more intense and the amide I band has an elevated component near 1625 cm$^{-1}$ in the IR spectra of LC (trace D). A negative difference band indicates that the intensity near 1400 cm$^{-1}$ is smallest in IR spectra of CC (trace E−B) and that the intensity at 1735 cm$^{-1}$ is smallest in IR spectra of BC (trace F−B). The difference spectrum (trace C−B) shows that the high wavenumber difference band is centered near 3350 cm$^{-1}$, which is typical for OH groups in glycogen.

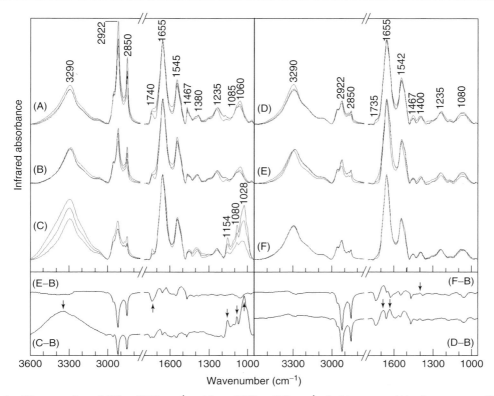

**Figure 6.** IR spectra from 3600 to 2750 cm$^{-1}$ and from 1800 to 950 cm$^{-1}$ of white matter (A), of gray matter (B) found in normal brain tissue and of brain metastases from renal cell carcinoma (C), from lung cancer (D), from colorectal cancer (E), and from breast cancer (F). Difference spectra are included to demonstrate the variations between metastases and gray matter. For comparison, spectra are baseline corrected, normalized, and shifted to avoid overlap. Three spectra from each tissue type demonstrate the intraclass variability.

### 3.3.2 Supervised classification of infrared spectra from brain metastases

Two independent classification models were developed for FT-IR images to distinguish normal brain tissue from brain metastases and to identify the primary tumor of brain metastases. The first model used the LDA algorithm,[57] which was applied to malignant gliomas in Section 3.1.2. The second model used the SIMCA algorithm.[58] The principles of both algorithms are described in Section 3.5.

The 18 IR spectra shown in Figure 6 were used to train the LDA model. After preprocessing the IR spectra, including normalization to the band at 1655 cm$^{-1}$, as described in Section 3.1.1,

the following band intensities were determined as molecular descriptors: $(1030 + 1153)$, 1080, 1400, 1542, 1629, 1735, 2850, and 3290 cm$^{-1}$. Because the spectral differences between the six classes (RCC, LC, CC, BC, and white matter and gray matter of normal brain tissue) are small, eight molecular descriptors were required as variables for the LDA model to classify six tissue types. For comparison, three molecular descriptors were sufficient for the LDA model to identify malignant gliomas based on more intense spectral differences.

To train the SIMCA algorithm, spectra representing each class were selected from FT-IR images with confirmed diagnosis. It is important that the training spectra encompass the specific spectral fingerprint of the tissue class and its inherent intrasample variances. The SIMCA model for the identification of the primary tumor of brain metastases consists of 10 PCs in the wavenumber interval $1800-950\,cm^{-1}$. The same color codes were used for display of the results as for the results of the LDA model.

### 3.3.3 Application of infrared-based classification models to brain metastases

The classification models (discussed above in Section 3.3.2) were applied to three normal control brain specimens and 17 brain metastases. Before the results are summarized, sample RCC7 deserves special attention, as the primary tumor could not be identified unambiguously by standard methods. The primary tumor was tentatively diagnosed as "suspicious to RCC or LC". Photomicrographs of unstained and H&E stained cryosections (Figure 7a and d) of brain metastasis RCC7 are compared with the classification results. The SIMCA classification of the FT-IR image segmented the tissue section into three main parts (Figure 7b): normal brain tissue consisting of white (3%) and gray matter (4%), RCC (66%), and others (CC 15%, LC 11%, BC 1%) in-between regions of normal tissue and RCC. The LDA classification gave a similar segmentation (Figure 7c) with slightly more normal tissue (white matter 10%, gray matter 11%), less RCC (46%), and more other tissue (CC 7%, LC 3%,

BC 22%). The class assignments deviate in particular at the transitions between tissue types. The detection of glycogen in the IR spectra and the majority classification rates by the SIMCA and the LDA model strongly support that the primary tumor of the brain metastasis is RCC. The fact that the classification rates are less than 100% can be explained by the presence of normal tissue and by the presence of tissue that was not included in the training sets. Histopathological inspection confirmed normal brain tissue at the bottom region and revealed a mixture of necrosis, cerebral gliosis, and isolated tumor cells from RCC without accumulation of glycogen in an intermediate zone between RCC and normal tissue. Such tissue types have not been included in the classification models yet. In contrast to the classification model for gliomas (Section 3.1.2), the classification models for brain metastases did not contain a class "not assigned" or "other tissue". Therefore, the unknown tissue types were incorrectly assigned to LC, CC, and BC.

The diagram in Figure 8 summarizes the percentage of correct assignments by SIMCA and by the LDA model. If the criterion is defined that the majority of spectral classification determines the diagnosis, then all normal brain tissue samples and all brain metastases were correctly classified, except CC3, which the LDA model incorrectly assigned to BC. Overall, the results agree qualitatively well for both classification models. Both models seem to be able to describe the diagnostic information in the IR spectra and to utilize it for

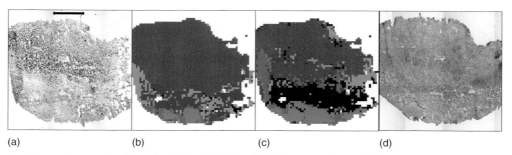

(a)  (b)  (c)  (d)

**Figure 7.** Photomicrographs of unstained (a) and H&E stained (d) cryosections of brain metastasis RCC7. Color-coded classification of a FT-IR image by the SIMCA algorithm (b) and by the LDA algorithm (c). Color code: white matter (light green) and gray matter (dark green) of normal brain tissue, brain metastases of renal cell carcinoma (red), of lung cancer (blue), of colorectal cancer (cyan), and of breast cancer (brown). Bar = 1 mm.

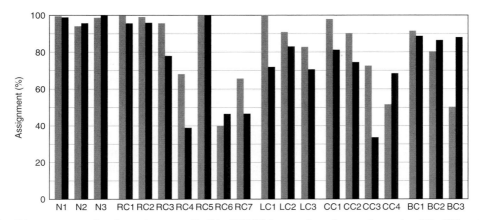

**Figure 8.** Diagram comparing the assignments (in %) of FT-IR images from 3 normal samples (N1–N3) and from 17 brain metastases of renal cell carcinoma (RCC1–RCC7), of lung cancer (LC1–LC3), of colorectal cancer (CC1–CC4), and of breast cancer (BC1–BC3) by the SIMCA model (gray bars) and by the LDA classification model (black bars).

tissue identification. Quantitatively, the numbers differ because the training data and the classification algorithms are different. In 12 out of 17 brain metastases, the percentage of IR spectra that were assigned to the correct primary tumor was higher for the SIMCA classification than for the LDA classification. This comparison suggests that the SIMCA model is able to describe the diagnostic spectral features more comprehensively. This conclusion is restricted by the fact that RCC1, LC1, CC1, and BC1 were selected for training of the SIMCA model. Therefore, these FT-IR images are not independent and the values just indicate the reclassification rates, which are usually very high. The different percentages of both classification approaches also represent the measurable error due to spectral noise and model deficiencies. Therefore, both models require refinement by including more tissue classes, other frequent primary tumors of brain metastases, and optimization of the training data. Furthermore, more independent FT-IR images have to be recorded for validation.

## 3.4 Raman and FT-IR imaging of a murine brain metastasis model

An important step toward the development of new diagnostic tools is their application to animal

models to further optimize instrumentation and data analysis strategies. In the first animal model example for malignant brain tumors, called *C6 cell glioma*, tumor cells are directly inoculated in adult rat brains, which form a tumor of 2.5 to 4 mm diameter within 2 weeks after implantation. As the induced tumors exhibit similar characteristics to those in human GBM, the model is often used for experiments during the preclinical stage to test novel therapeutic approaches. Tissue sections obtained from such rat brains were recently studied by FT-IR imaging[59,60] and Raman imaging.[61] Cluster analyses of FT-IR images identified anatomical structures in control rat brain without tumor and a solid tumor in rat brain after inoculation of glioma cells.[59] The morphology of the brains and the detection of tumors obtained by cluster analysis of FT-IR images were similar in a second study, although the instrumentation, pixel resolution, and clustering algorithm differed.[60] The IR spectra characteristic of the chemical signature of each anatomical structure differed significantly between the studies because flash-frozen sections were used in one, and paraffin-embedded sections in the other study. The sample preparation steps in the latter method washed out virtually all lipids that constitute the main compositional difference between normal brain tissue and gliomas, and the different data collection modes caused physical dispersive artifacts in IR spectra, which complicated data analysis. Raman imaging showed a clear distinction in rat brain thin sections between normal, tumoral, necrotic, and edematous tissues.[61]

Another murine brain tumor model was established that develops brain tumors by injection of tumor cells into the carotid artery.[62,63] A variety of brain tumors such as gliomas or brain metastases can be induced in this way. An advantage compared to the above-mentioned rat model is that this procedure mimics hematogenous tumor development preferentially in one brain hemisphere, which is a more natural way than intracerebral implantation of tumor cells. The unaffected brain hemisphere can be used as a control for normal murine brain tissue. Human brain tumor specimens collected during surgery usually do not contain normal tissue since this is spared as much as possible during neurosurgical procedures. This murine tumor model was studied by FT-IR imaging and fiber-optic Raman imaging.[64]

Ten-micrometer-thick dried tissue sections with and without H&E staining and 2-mm-thick pristine tissue sections were prepared consecutively in order to compare the results of FT-IR imaging, Raman imaging, and histopathology as the gold standard. The instrumentation differed slightly compared with the instrumentation in the previous sections. FT-IR images were collected at a resolution of $25 \, \mu m \times 25 \, \mu m$ per pixel using the FT-IR imaging spectrometer Spotlight 300 (Perkin-Elmer, UK) with a 16-element array for detection. The sample was scanned over the detector array in a raster pattern with 2 scans per step at a spectral resolution of $8 \, cm^{-1}$. Approximately 500 IR spectra could be collected per minute, which is comparable with the acquisition rate of a $64 \times 64$ FPA detector. Raman images were collected by coupling the Raman spectrometer HoloSpec (Kaiser Optical Systems, USA) with a fiber-optic probe (Inphotonics, USA) and a motorized stage. The fiber-optic probe focused the laser to a spot of $60 \, \mu m$ diameter, which is identical to the laser spot size using a microscope and an objective with $10\times$ magnification (see Section 3.2). Raman

images were accumulated with an exposure time of 8 s per spectrum and a step size of $120 \, \mu m$.

Figure 9 displays first results for two specimens. The Raman images (a, d) were obtained from the 2-mm-thick pristine tissue sections. Boxes in the consecutive H&E stained 10-$\mu m$ tissue sections (c, f), which are included as references, indicate the approximate position of the Raman images. The FT-IR images (b, e) were obtained from the consecutive 10-$\mu m$-thick unstained tissue sections. Higher magnification views indicate metastatic cells of a murine K1735 malignant melanoma along the arachnoid cistern and brainstem in the first specimen (inset of c), and metastatic cells of the human A375 malignant melanoma within the temporal lobe of the second specimen (inset of f). Cluster analysis of the Raman images grouped the spectra into six clusters. Shape deformations due to the transfer of the soft pristine brain tissue to the sample compartment were small enough, and the main morphological structures were sufficiently large, so that the Raman images coincided well with the FT-IR images of the unstained tissue section and with the photomicrographs of the H&E stained tissue section. Whereas thin sections were prepared from frozen tissue, brain tissue at room temperature is extremely soft due to the high water content of 70–80%. Upon comparison with the consecutive tissue sections, the clusters are assigned to brain metastases with high (gray) and medium (black) concentration of tumor cells and to normal brain tissue with high (orange), medium high (yellow), medium low (cyan), and low (blue) lipid-to-protein ratios. The brain metastases (gray) are always surrounded by a ring of tissue (black), which is consistent with the transition zone of lower tumor cell concentration bordering normal tissue. A small tumor of just four pixels was also detected in the center of Raman map (d).

The regions that were probed by FT-IR imaging were expected to contain tumor cells. However, cluster analyses only distinguished morphological features of normal mouse brain tissue. Analyses by other chemometric methods were also unsuccessful in detecting tumor cells. The general morphology in both tissue sections

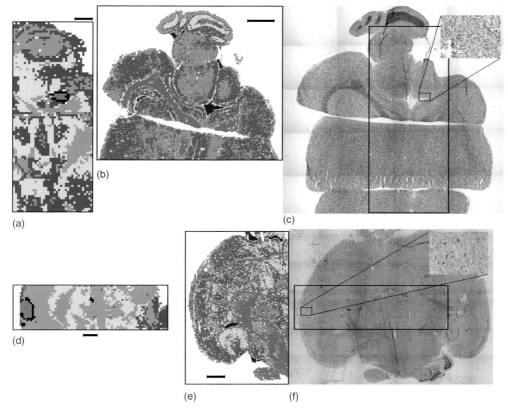

**Figure 9.** Raman images of pristine mouse brains (a, d), FT-IR images of consecutive dried cryosections (b, e), and consecutive H&E stained cryosections (c, f). Metastatic cells of brain metastases from malignant melanomas are identified in the Raman images (black, gray clusters) and confirmed at higher magnification in H&E stained cryosections (insets of c, f). Bar = 1 mm.

differs because they were cut at different levels. Arachnoid cisterns (brown), white matter fiber tracts (orange), cerebellar gyri (yellow), and a ventricular wall in the right hemisphere (yellow/brown) are resolved in Figure 9(b) and (e). The position of the cistern and the shape of the curved white matter fiber tracts in the FT-IR image (b) coincide with the Raman map (a). The orange cluster in the left brain hemisphere of the FT-IR image (e) also coincides remarkably well with the Raman map (d).

The Raman spectra of white matter (Figure 10, trace A) and gray matter (trace B) in normal murine brain tissue are similar to the spectra of normal human brain tissue in Figure 3. However, the Raman spectra of brain metastases from

malignant melanomas show additional spectral contributions in the range 1800–400 cm$^{-1}$ (traces C, D). Raman bands at 457, 599, 727, 976, 1124, 1400, and 1592 cm$^{-1}$ can be assigned to the pigment melanin. Their intensities are resonance enhanced with 785-nm excitation because the pigment has electronic absorption at the excitation wavelength. The absorption spectrum of melanin is included in Figure 10 (inset). The absorption peak near 335 nm is consistent with the protective function of melanin against UV radiation. High absorption occurs throughout the visible range, which also extends toward the NIR range. The most intense Raman bands of melanin are found in the spectrum of the gray-colored cluster (trace D), which obscure other spectral contributions from

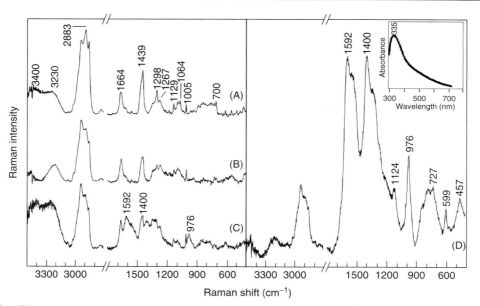

**Figure 10.** Cluster-averaged Raman spectra of normal murine brain tissue (traces A, B) and of brain metastases from malignant melanomas with low (trace C) and high melanin content (trace D) from 3500 to 2700 cm$^{-1}$ and from 1800 to 400 cm$^{-1}$. Absorption spectrum of melanin (inset).

lipids and proteins in the region 1800–400 cm$^{-1}$. In the Raman spectrum of the black-colored cluster (trace C), Raman bands of lipids and proteins have almost the same intensities as bands of the pigment. In the interval 3550–2700 cm$^{-1}$, the Raman spectra of brain metastases (traces C, D) are almost identical to the Raman spectrum of gray matter (trace B). The CH, NH, and OH bands in this spectral range were not resonance enhanced because these bonds are not associated with the chromophore, giving rise to the electronic transitions. Analyzing this high wavenumber range did not permit distinction between normal tissue and tumors.

Beside the deeper penetration of radiation into nondried tissue, another advantage of Raman spectroscopy is the possibility to utilize enhancement effects. Here, the enhanced spectral contributions of melanin could be used as tumor markers and facilitated for the sensitive detection of metastatic cells of malignant melanomas. The absorption properties might not only be used for diagnosis but also offer the option for a selective

treatment. Prolonged exposure or higher intensities of the laser radiation induce irreversible damages in these tumor cells, whereas nonabsorbing normal cells remain intact. Tumor cells could not be detected in 10-μm-thick cryosections by FT-IR imaging. A first possible explanation is that the contribution of the pigment melanin to the infrared spectrum is not resonance enhanced so the differences between IR spectra of normal tissue and metastases of malignant melanomas are too small. A second possible explanation is that tumor cells did not form a solid tumor and the resolution of 25 μm × 25 μm per pixel is too large for detection of small, single, and scattered tumor cells. Further studies will be performed to validate these conclusions.

## 3.5 Algorithms for unsupervised segmentation and supervised classification

Unsupervised algorithms are used as exploratory tools to display variations or similarities within spectroscopic data sets. In contrast to supervised algorithms, they do not require a priori information. In this chapter, principal component analysis (PCA) and $k$-means cluster analysis were applied as unsupervised algorithms for segmentation. Supervised algorithms are used to classify spectroscopic data. They require training data with well-known properties. In this chapter, LDA and SIMCA were applied for supervised classification. The principles of these algorithms are briefly described in this section. Mathematical descriptions are beyond the scope of this chapter and can be found elsewhere.[65]

In contrast to the IR-based classification of microbial samples,[66] derivatives of IR spectra were not calculated to compensate different baseline slopes before analysis. IR spectra of bacterial microcolonies are usually collected with high SNRs and a so-called zero filling factor of 4 in the Fourier transformation. This parameter increases the number of data points by a factor of 4 by interpolation without adding chemical information. Derivatives with sufficient SNR can be calculated from these spectra. However, derivatization algorithms often include a smoothing step, which improves the SNR at the expense of spectral resolution. The IR and Raman spectra in images of brain tissue have an inferior SNR to the previously reported spectra of bacteria. In addition, in order to keep the number of data points low and each imaging data set size to a workable level (approximately 20 MB), the zero filling factor was set to 1. We have shown that derivatives of these spectra are strongly affected by noise.[57] Extensive smoothing would change the spectral information more strongly than the baseline correction described in Section 3.1.1. Therefore, better results for unsupervised segmentation and supervised classification of vibrational spectroscopic images from brain tissue were achieved without calculating derivatives.

### 3.5.1  Principal component analysis

PCA has been used widely with large multidimensional data sets such as vibrational spectroscopic images. The use of PCA allows the number of variables in a data set to be reduced, while retaining as much as possible of the variance present in the data sets. This reduction is achieved by decomposing the data set into eigenvectors, also called *principal components* or factors, which are uncorrelated or orthogonal to each other. The contribution of each eigenvector to a given spectrum is given by its score. PCs are ordered so that PC1 exhibits the greatest amount of variation, PC2 the second greatest amount of variation, PC3 the third greatest amount of variation, and so on. That is $\text{var(PC1)} \geq \text{var(PC2)} \geq \text{var(PC3)} \geq \cdots$, where var(PCi) expresses the variance of PCi. Var(PCi) is the eigenvalue of PCi. When using PCA, it is hoped that the eigenvalues of only few PCs are of significance, while all subsequent PCs will be so low as to be virtually negligible. For spectroscopic data, the latter PCs are dominated by noise. Where this is the case, the variation in the data set can be adequately described by means of the few significant PCs, where the eigenvalues are not negligible. For spectroscopic data, the PCs represent spectroscopic features. Plotting the scores for each spectrum as in Figure 4 represents the "intensity" of the PC. PCA in this work was performed using the PLS Toolbox (Eigenvector Research, USA) under a Matlab platform (The Mathworks, USA).

### 3.5.2  k-means cluster analysis

In general, clustering is the partitioning of a data set into subsets, called *clusters*, so that the differences between data within each cluster are minimized and the differences between clusters are maximized according to some defined distance measure. The $k$-means algorithm assigns each point to the cluster whose center is nearest. The center is the average of all the points in the cluster. The algorithm steps are as follows: (i) choosing the number of clusters, $k$, (ii) selecting randomly $k$ spectra as initial cluster centers, (iii) assigning each spectrum to the nearest cluster center, (iv) recomputing the new cluster centers, and (v) repeating the two previous steps until the solution converges. The results of the $k$-means cluster analysis are the centers of each cluster and the cluster membership map. The main advantages of this algorithm are its simplicity and speed, which allow it to run on large data sets. Its disadvantage is that it might not yield the same results with each run, since the results depend on the initial random assignment. In the current implementation, data were normalized using multiplicative signal correction before cluster analysis and a Euclidean distance metric was used.

### 3.5.3 Linear discriminant analysis

LDA is also presented in another chapter (see **Raman Spectroscopy as a Potential Tool for Early Diagnosis of Malignancies in Esophageal and Bladder Tissues** by Stone, Kendall and Barr). Briefly, a number of objects belong exactly to one out of $k$ similar classes and the class membership is known for each object. Furthermore, each object is defined by characteristic parameters. LDA uses this information to calculate $(k - 1)$ linear discriminant functions that optimally separate $k$ classes. LDA uses these functions to assign unknown objects to classes. In the present vibrational spectroscopic context, the objects are IR or Raman spectra, the classes are tissue types, and the characteristic parameters or features are band intensities or intensities ratios. In case of imaging data sets, the resulting class memberships can be represented by colors, and color-coded images are assembled for visualization, which are interpreted in analogy to histopathologically stained tissue sections. This current work used the LDA algorithm of the Discriminant Analysis Toolbox developed by M. Kiefte (downloadable under http://www.mathworks.com).

### 3.5.4 Soft independent modeling of class analogies

Analogous to LDA, SIMCA requires a training set of objects with a set of attributes and their class membership. In contrast to LDA, the term soft refers to the fact that the classifier can identify objects as belonging to multiple classes and not necessarily producing a classification of objects

into nonoverlapping classes. In order to build the classification models, the objects belonging to each class need to be analyzed using PCA. Only the significant PCs are retained. For a given class, the resulting model then describes a line (for one PC), plane (for two PCs), or hyper-plane (for more than two PCs). For each modeled class, the mean orthogonal distance of training data objects from the line, plane, or hyper-plane is used to determine a critical distance for classification. New unknown objects are projected into each PC model and the residual distance is calculated. An object is assigned to the model class when its residual distance from the model is below the statistical limit for the class. SIMCA in this work was performed using the PLS Toolbox (Eigenvector Research, USA). Here, the class prediction was not soft, which means the object was assigned to the one class to which the sample was closest.

## 4 CONCLUSIONS

According to a recent review,[39] IR and Raman spectroscopy of brain tissue is the third most frequent biomedical application of these methods after bone and skin studies. The main reasons might be that there is a need to complement currently applied standard methods by rapid and nondestructive methods and that the specific compositional and structural changes of proteins and lipids in brain tumors can be sensitively detected by vibrational spectroscopy. Although morphological changes characteristic of disease processes can be observed via staining and microscopic examination of thin tissue sections, vibrational spectra have the potential to offer additional information as modifications can be assessed on the molecular level before visual abnormalities become apparent. Three potential applications have been demonstrated in this chapter: identification of malignant gliomas, determination of the primary tumor of brain metastases, and studies on murine brain tumor models. Advantages of Raman spectroscopy with NIR excitation include in vivo applications after coupling to fiber-optic probes, higher lateral resolution, and utilization of enhancement effects. Instrumentation for IR spectroscopy is usually less complex than for Raman spectroscopy, and state-of-the-art FT-IR imaging spectrometers offer advantages for rapid assessment of tissue sections. Both factors contribute that applications on brain tissue using IR spectroscopy have been reported three times more frequently than studies using Raman spectroscopy.[39] The general concept behind FT-IR imaging of primary and secondary brain tumors was to measure profiles from tissue types, to identify relevant spectral markers, to organize them into a prediction algorithm, and to validate the model by classification of independent specimens. The high success rates demonstrate that a combination of FT-IR imaging with multivariate feature extraction and classification methods has the clear potential of providing the pathologist with a new tool for the diagnosis of brain tumors. This concept can be transferred to other biomedical applications making FT-IR imaging a versatile analytical technique in this field.

## ACKNOWLEDGMENTS

This work was part of the project "Molecular Endospectroscopy", which was supported by the Volkswagen Foundation within the program junior research groups at universities. We thank Prof. G. Schackert, Dr S.B. Sobottka, Dr M. Kirsch (Clinic of Neurosurgery, University Hospital Dresden) and PD Dr K.D. Geiger (Department of Neuropathology, University Hospital Dresden) for cooperation and support.

## ABBREVIATIONS AND ACRONYMS

| | |
|---|---|
| ALA | 5-Aminolevulinic Acid |
| ATP | Adenosine Triphosphate |
| AU | Absorbance Units |
| BC | Breast Cancer |
| BLI | Bioluminescence Imaging |
| CC | Colorectal Cancer |
| CCD | Charge Coupled Device |
| CE | Cholesterol Ester |
| CT | Computed Tomography |
| [$^{18}$F]FDG | 2-Deoxy-2-[$^{18}$F]fluoro-D-glucose |
| FPA | Focal Plane Array |
| FT | Fourier Transform |
| FT-IR | Fourier Transform Infrared |
| GBM | Glioblastoma Multiforme |
| H&E | Hematoxylin and Eosin |
| IOUS | Intraoperative Ultrasound |
| IR | Infrared |
| LC | Lung Cancer |
| LDA | Linear Discriminant Analysis |
| mid-IR | Mid-Infrared |
| MRI | Magnetic Resonance Imaging |
| NIR | Near-Infrared |
| NIRF | Near-Infrared Fluorescence |
| NMR | Nuclear Magnetic Resonance |
| OCT | Optical Coherence Tomography |
| PC | Principal Component |
| PCA | Principal Component Analysis |
| PEG | Polyethylene Glycol |
| PET | Positron Emission Tomography |
| QDs | Quantum Dots |
| RCC | Renal Cell Carcinoma |
| RGD | Arginine–Glycine–Aspartic Acid |
| SD | Spectral-Domain |
| SIMCA | Soft Independent Modeling of Class Analogies |
| SNR | Signal-to-Noise Ratio |
| TD | Time Domain |
| UHR | Ultra-High Resolution |
| WHO | World Health Organization |

## REFERENCES

1. L.M. Miller, Q. Wang, T.P. Telivala, R.J. Smith, A. Lanzirotti and J. Miklossy, *J. Struct. Biol.*, **155**, 30–37 (2006).

2. M. Szczerbowska-Boruchowska, P. Dumas, M.Z. Kastyak, J. Chwiej, M. Lankosz, D. Adamek and A. Krygowska-Wajs, *Arch. Biochem. Biophys.*, **459**, 241–248 (2007).

3. A. Kretlow, Q. Wang, J. Kneipp, P. Lasch, M. Beekes, L. Miller and D. Naumann, *Biochim. Biophys. Acta*, **1758**, 948–959 (2006).

4. J.M. Legler, L.A. Ries, M.A. Smith, J.L. Warren, E.F. Heinemann, R.S. Kaplan and M.S. Linet, *J. Natl. Cancer Inst.*, **91**, 1382–1390 (1999).

5. P. Kleihues and W.K. Cavenee, 'Pathology and Genetics of Tumors of the Nervous System', IARC Press, Lyon (2000).

6. R.O. Mirimanoff, D.E. Dosoretz, R.M. Linggood, R.G. Ojemann and R.L. Martuza, *J. Neurosurg.*, **62**, 18–24 (1985).

7. K.S. Polyzoidis, G. Miliaras and N. Pavlidis, *Cancer Treat. Rev.*, **31**, 247–255 (2005).

8. P.L. Madsen and N.H. Secher, *Prog. Neurobiol.*, **58**, 541–560 (1999).

9. N. Fujiwara, K. Sakatani, Y. Katayama, Y. Murata, T. Hoshino, C. Fukaya and T. Yamamoto, *Neuroimage*, **21**, 1464–1471 (2004).

10. S. Asgari, H.J. Röhrborn, T. Engelhorn and D. Stolke, *Acta Neurochir.*, **145**, 453–460 (2003).

11. D. Goujon, M. Zellweger, A. Radu, P. Grosjean, B.C. Weber, H. van den Bergh, P. Monnier and G. Wagnieres, *J. Biomed. Opt.*, **8**, 17–25 (2003).

12. W.C. Lin, S.A. Toms, M. Johnson, E.D. Jansen and A. Mahadevan-Jansen, *Photochem. Photobiol.*, **73**, 396–402 (2001).

13. S.A. Toms, W.C. Lin, R.J. Weil, M.D. Johnson, E.D. Jansen and A. Mahadevan-Jansen, *Neurosurgery*, **57**, 382–381 (2005).

14. J. Leppert, J. Krajewski, S.R. Kantelhardt, S. Schlaffer, N. Petkus, E. Reusche, G. Hüttmann and A. Giese, *Neurosurgery*, **58**, 759–767 (2006).

15. M.M. Haglund, M.S. Berger and D.W. Hochman, *Neurosurgery*, **38**, 308–317 (1996).

16. M. Kabuto, T. Kubota, H. Kobayashi, T. Nakagawa, H. Ishii, H. Takeuchi, R. Kitai and T. Kodera, *Neurol. Res.*, **19**, 9–16 (1997).

17. P. Kremer, A. Wunder, H. Sinn, T. Haase, M. Rheinwald, U. Zillmann, F.K. Albert and S. Kunze, *Neurol. Res.*, **22**, 481–489 (2000).

18. W. Stummer, S. Stocker, S. Wagner, H. Stepp, C. Fritsch, C. Goetz, A. Goetz, R. Kiefmann and H.J. Reulen, *Neurosurgery*, **42**, 518–526 (1998).

19. W. Stummer, U. Pichlmeier, T. Meinel, O.D. Wiestler, F. Zanella and H.J. Reulen, *Lancet Oncol.*, **7**, 392–401 (2006).

20. X. Chen, P.S. Conti and R.A. Moats, *Cancer Res.*, **64**, 8009–8014 (2004).

21. Z. Cheng, Y. Wu, Z. Xiong, S.S. Gambhir and X. Chen, *Bioconjugate Chem.*, **16**, 1433–1441 (2005).

22. Z. Cheng, J. Levi, Z. Xiong, O. Gheysens, S. Keren, X. Chen and S.S. Gambhir, *Bioconjugate Chem.*, **17**, 662–669 (2006).

23. X. Gao and S. Nie, *Methods Mol. Biol.*, **303**, 61–71 (2005).

24. W. Cai, D.W. Shin, K. Chen, O. Gheysens, Q. Cao, S.X. Wang, S.S. Gambhir and X. Chen, *Nano Lett.*, **6**, 669–576 (2006).

25. S.A. Toms, P.E. Konrad, W.C. Lin and R.J. Weil, *Technol. Cancer Res. Treat.*, **5**, 231–238 (2006).

26. M.F. Kircher, U. Mahmood, R.S. King, R. Weissleder and L. Josephson, *Cancer Res.*, **63**, 8122–8125 (2003).

27. R. Trehin, J.L. Figueiredo, M.J. Pittet, R. Weissleder, L. Josephson and U. Mahmood, *Neoplasia*, **8**, 302–311 (2006).

28. O. Veiseh, C. Sun, J. Gunn, N. Kohler, P. Gabikian, D. Lee, N. Bhattarai, R. Ellenbogen, R. Sze, A. Hallahan, J. Olson and M. Zhang, *Nano Lett.*, **5**, 1003–1008 (2005).

29. K. Shah and R. Weissleder, *NeuroRx*, **2**, 215–225 (2005).

30. A. Rehemtulla, L.D. Stegman, S.J. Cardozo, S. Gupta, D.E. Hall, C.H. Contag and B.D. Ross, *Neoplasia*, **2**, 491–495 (2000).

31. J.S. Burgos, M. Rosol, R.A. Moats, A. Khankaldyyan, D.B. Kohn, J.M.D. Nelson and W.E. Laug, *Biotechniques*, **34**, 1184–1188 (2003).

32. L. Uhrbom, E. Nerio and E.C. Holland, *Nat. Med.*, **10**, 1257–1260 (2004).

33. K. Shah, Y. Tang, X. Breakefield and R. Weissleder, *Oncogene*, **22**, 6865–6872 (2003).

34. K. Shah, E. Bureau, D. Kim, K. Yang, Y. Tang, R. Weissleder and X.O. Breakefield, *Ann. Neurol.*, **57**, 34–41 (2005).

35. H.J. Böhringer, D. Boller, J. Leppert, U. Knopp, E. Lankenau, E. Reusche, G. Hüttmann and A. Giese, *Laser Surg. Med.*, **38**, 588–597 (2006).

36. K. Bizheva, A. Unterhuber, B. Hermann, B. Povazay, H. Sattmann, A.F. Fercher, W. Drexler, M. Preusser, H. Budka, A. Stingl and T. Le, *J. Biomed. Opt.*, **10**, 11006 (2005).

37. X.D. Li, S.A. Boppart, J. Van Dam, H. Mashimo, M. Mutinga, W. Drexler, M. Klein, C. Pitris,

M.L. Krinsky, M.E. Brezinski and J.G. Fujimoto, *Endoscopy*, **32**, 921–930 (2000).

38. C. Krafft, *Anal. Bioanal. Chem.*, **378**, 60–62 (2004).

39. C. Krafft and V. Sergo, *Spectrosc. Int. J.*, **20**, 195–218 (2006).

40. C. Krafft, S.B. Sobottka, G. Schackert and R. Salzer, *Analyst*, **129**, 921–925 (2004).

41. R. Campanella, *J. Neurosurg. Sci.*, **36**, 11–25 (1992).

42. P.E. Sijens, P.C. Levendag, C.J. Vecht, P. van Dijk and M. Oudkerk, *NMR Biomed.*, **9**, 65–71 (1996).

43. G. Steiner, A. Shaw, L.P. Choo-Smith, M.H. Abuid, G. Schackert, S. Sobottka, W. Steller, R. Salzer and H.H. Mantsch, *Biopolymers*, **76**, 464–471 (2003).

44. C. Krafft, K. Thümmler, S.B. Sobottka, G. Schackert and R. Salzer, *Biopolymers*, **82**, 301–305 (2006).

45. C. Beleites, G. Steiner, M.G. Sowa, M.G. Baumgartner, S. Sobottka, G. Schackert and R. Salzer, *Vib. Spectrosc.*, **38**, 143–149 (2005).

46. C. Krafft, S.B. Sobottka, K.D. Geiger, G. Schackert and R. Salzer, *Anal. Bioanal. Chem.*, **387**, 1669–1677 (2007).

47. C. Krafft, S.B. Sobottka, G. Schackert and R. Salzer, *Analyst*, **130**, 1070–1077 (2005).

48. C. Krafft, L. Neudert, T. Simat and R. Salzer, *Spectrochim. Acta A*, **61**, 1529–1535 (2005).

49. B.R. Wood and D. McNaughton, *J. Raman Spectrosc.*, **33**, 517–523 (2002).

50. J.F. Brennan, T.J. Romer, R.S. Lees, A.M. Tercyak, J.R. Kramer and M.S. Feld, *Circulation*, **96**, 99–105 (1997).

51. K.E. Shafer-Peltier, A.S. Haka, M. Fitzmaurice, J. Crowe, J. Myles, R.D. Dasari and M.S. Feld, *J. Raman Spectrosc.*, **33**, 552–563 (2002).

52. N. Stone, M.C. Hart Prieto, P. Crow, J. Uff and A.W. Ritchie, *Anal. Bioanal. Chem.*, **387**, 1657–1668 (2007).

53. C. Krafft, S.B. Sobottka, G. Schackert and R. Salzer, *J. Raman Spectrosc.*, **37**, 367–375 (2006).

54. C. Nygren, H. von Holst, J.E. Mansson and P. Fredman, *Br. J. Neurosurg.*, **11**, 216–220 (1997).

55. S. Koljenovic, L.P. Choo-Smith, T.C. Bakker Schut, J.M. Kros, H.J. van den Berge and G.J. Puppels, *Lab. Invest.*, **82**, 1265–1277 (2002).

56. S. Koljenovic, T. Bakker Schut, A. Vincent, J.M. Kros and G.J. Puppels, *Anal. Chem.*, **77**, 7958–7965 (2005).

57. C. Krafft, L. Shapoval, S.B. Sobottka, G. Schackert and R. Salzer, *Technol. Cancer Res. Treat.*, **5**, 291–298 (2006).

58. C. Krafft, L. Shapoval, S.B. Sobottka, K.D. Geiger, G. Schackert and R. Salzer, *Biochim. Biophys. Acta*, **1758**, 883–891 (2006).

59. N. Amharref, A. Beljebbar, S. Dukic, L. Venteo, L. Schneider, M. Pluot, R. Vistelle and M. Manfait, *Biochim. Biophys. Acta*, **1758**, 892–899 (2006).

60. K.R. Bambery, E. Schültke, B.R. Wood, S.T. Rigley MacDonald, K. Ataelmannan, R.W. Griebel, B.H.J. Juurlink and D. McNaughton, *Biochim. Biophys. Acta*, **1758**, 900–907 (2006).

61. N. Amharref, A. Beljebbar, S. Dukic, L. Venteo, L. Schneider, M. Pluot and M. Manfait, *Biochem. Biophys. Acta*, **1768**, 2605–2615 (2007). DOI 10.1016/j.bbamem.2007.06.032.

62. G. Schackert and I.J. Fidler, *Int. J. Cancer*, **41**, 589–594 (1988).

63. M. Kirsch, P. Weigel, T. Pinzer, R.S. Carrol, P. Black, H.K. Schackert and G. Schackert, *Clin. Cancer Res.*, **11**, 1259–1267 (2005).

64. C. Krafft, M. Kirsch, C. Beleites, G. Schackert and R. Salzer, *Anal. Bioanal. Chem.*, **389**, 1133–1142 (2007). DOI 10.1007/s00216-007-1453-2.

65. M. Otto, 'Chemometrics: Statistics and Computer Application in Analytical Chemistry', 2nd edition, Wiley-VCH, Weinheim (2007).

66. D. Naumann, 'Infrared Spectroscopy in Microbiology' in "Encyclopedia of Analytical Chemistry", ed R.A. Meyers, John Wiley & Sons Ltd, Chichester, 102–131 (2000).

# Resonance Raman Spectroscopy of Erythrocytes

## Bayden R. Wood and Don McNaughton

*Monash University, Melbourne, Victoria, Australia*

## 1  INTRODUCTION

Hemoglobin (Hb) is possibly the most thoroughly investigated metalloprotein and the subject of innumerable kinetic, thermodynamic, and spectroscopic studies. The elegant molecular symmetry and unique electronic structure of the heme group in Hb and other heme-based metalloproteins underpin its importance in the most fundamental of biological processes including respiration, energy transfer (ET), and catalysis of a myriad of biochemical reactions. These intrinsic molecular properties provide exquisite resonance Raman (RR) spectra rich in information that have enabled a deeper understanding of the structure and function of these dynamic heme-based molecules. The unique spectral properties of hemes were realized and exploited in the 1970s by Spiro,[1-11] Brunner,[12-14] Yu,[15,16] Rousseau,[17] Shelnutt[17,18] Asher,[19-21] Kitagawa,[19,20,22-28] Armstrong[29] and Champion,[30-35] along with many others, laying the foundation for future applications of the Raman technique in erythrocyte analysis. The spectroscopic measurements performed by these early pioneers occurred mainly in the solution phase, taking advantage of the RR effect, which occurs when the incident laser light frequency is in the vicinity of an electronic absorption band of a chromophore. Under these conditions the intensities of certain Raman bands, those associated with the chromophore, may be enhanced by factors of $10^3 - 10^5$. The application of pulsed lasers to dynamical studies of Hb began in the early 1980s with extensive studies by Friedman and coworkers.[36] These studies showed it was possible to kinetically isolate transients on the picometers time scale by flowing samples rapidly past a tightly focused laser.[36] In these experiments an initial laser pulse photodissociates Hb, removing the ligand from the sixth coordination site. This is followed by a temporally delayed probe laser pulse that generates the Raman spectrum of the photoproduct.[36] With these systems, information on the effect of ligand binding on tertiary structure could be deduced, based on the frequency shift of the sixth coordination site bond that links the heme group to the protein, namely the iron–histidine stretching vibration designated $\nu_{Fe-His}$.[36]

The coupling of a Raman spectrometer to a microscope, along with the advent of back-illuminated charge coupled device (CCD) detectors

*Vibrational Spectroscopy for Medical Diagnosis.* Edited by Max Diem, Peter R. Griffiths and John M. Chalmers.
© 2008 John Wiley & Sons, Ltd. ISBN 978-0-470-01214-7.

and holographic notch filters enabled biospectroscopists to probe the molecular architecture of hemes in single cells, opening up in vivo analysis of the fundamental units of life and yielding spectacular results. Raman techniques to probe single living blood cells were developed in the late 1990s by Puppels *et al.*[37,38] investigating lymphocytes[38,39] and granulocytes.[38,40,41] They identified eosinophil peroxidase (EPO) and myeloperoxidase (MPO) in eosinophils and neutrophils, respectively,[38] and confirmed the high-spin six-coordinated structure of these hemes. Otto's group investigated the activation of nicotinamide adenine dinucleotide phosphate (NADPH) oxidase by both soluble and particulate activators by recording RR spectra from the cytoplasm of living neutrophils using a confocal Raman microspectrometer.[41,42] Consistent with earlier absorption spectroscopic studies, the enzymes cyt b558 and MPO were partly reduced upon activation in these cells, whereas activated cyt b558-deficient neutrophils from patients with chronic granulomatous disease (CGD) did not show any MPO reduction. CGD is characterized by a high susceptibility of affected patients to bacterial and fungal infections.

Functional erythrocytes make an excellent subject for RR microspectroscopy and imaging because of the high concentration of Hb (22 mM) and interesting morphology. In this chapter, we will convey the central findings of the application of RR spectroscopy to erythrocyte analysis. We begin with a basic overview of the structure and function of Hb before commencing a discussion on the electronic structure to provide a basis to interpret the RR spectrum of heme molecules. The next section will introduce the instrumentation and methodologies for single cell analysis followed by a detailed analysis of the RR excitation profile of the normal erythrocyte. The importance of determining the effects of temperature, laser power, and laser exposure on the cells will also be discussed and examples of these effects presented. The final sections will focus on applications of the technique in medicine with particular emphasis on malaria and sickle cell disease.

# 2 STRUCTURE AND FUNCTION OF HEMOGLOBIN

## 2.1 Allosteric nature of hemoglobin

The heme moiety behaves as an enzyme, catalyzing important oxygen binding reactions in Hbs, cytochromes P450, and peroxidases. The tetrameric structure of Hb consists of four heme groups bound to one pair of $\alpha$-like globin chains ($\alpha$ and $\xi$), encoded by a cluster of genes on the short arm of chromosome 16, and another pair of $\beta$-like chains ($\beta$, $\varepsilon$, $\gamma$, and $\delta$) encoded by a cluster of genes on the terminal portion of the short arm of chromosome 11. Both $\alpha$- and $\beta$-globin proteins share similar secondary and tertiary structures, each with eight helical segments (labeled helix A–G). When fully oxygenated two of the heme groups in the Hb tetramer move approximately 100 pm toward one another, while the other two separate by approximately 700 pm.[43] This movement was explained by Perutz[44] who suggested that the binding of oxygen to the sixth coordination position on the Fe atom in the high-spin ferrous state triggers a biochemical cascade that characterizes the T (tense-deoxygenated) to R (relaxed-oxygenated) state transition. In the T state, the sixth coordination site of all four Fe atoms is unligated and the proximal histidine is constrained by intersubunit interactions, thus resisting movement into the porphyrin plane and diminishing the oxygen binding constant. When two oxygen molecules bind to the Hb molecule, the quaternary structure "relaxes" to the R state, facilitating $O_2$ binding to the two remaining subunits. The high- and low-spin trigger mechanism was first suggested by Hoard[45] and demonstrates how a simple chemical change can initiate a complex biochemical cascade. Figure 1 illustrates the structural and spin state changes that occur upon the binding of $O_2$ to heme. In the deoxygenated state (Figure 1b) the Fe atom is in a ferric high-spin state ($S = 2$) and lies approximately 40 pm out of the porphyrin plane.[46] In the oxygenated state (Figure 1a) the Fe atom is in a ferrous low-spin state ($S = 1/2$) and resides much closer to the plane of the porphyrin. As

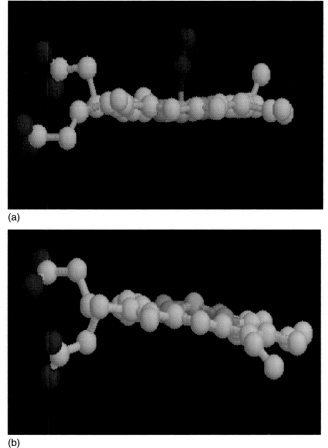

**Figure 1.** Structural change of the heme group of hemoglobin. (a) Oxygenated cell with Fe atom translocated into porphyrin plane. (b) Deoxygenated hemoglobin with Fe atom out of porphyrin plane. The red is oxygen, orange is iron, blue is nitrogen, and gray is carbon. The hydrogen atoms are not represented. [Reproduced from Wood and McNaughton.[88] © Nova Science Publisher, 2006.]

the Fe atom moves into the porphyrin plane it pulls the proximal F8 histidine coordinated to the Fe in the fifth position, which shifts the F helix.[47] The resulting conformational change is transmitted to the subunit interfaces, breaking salt bridges between terminal amino acid groups on the subunits and leading to the R state. Consequently, oxygen binding at one heme site is communicated to neighboring heme sites.[47] The Raman technique is particularly sensitive to the porphyrin core vibrations and hence it is possible to monitor the core expanding, ruffling, and

contracting as the Fe atom translocates in and out of the porphyrin plane.

## 2.2 Electronic structure of heme moieties

Extensive studies utilizing polarized absorption spectroscopy and linear dichroism by Eaton and colleagues[48,49] have led to a robust understanding of the types of electronic transitions observed in Hb and its derivatives. The extended, conjugated system of the porphyrin along with the small energy gap between its valence and conduction band ($\sim$2 eV) enables metalloporphyrins to absorb light strongly in the visible and near-ultraviolet (UV) region of the spectrum. The absorption spectra of all Hb complexes exhibit intense bands known as the *Soret* (or B *state*) *bands* between 400 and 500 nm and two Q bands resulting from $\pi \rightarrow \pi^*$ transitions. The higher frequency Q band, designated $Q_v$ (also called $\beta$ or Q(0,1)), corresponds to a series of upper vibrational levels of the Q transition, while the lower frequency Q band, designated $Q_0$ (also called $\alpha$ or Q(0,0)), corresponds to a vibrationless transition of the Q transition. The $Q_0$ band has a distinct peak in low-spin ferrous complexes and appears as a shoulder on the $Q_v$ band in other heme complexes,[49] the exception being oxyhemoglobin, which exhibit a distinct $Q_0$ band but is in the ferric low-spin state according to the Weiss model.[50] In addition, there are two broad bands between 200 and 400 nm denoted the N and L bands. Figure 2 shows the UV/visible spectrum of Hb highlighting the major electronic transitions predicted by Gouterman's four-orbital

model along with the wavelengths used to excite red blood cells (RBCs) in our studies. In this model the Soret and Q bands result from electronic excitation from a nondegenerate ground state to two doubly degenerate excited states of $E_u$ symmetry.[51] In the following discussion, the symmetry of the heme group is assumed to be $D_{4h}$. More details of the symmetry of differently ligated species will be presented in Section 5.1. The Soret band and quasi-forbidden Q states result from configuration interaction between electron excitations from a nearly degenerate pair of $\pi$ orbitals ($a_{1u}, a_{2u}$) into a doubly degenerate pair of unoccupied $\pi^*$ orbitals ($e_g$). The B and Q states constitute in-phase and out-of-phase combinations of the $\pi-\pi^*$ excitations with approximately equal transition dipole moments. Because the filled orbitals are similar in energy there is a large configuration interaction between the orbital excitations $a_{1u} \rightarrow e_g$ and $a_{2u} \rightarrow e_g$. The Soret band results when the two transition dipoles add together giving the intense band observed at $\sim$400 nm, while for the weaker $Q_0$ band the transition dipoles almost cancel one another out.

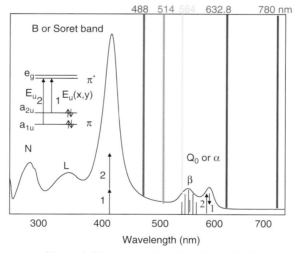

**Figure 2.** An electronic spectrum of hemoglobin showing the N and L bands along with the major $\pi \rightarrow \pi$ transitions including the Soret band and Q bands along with the wavelengths used in our resonance Raman studies on erythrocytes. **Inset**: Gouterman's four-orbital energy level diagram that is used to explain the major $\pi \rightarrow \pi$ transitions from the porphyrin. For the Soret band, the two transition moments, 1 and 2, add, while for the $Q_0$ band the transition moments nearly cancel one another out. The remainder of the intensity from the $Q_0$ band is regained through vibronic mixing with the Soret band resulting in the $Q_v$ side band.

The remainder of the intensity from the $Q_0$ band is regained through vibronic mixing with the Soret band resulting in the $Q_v$ side band, which is an envelope of vibrational bands built on the side of the $Q_0$ band.[52] Crystal field theory enables the determination of the allowed d $\rightarrow$ d transitions and possible charge-transfer transitions. The intense Raman spectra obtained from metalloporphyrin complexes have been interpreted on the basis of vibronically induced scattering from the B (Soret) or Q states from the porphyrin macrocycle.[52]

Laser excitation into the near-IR transitions of heme moieties provides an avenue to explore RR enhancement associated with charge transfer and d–d electronic transitions of heme moieties. It is important to note that the oxyhemoglobin ground state does not classify according to the simple crystal field model. Oxyhemoglobin exhibits a very broad band centered at 925 nm in solution absorption spectra. Magnetic circular dichroism (MCD) and natural circular dichroism (CD) measurements show this band to be composed of four separate bands designated by Eaton *et al.*[48] as bands I to IV. Other bands designated V and VI are observed in the CD spectrum and are thought to result from d $\rightarrow$ d promotions, while two more bands designated VI and VII are observed only in the single crystal spectrum because of their $z$ polarization. Extended Hückel calculations on

the ferrous porphine complexed to imidazole and oxygen indicate that while the $d_{xy}$ orbital remains relatively pure, both the $d_{xz}$ and $d_{yz}$ are strongly mixed with oxygen $\pi_g$ orbitals resulting in four closely spaced molecular orbitals.[49] On the basis of CD spectroscopy, bands I and II, which appear at 1300 and 1150 nm, have been assigned to the magnetic dipole allowed $d_{yz} + O_2(\pi_g) \rightarrow d_{xz} + O_2(\pi_g)$ and $d_{x^2-y^2} \rightarrow d_{xz} + O_2(\pi_g)$, respectively. On the basis of curve fitting MCD spectral data of oxyhemoglobin, bands III ($x$ polarized) and IV ($y$ and $z$ polarized), which appear at 980 and 780 nm, were assigned to $a_{2u}(\pi) \rightarrow d_{xz} + O_2(\pi_g)$ and $a_{1u}(\pi) \rightarrow d_{xz} + O_2(\pi_g)$, respectively. Band VII has been assigned to $O_2(\pi_u) \rightarrow d_{xz} + O_2(\pi_g)$ transition but the assignment is in doubt.[49] Figure 3 shows the band assignments along with the calculated and observed frequencies and magnetic and dipole selection rules for an oxy-heme complex based on extended Hückel calculations performed by Eaton and coworkers.[48]

In the case of deoxygenated Hb, Eaton *et al.*[49] assigned a $z$-polarized band appearing at 917 nm to the $d_{xz} \rightarrow e_g(\pi^*)$ transition and designated it as band I. A band at 813 nm, designated band II, was assigned to the mostly $z$-polarized transition $d_{xz} \rightarrow d_{z^2}$. A very weak charge transfer (CT) band, centered at 758 nm and designated band III, is observed for deoxygenated Hb and

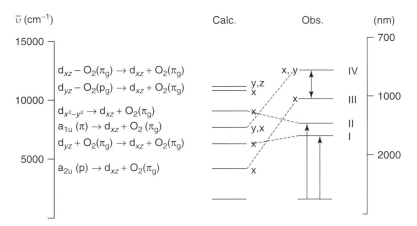

**Figure 3.** Predicted energy levels for the complex of iron–porphine with imidazole and oxygen, and observed wavenumber values for oxyhemoglobin based on the calculations by Eaton and coworkers.[49]

deoxymyoglobin. This band is more than a 1000 times less intense than the Soret band,[53] and is quite narrow. Models that fit the line shape are reported to exhibit only weak coupling to a low frequency vibrational mode.[54] Band III has attracted considerable attention due to its sensitivity to protein conformational changes in response to bond breaking and recombination.[53] The position of band III is assumed to be related to the out-of-plane position of the Fe. On the basis of B-term MCD spectrum, band III is assigned to the $a_{1u}(\pi) \rightarrow e_g(d_{yz})$ transition.[48]

## 3  RAMAN SCATTERING

For a detailed discussion of basic Raman theory and RR theory the reader is referred to, for example, chapters in the Handbook of Vibrational Spectroscopy, volume 1: Raman Spectroscopy:

Theory by G Keresztury (p71–87) and Resonance Raman Spectroscopy by J. L. McHale (p534–556),[55] and only a brief discussion of the main points relevant to heme analysis is presented here. Figure 4 compares infrared (IR) absorption, Rayleigh scattering, Raman scattering, and fluorescence. When the energy ($h\nu_0$) of the incident light is less than the energy of the scattered light, the effect is referred to as *Stokes* scattering, and results from an incident photon interacting with the molecule in the $\nu = 0$ vibrational level of the ground electronic state. When $h\nu_0$ is greater than $h\nu_s$, the effect is referred to as *anti-Stokes* scattering, and is a consequence of the incident photon interacting with the molecule in the $\nu = 1$ vibrational level of the ground electronic state. Away from the laser exciting line, Stokes scattering is much more intense than anti-Stokes scattering due to the relative populations of molecules in the

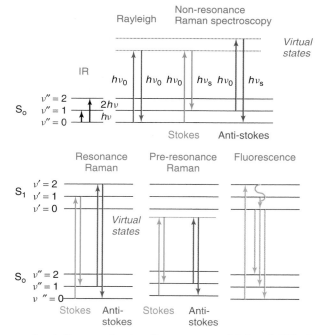

**Figure 4.** Qualitative energy level diagram showing the processes of infrared (IR) absorption, Raman scattering including Stokes and anti-Stokes, Rayleigh scattering and fluorescence. The $S_0$ is the ground electronic state, $S_1$ the excited electronic state, $\nu''$ denotes the vibrational quantum numbers for the ground electronic state, $\nu'$ denotes the vibrational quantum numbers for the excited electronic state, $h\nu_0$ is the energy of the incident photon, and $h\nu_s$ is the energy of the scattered photon.

ground and excited state, which is governed by the Boltzmann distribution.

The scattered light intensity ($I_s$) of a Raman transition between a ground (g) and final (f) state is given by

$$I_s = \frac{8\pi \nu_s^4}{9c^4} I_0 \sum |(\boldsymbol{\alpha}_{\rho\sigma})_{gf}|^2 \qquad (1)$$

where $I_0$ is the incident intensity of the laser at frequency $\nu_0$, $\nu_s$ is the scattering frequency, $c$ is the speed of light, and $(\boldsymbol{\alpha}_{\rho\sigma})_{gf}$ is the transition polarizability tensor with $\rho$ and $\sigma$ the incident and scattered polarizations. The intensity depends on the fourth power of the scattered frequency and $|(\boldsymbol{\alpha}_{\rho\sigma})_{gf}|^2$, which is the Raman polarizability tensor for the vibration $\nu_s$. Thus, $(\boldsymbol{\alpha}_{\rho\sigma})_{gf}$ specifies the coupling between the molecular vibration and the molecular polarizability at frequency $\nu_0$.

# 4 RESONANCE RAMAN ENHANCEMENT

Laser excitation with wavelengths that closely approach an electronic absorption peak of a molecule can greatly enhance the Raman intensities of specific vibrational modes through a mechanism known as *resonance Raman (RR) scattering*. The enhancement can been used to detect analyte concentrations down to $10^{-5}$ M, whereas under nonresonance Raman excitation is usually limited to concentrations greater than $10^{-1}$ M.[56] Figure 4 also illustrates the energy changes associated with RR scattering and fluorescence. In RR scattering, an electron is promoted to an excited electronic state, which is immediately followed by relaxation to a vibrational level in the electronic ground state. Fluorescence, on the other hand, differs from RR scattering in that the relaxation to the electronic ground state ($S_0$) is preceded by radiation-less relaxation of the molecule to the lowest vibrational level in the electronic excited state ($v' = 0$). This process is characterized by a "vibrational cascade", where energy in the form of heat is released. The molecules eventually relax to vibrational levels in the ground electronic state

and in the process they emit photons at various frequencies. This manifests in the spectrum as a broad baseline feature that sometimes can swamp the Raman signal of the analyte under investigation. To avoid fluorescence, samples can be cooled to reduce the bandwidth of the fluorescence or alternatively by selecting excitation wavelengths well away from major electronic transitions of the analyte under investigation.

A number of theories have been espoused to account for the large band enhancement observed when exciting molecules in the vicinity of electronic transitions.[57–64]

The overall polarizability is given by the Kramers–Heisenberg equation based on second-order perturbation theory:

$$(\boldsymbol{\alpha}_{\rho\sigma})_{gf} = \frac{1}{h} \sum_e \frac{\langle f|\mu_\rho|e\rangle\langle e|\mu_\sigma|g\rangle}{\nu_{eg} - \nu_0 + i\Gamma_e}$$
$$+ \frac{\langle f|\mu_\sigma|e\rangle\langle e|\mu_\rho|g\rangle}{\nu_{ef} + \nu_0 + i\Gamma_e} \qquad (2)$$

where $h$ is Planck's constant, $\mu_\rho$ and $\mu_\sigma$ are dipole moment operators, $|g\rangle$ and $\langle f|$ are the initial and final-state wave functions, $|e\rangle$ is the excited state wave function, $\Gamma_e$ is the half-width life time of the excited state, $\nu_{eg}$ and $\nu_{ef}$ are transition frequencies. As $\nu_0$ approaches $\nu_{eg}$, the denominator of one of the terms in the first summation becomes very small so one of these terms becomes very large and is responsible for resonance effects.

The Born–Oppenheimer approximation enables the wave functions of the electronic and vibronic parts to be separated giving

$$\langle f|\mu|e\rangle = \langle j|\mathrm{M}_e|v\rangle \qquad (3)$$

and

$$\langle e|\mu|g\rangle = \langle v|\mathrm{M}_e|i\rangle \qquad (4)$$

where $|i\rangle$ and $\langle j|$ are the initial and final wave functions of the ground electronic state and $|v\rangle$ is the vibrational wave function of the excited electronic state e. The pure electronic transition moment between g and e is $M_e$, which may be expanded in a Taylor's series

$$M_e = M_e^0 + \left(\frac{\partial M}{\partial Q}\right)^0 Q + \cdots \quad (5)$$

where $Q$ is a given normal mode of the molecule.

The first term in the Taylor's series (A term) represents the scattering amplitude, which is directly proportional to the strength of the electronic transition and inversely proportional with its bandwidth, and also depends on the magnitude of the Franck–Condon (FC) integral products $\langle j|v\rangle\langle v|i\rangle$. The magnitude of the FC products reflects the extent of displacement of the excited state potential along the normal coordinate:

$$A = (M_e^0)^2 \frac{1}{\hbar} \sum_v \frac{\langle j|v\rangle\langle v|i\rangle}{\Delta v_v + i\Gamma_v} \quad (6)$$

where $|i\rangle$ and $|j\rangle$ are once again the initial and final vibrational wave functions of the ground electronic state, $v$ represents the intermediate vibrational levels in the resonant excited state, $\Delta v_v$ is the difference between the frequency of laser excitation wavelength and the frequency of the vibration level $v$, $\Gamma_v$ is the half-width of the vibrational wave function for the excited state, and $\hbar$ is Plank's constant. The A term depends on the square of the transition dipole moment, $M_e$, and Franck–Condon integral products. From the equation we can see that as $\Delta v_v$ approaches zero, which occurs when the frequency of the laser matches the energy required to put the molecule into the vibronically excited state, the denominator of the first term reduces to $\Gamma_v$, which is a small energy correction factor for the lifetime of the molecule in the excited state. Thus, under resonant conditions the denominator becomes very small leading to the first term of the Taylor's expansion becoming very large, increasing polarizability, and giving much greater Raman scattering. The A term enhances vibrational modes whose atomic displacements occur along the molecular coordinates that differ between the ground and excited states. To obtain intense scattering, the transition should start from a region of high electron density in the ground state and go to a state where the wave function is such that once populated would also contain significant electron density. This is referred to as *overlap* because of the vertical nature of the transitions and is illustrated in Figure 5.

The second term in the Taylor's series (B term) represents the scattering amplitude and is dependent on the transition moments and electronic

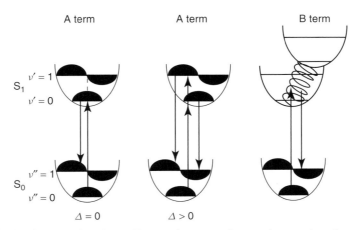

**Figure 5.** Schematic showing type A and type B scattering terms. In type A scattering, the most intense transitions occur when there is significant overlap between the wavefunctions of the ground and excited vibronic states as a function of the displacement of the normal coordinate ($\Delta$). Type B scattering occurs when the vibronic mode couples to a nearby excited electronic transition of similar energy.

wavefunctions of the normal coordinate, $Q$:[65]

$$
\begin{aligned}
B = M_e^0 &\left(\frac{\partial M_e}{\partial Q}\right)^0 \frac{1}{\hbar} \\
&\times \sum_v \frac{\langle j|Q|v\rangle\langle v|i\rangle + \langle j|v\rangle\langle v|Q|i\rangle}{\Delta\nu_v + i\Gamma_v}
\end{aligned}
\tag{7}
$$

The contribution of the B term to the resonance scattering depends on the extent the normal coordinate perturbs the electronic transition moment.[65] Type B modes mix the resonant electronic transition to one of higher energy. The B term depends linearly on the extinction coefficient when one state is weakly dipole allowed.[66] In the B-term scattering, two excited states are mixed through the coordinate operator, $Q$. Consequently, any displacement in the molecule results in a geometry change that is needed to remix the electronic states to obtain new molecular states.[20] Thus, if two transitions are close together in the visible region, like the Q and B (Soret) bands in Hb, then the coordinate operator will help mix these two transitions together. B-term enhancement is only strong from the zero and first vibronic states of the excited state, whereas there is no restriction on the excited vibronic states giving rise to A-term enhancement, hence overtones are allowed by this mechanism.[20]

Since the excited states generated by the absorption in the Soret and Q bands are of $E_u$ symmetry, the allowed vibrational symmetries are $\Gamma_{E_u} \times \Gamma_{E_u} = A_{1g} + A_{2g} + B_{1g} + B_{2g}$. The vibrational symmetries result in depolarization ratios of $\rho = <0.75, \infty, 0.75, 0.75$, respectively. Depolarization ratios of $\rho = \infty$ can be achieved through "inverse" polarization of Raman scattering, first experimentally demonstrated by Spiro and Strekas[1] and originally predicted by Placzek.[67]

Inverse polarization can take place in systems having a threefold or higher symmetry axis and degenerate resonant electronic levels. When the symmetry is lowered, the state of polarization is collapsed to "anomalous" polarization giving values of $3/4 < \rho < \infty$.

Generally, the vibrational modes enhanced by A-term scattering are the totally symmetric $A_{1g}$ modes, which are highly polarized with low depolarization ratio. A-term enhancement of heme macrocyclic vibrational modes generally occurs in the Soret band. Type B modes may have any symmetry depending on the direct product of the two electronic transition representations e.g., $B_{1g}$, $B_{2g}$, and $A_{2g}$ for $D_{4h}$.

Enhancement of normal modes can also occur in the CT region of hemes because of vibronic coupling mechanisms. The theory of vibronic coupling developed by Albrecht[68] is based on a one-electron perturbation model and forms the basis to understanding the type B scattering mechanism. One-electron perturbations of $A_{1g}$, $A_{2g}$, $B_{1g}$, and $B_{2g}$ symmetries can couple to Q transitions.[69] In the 16-membered cyclic polyene model, only nontotally symmetric vibrations $A_{2g}$, $B_{1g}$, and $B_{2g}$ symmetries are vibronically active.[70] The $A_{1g}$ mode is inactive in this model, although theoretically possible if a porphyrin is considered rather than a polycyclic ring.[69] Vibronic coupling is also possible between CT bands that overlap and sometimes mix with the $\pi$ and $\pi^*$ transitions. The charge can be transferred from the porphyrin to the metal and vice versa or from the metal to axial ligands.[66] CT states with $e_g$ symmetry can mix strongly with B and Q states resulting in the enhancement of metal–pyrrole and porphyrin normal modes.[66] Consequently, CT can take place from occupied d orbitals to the vacant porphyrin orbitals, $e_g^*$. Although this process is parity-forbidden, the restriction can be removed by vibronic mixing with intense $\pi$–$\pi^*$ transitions via out-of-plane $E_g$ mixing modes. RR studies on metalloporphyrins have revealed the existence of other scattering mechanisms including nonadiabatic effects and Jahn–Teller distortions. Nonadiabatic effects result from a breakdown of the separability of nuclear and electronic wavefunctions in the Born–Oppenheimer approximation. Jahn–Teller-active modes of $B_{1g}$ and $B_{2g}$ symmetry of variable intensity can be observed, if the excited state is degenerate.[25] It has been shown that the strength of these modes correlates with the strength of the vibronic $Q_0$ band in metalloporphyrins.[16]

# 5 RAMAN SPECTROSCOPY OF HEME PROTEINS

## 5.1 Vibrational assignments of heme

The heme moiety central to Hb, known as *protoporphyin IX*, is depicted in Figure 6. The chromophore produces strong RR scattering with a variety of excitation laser wavelengths due to its high symmetry, large Raman cross-section, and electronic structure. Pioneering RR studies on Hb solutions by the Spiro group[1,4,6,52] and independently by Brunner and coworkers[12,13] led

to an understanding of the resonance effects and ultimately the assignment of the Hb spectrum. The symmetry of the macrocyclic core of the protoporphyrin IX is $D_{2h}$; however, the symmetry of the core without the functional groups is $D_{4h}$. In the high-spin state, when the Fe atom is out of the porphyrin plane the effective symmetry is $C_{4v}$. The point group table for $D_{4h}$ indicates that only the g (gerade) modes (apart from $A_{2g}$, which is both Raman and IR inactive) are Raman active, while the $E_u$ modes are only IR active. Deviation from $D_{4h}$ symmetry results in other modes being weakly allowed. The porphyrin consists of four pyrrole rings connected by four unsaturated methine carbons. If the methyl, ethyl, and propyl groups are treated as point masses, then the 37-atom molecule has 71 in-plane modes

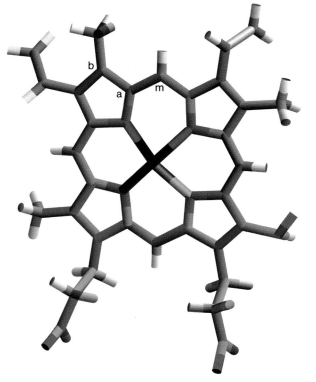

**Figure 6.** Structure of iron protoporphyrin IX showing oxygen atoms (orange), nitrogen atoms (blue), hydrogen atoms (white), carbon skeleton (gray), and iron atom (red). The nomenclature "a" and "b" and "m" are carbon positions used for the assignments in Table 1 with the 'm' referring to the *meso-* carbon atom. [Reproduced from Buller *et al.*[126] © American Chemical Society, 2002.]

(2N − 3), which constitute the following irreducible representation:

$$\Gamma_{inplane} = 9A_{1g} + 8A_{2g} + 9B_{1g} + 9B_{2g} + 18E_u \tag{8}$$

The "g" modes are Raman active, while the "u" modes are IR active. Extensive isotopic studies by Kitagawa and coworkers[28] involving $^{2/1}$H, $^{15/14}$N, and $^{13/12}$C substitution into model metalloporphyrins, including the Ni(II) complexes of octaethylporphyrin (OEP), porphine, and tetraphenylporphyrin (TPP), enabled the first relatively complete set of assignments. The Ni(II) complexes provided excellent model compounds because they are not complicated by the effects of axial ligation. The assigned modes based on these studies are numbered as follows (rather than by normal modern accepted convention) $\nu_1 - \nu_9$ for $A_{1g}$, $\nu_{10} - \nu_{18}$ for $B_{1g}$, $\nu_{19} - \nu_{26}$ for $A_{2g}$, $\nu_{27} - \nu_{35}$ for $B_{2g}$, and $\nu_{36} - \nu_{55}$ for $E_u$.

Out-of-plane modes can also be observed in the Raman spectra of metalloporphyrins. There are 34 out-of-plane vibrations, which classify as follows:

$$\Gamma_{out-of-plane} = 3A_{1u} + 6A_{2u} + 5B_{1u} \\ + 4B_{2u} + 8E_g \tag{9}$$

The $\gamma$ symbol is often used to designate out-of-plane modes of which there are a number of categories.[72] These include (i) out-of-plane wagging modes, (ii) tilting, and (iii) internal folding of the pyrrole rings. Most out-of-plane modes are located at the low wavenumber end of the spectrum, below 800 cm$^{-1}$. The exceptions include hydrogen-wagging modes, which are located between 1100 and 800 cm$^{-1}$.[52] The pyrrole tilting modes are believed to play a role in the dynamics of ligand dissociation in heme proteins.[52] The out-of-plane modes are very weak in the spectra of RBCs and are not considered in the ensuing discussion.

## 5.2 Core size marker bands

RR studies on heme moieties undertaken in the mid 1970s showed a correlation between particular bands in the 1650–1450 cm$^{-1}$ region and the spin state of the heme complex.[2,3,10,52,73] Other studies showed a consistent correlation between the size of the porphyrin cavity and the anomalous polarized band designated $\nu_{19}$ (1590–1550 cm$^{-1}$) and the depolarized band $\nu_{10}$ (1645–1605 cm$^{-1}$).[73,74] It became apparent that the wavenumber values of these vibrational modes depended on the spin state of the Fe atom and are not affected by the spin state density (oxidation state), particularly for $\nu_{19}$.[10] The correlation between porphyrin cavity size and spin state can be explained by the fact that the size of the Fe atom depends on its spin state. The larger the Fe atom the more it expands the porphyrin macrocycle. In the low-spin state both the six electrons of the Fe$^{2+}$ and the five electrons of the Fe$^{3+}$ are accommodated in the d$_\pi$ orbitals leaving the d$_{z^2}$ and d$_{x^2-y^2}$ orbitals empty.[52] Because the antibonding orbitals are vacant the bond order is higher and thus the Fe−ligand bonds are shorter and the core atom smaller, hence the Fe atom is translocated more into the porphyrin plane.[52] When the heme complex is in a high-spin state the antibonding orbitals are occupied and thus the bond order is reduced, lengthening the bond and increasing the core size occupied by the Fe atom and expanding the porphyrin macrocycle. The core size difference between six coordinated Fe$^{2+}$ and Fe$^{3+}$ is only ∼6 pm; however, a core size change of only 1 pm can give rise to a 5–6 cm$^{-1}$ wavenumber shift.[52] The coordination number is also an important factor in determining the core size.[52] In high-spin five-coordinated heme complexes, the Fe atom can be translocated ∼50 pm out of the heme plane, while for six-coordinated hemes the Fe atom is more or less in the heme plane.[52] The correlation with core size extends to other metalloporphyrins besides iron, including the metals Zn, Mn, Co, Ru, and Ni. OEP[75] and TPP[76,77] give core size marker band wavenumber values similar to protoporphyrins except that the E$_u$ core size bands are not seen for OEPs because their activation requires the symmetry lowering effect of the protoporphyrin vinyl groups.[52,75]

Choi explained large deviations from core size plots for $(ImH)_2Fe(II)PP$ by the effects of back donation of the $d_\pi$ electrons of the low-spin $Fe^{2+}$ to empty $\pi^*$ orbitals of the porphyrin.[75] Choi also reported significant deviations from core size plots for the high-spin complex $(2-MeImH)Fe(II)PP$, which were attributed to porphyrin doming.[75] The doming manifests in tilting of the pyrrole rings with the pyrrole N atoms following the out-of-plane motion. This is expected to reduce the $\pi$ conjugation of the methine bridges accounting for the deviations in the core size plots.[52] As a result of doming, the $C_a–C_m$ stretching modes $\nu_3$, $\nu_{10}$, and $\nu_{19}$ are principally affected and are shifted to lower wavenumber.[78]

## 5.3  $\pi$ Electron density marker band

One band that has received considerable attention, due to its apparent sensitivity to electron distribution in the $\pi$ orbitals of the porphyrin macrocycle, is the band appearing between 1376 and 1355 $cm^{-1}$, assigned to a $C–N$ pyrrole-breathing mode designated $\nu_4$,[28] and usually termed the $\pi$ electron density marker band or the oxidation state marker band. In the $Fe^{3+}$ state this band is found near 1375 $cm^{-1}$, while in the $Fe^{2+}$ state it is found at 1355 $cm^{-1}$ [4,79] relatively independent of the spin state. However, when $\pi$ acid ligands such as CO, NO, and $O_2$ are bound to the $Fe^{2+}$, the wavenumber can shift toward the $Fe^{3+}$ region due to the withdrawal of $d_\pi$ electrons from the porphyrin $\pi^*$ orbitals resulting from competition with the $\pi^*$ orbitals of the axial ligands.[4,26,52] $\nu_4$ also correlates with the porphyrin core size, although the slope of core size plots is much smaller than that for the higher wavenumber core size marker bands.[52] The observation that the position of $\nu_4$ correlated well with the core size of heme isolates in the ferric state, led to the suggestion by Yamamoto and Palmer[79] that the RR bands of oxygenated Hb (oxyHb) supported the Weiss model[50] as to the nature of the iron–oxygen bond upon ligation.[9] In this model, the oxygen molecule upon binding to deoxygenated Hb (deoxyHb) withdraws an electron from the ferrous ion, leaving the metal in a low-spin ferric state with the oxygen converted to superoxide $(O_2^-)$.[80] The subsequent diamagnetism implies an antiferromagnetic exchange interaction between the low-spin ferric ion ($S = \frac{1}{2}$) and the $O_2^-$ ($S = \frac{1}{2}$) of the ligand.[81]

## 5.4  Peripheral substituent groups

The enhancement of the internal modes of non-chromophoric substituent groups such as vinyl and phenyl groups depends on kinematic or electronic coupling to the porphyrin core.[52] Electronic coupling involves conjugation with the porphyrin $\pi$ system in either the ground or excited states.[52] In this case, the porphyrin $\pi \to \pi^*$ transition leads to a geometry change of the substituent because of ground or excited state delocalization, thus the normal modes involved in the geometry change are directly enhanced.[52] Substituent modes can also gain intensity from kinematic mixing with porphyrin skeletal modes.[52] Another mechanism that can cause enhancement of peripheral modes involves the inclusion of electronic effects in the excited state.

# 6  RAMAN INSTRUMENTATION AND METHODOLOGIES FOR SINGLE ERYTHROCYTE ANALYSIS

## 6.1  Raman instrumentation

Highly sensitive Raman spectrometers coupled to microscopes are now available as bench top instruments and are sure to become as common as the traditional general laboratory workhorse, the mid-IR spectrometer. Detailed information on the instrumentation and methods for both microspectroscopy can be found in the chapter by Dhamelincourt in, for example, the *Handbook of Vibrational Spectroscopy*.[82] The components in a modern Raman microspectrometer are vastly different from those of the original experiments carried out by C.V. Raman in 1928: the excitation energy is now provided by a highly stable continuous-wave (cw) laser, the dispersing element a holographic grating, the detector a multichannel CCD, and the Rayleigh filter either a holographic notch filter or a highly efficient edge filter. The CCD multichannel detector allows for the collection of data on a time scale unheard of 20 years ago with sampling times typically less than 10 s typical. The filters allow for the collection of Raman scattered light very close to the laser line with a low energy, $\sim 50\,cm^{-1}$, cutoff achievable and typical cutoffs being ca. $200\,cm^{-1}$. For work close to the Rayleigh line, near-edge excitation filters are now available with cutoffs as close as $10\,cm^{-1}$. The excitation laser can range from the UV through the visible and into the near-IR regions and is mainly limited by the CCD detector response curve that falls off rapidly above 1000 nm. Hence, for near-IR lasers with wavelengths longer than $\sim 800\,nm$ the spectral range is reduced to typically $50-2000\,cm^{-1}$. To maximize the spectral range for near-IR systems a back-illuminated, deep depletion CCD detector is employed. With UV and visible lasers the high energy cutoff results in a spectral range similar to that of a mid-IR spectrometer at ca. $200-4000\,cm^{-1}$. Recently, fast confocal Raman imaging systems, relying on rapid read out CCDs and raster scanning at sub second exposures or line scanning of a linearized laser beam, have vastly increased the usefulness of Raman imaging.

One of the major drawbacks of dispersive Raman instruments is that for many colored materials, fluorescing samples and samples containing even minute amounts of fluorescent material, the high photon energy sources (typically 488, 514, 532, 547 or 633 nm) give rise to strong broad fluorescence that can easily swamp the weak Raman signals. Using a low photon energy laser ($>700\,nm$), will often, but not always, reduce or eliminate the fluorescence problem but with a consequent restriction in spectral range and at the expense of signal due to the $\nu_4$ dependence of the Raman scattered intensity. For samples such as biological specimens that are easily burnt or damaged, the longer wavelength lasers are often preferred. For these reasons, most modern dispersion-based systems are equipped with multiple lasers, gratings, and filters.

Despite the use of near-IR lasers, many samples are damaged or still give rise to extreme fluorescence that precludes Raman spectroscopy. The other system available, the Fourier transform (FT) Raman spectrometer reduces the fluorescence by using an Nd/YAG laser at 1064 nm for excitation. For a description of these systems, see the chapter by Chase in Volume 1 of the Handbook of Vibrational Spectroscopy.[83] At this wavelength the CCD detector is of no use and the signal is recorded as an interference pattern and Fourier transformed into the spectral domain in the same manner as for FT-IR spectroscopy. Microscopy systems are available but the disadvantage of such systems is that high laser powers need to be used rather than the $1-20\,mW$, typical of the dispersive systems for microscopy. Such systems are typically 3 orders of magnitude less sensitive than macrodispersive systems and often require hundreds of coadded scans, and hence a long time, to record a useful spectrum from a biological sample. The high power can also lead to sample damage. Recently, new linear array detectors operating in the near-IR have become available and are being coupled to dispersive-based microscopy systems operating with 1064 nm lasers. These promise an order of magnitude increase in sensitivity compared to FT-Raman enabling the use of lower laser powers.

In our work, we have used dispersive Renishaw Raman system 2000 and inVia microscopes (Renishaw plc, Wotton-under-Edge, UK) for single cell work and a variety of other manufacturer systems, including FT-Raman for the investigation of isolated heme products. The lasers and excitation wavelengths that can be used to study erythrocytes include

1. HeNe: 632.8 nm
2. Argon ion: 454.6, 457.9, 465.8, 476.5, 488.0, 496.5, 501.7, 514.5, and 528.7 nm
3. Krypton ion: 406.7, 413.1, 415.4, 468.0, 476.2, 482.5, 520.8, 530.9, 568.2, 647.1, and 676.4 nm.

4. Mixed argon/krypton lasers can be used to produce the wavelengths listed in 2 and 3
5. Near-IR diode lasers: 780, 830 nm and others
6. Ti:sapphire tunable between 675 and 1100 nm wavelengths.

## 6.2 Methods to record spectra of functional erythrocytes

Raman spectra of functional erythrocytes can be recorded using a water immersion objective and affixing cells with poly-L-lysine or Celltack™ to an aluminum-coated or quartz Petri dish. With this configuration, it is possible to perform measurements on functional erythrocytes in growth media, phosphate buffered saline (PBS), and isotonic saline.[84–89] It is also possible to maintain physiological temperatures by placing the Petri dish in a temperature-controlled, purpose-built unit designed to fit onto a microscope stage (Figure 7).[84] The Petri dish has a lid with openings for the water immersion

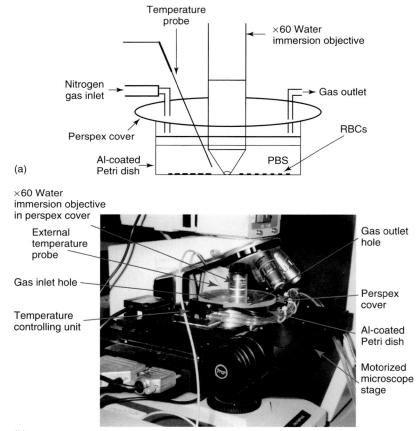

(a)

(b)

**Figure 7.** Device for recording Raman micro spectra of functional erythrocytes in phosphate buffered saline (PBS) using a water immersion objective. (a) Schematic showing the device sealed by a Perspex® cover with an opening for the water immersion objective and inlets and outlets for gas exchange. The Petri dish is coated with aluminum to help maximize collection of scattered photons. The red blood cells (RBCs) are immersed in PBS and attached to the bottom of the dish with poly-L-lysine. (b) The complete unit sitting in a temperature-controlled stage that fits onto the Raman microscope stage. [Reproduced from Wood *et al.*[84] © SPIE, 2004.]

objective and to enable gas exchange along with other insertion points for temperature, pH, and oxygen probes.[84]

Single cells can also be held in position using laser tweezers generated by a low-power diode laser (785 nm), which both traps the cell and simultaneously excites the spectrum.[90] In these experiments, the trapping required 2 mW of power while the spectrum was recorded using 20 mW of power.[90]

Ramser and colleagues used a microscope slide coated with poly-L-lysine and targeted single cells in 50 μL of buffer with a water immersion objective.[91] More recently they reported changes in the erythrocyte membrane of cells suspended in a droplet of PBS and affixed to a glass slide. They attributed these changes to a surface-induced effect mitigated by the poly-L-lysine. In our work, we found that by keeping the cells under physiological conditions in a Petri dish, the majority of erythrocytes retain their discoid morphology and show no evidence of Heinz bodies or perturbation in the exposed cell membrane.

### 6.3 Methods for micro-Raman spectroscopic imaging of functional erythrocytes

Raman images can be generated in a number of ways including raster mapping over the entire sample, which has the advantage of generating a full Raman spectrum at each point in the map but the disadvantage of long data acquisition times. The alternative of building an image through line scanning with the laser defocused to a line minimizes the acquisition and exposure time. For live erythrocytes the long acquisition times of these two techniques usually result in photo/thermal degradation of the Hb, even at low laser powers. Newer instruments are capable of obtaining rapid images at 70 ms per spectrum and may change this situation. Hitherto, we have used filter imaging as opposed to raster mapping to minimize laser exposure. To generate a Raman image, a strong and relatively isolated band that characterizes the chromophore of interest needs

to be selected and the required band-pass filter put in place. To calibrate the filter, a "filter spectrum" of the sample with the appropriate filter needs to be recorded. A filter spectrum enables one to verify where the band of interest lies when the filter is in position. The width of the band of interest determines the spectral resolution, which is usually 20–30 cm$^{-1}$. The image is captured by defocusing the laser spot using a beam expander so that the size of the focused beam matches that of the cell. To account for inhomogeneous illumination across the sample, a background image can be recorded on a silicon wafer and ratioed against the sample image.[91,92]

## 7  RESONANCE RAMAN SPECTROSCOPY OF HEMES IN SINGLE ERYTHROCYTES

### 7.1  Spectral band assignment of hemoglobin in erythrocytes

Strong, highly resolved bands have been reported in RBCs when using 488, 514, 568, 633, and 785 nm laser excitation. Furthermore, clear spectral differences have been observed between oxygenated and deoxygenated cells.[84–86,88,89,93] Most notable are changes in the "core size" or "spin state" marker band region between 1650 and 1500 cm$^{-1}$ between the two states. Other differences are observed in the methine deformation region between 1250 and 1200 cm$^{-1}$. Figures 8 and 9 depict Raman spectra of oxygenated and deoxygenated cells recorded at 488, 514, 564, 633, and 785 nm excitation, respectively. The band assignments for mainly in-plane modes observed in RBCs are listed in Table 1 and are based on the notation system originally proposed by Kitagawa and coworkers[28] for octaethylporphyrinato-Ni(II) with some modifications by Hu *et al.*[94] from their RR studies on myoglobin.

The spectrum taken using 785-nm excitation is significantly different from the other spectra, especially in the 1700–1500 cm$^{-1}$ region. The spectra using 488, 514, 568, and 633-nm excitation show the characteristic high-spin state

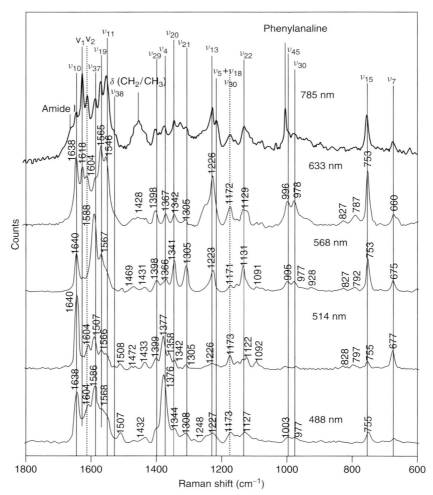

**Figure 8.** A comparison of recorded spectra of oxygenated cells using 785, 633, 568, 514, and 488 nm laser excitation lines showing the major band assignments in the $1800–600\,cm^{-1}$ region. $V_1$ and $V_2$ refer to the vinyl modes, while the $\nu$ indicates the in-plane modes based on Abe and co-workers[28] notation system (see Table 1). [Reproduced from Wood *et al.*[84] © SPIE, 2004.]

profile in the deoxygenated cells with three major bands at approximately 1608, 1581, and $1545\,cm^{-1}$. The spectrum collected at 785-nm excitation exhibits bands at 1650, 1620, 1563, and $1526\,cm^{-1}$ within the $1700–1500\,cm^{-1}$ domain and additionally shows bands from the protein components,[71] whereas the other excitation wavelengths exhibit only porphyrin modes. Protein bands observed with 785-nm excitation include the amide I band at $1650\,cm^{-1}$, the $CH_2$ and $CH_3$

deformation modes from primarily amino acid side chains at $1446\,cm^{-1}$, and the phenylalanine band at $1003\,cm^{-1}$. These bands, along with the band at $1608\,cm^{-1}$, also appear in the spectra of oxygenated cells, although the band at $1526\,cm^{-1}$ is much stronger in the deoxygenated state. Both 1620 and $1608\,cm^{-1}$ bands are assigned to vinyl modes of the heme moiety.[96]

It is interesting to note the position of $\nu_4$, which appears at $1356\,cm^{-1}$ under 785-nm excitation,

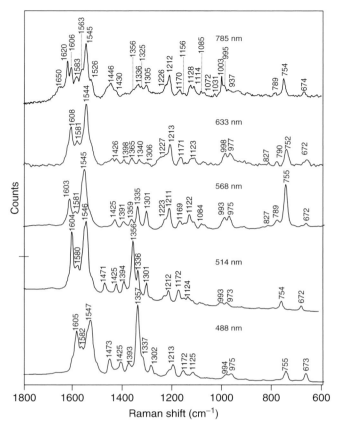

**Figure 9.** A comparison of recorded spectra of deoxygenated cells using 785, 633, 568, 514, and 488 nm laser excitation lines (see Figure 8). [Reproduced from Wood *et al.*[84] © SPIE, 2004.]

and which is consistent with the Yamamoto and Palmer[79] oxidation state marker band hypothesis for ferrous hemes. Yamomoto and Palmer[79] using 441.6-nm excitation (near the Soret band) noted that the strongest band in the spectra of heme proteins occurs between 1361 and 1356 cm$^{-1}$ for reduced (ferrous) heme proteins and between 1378 and 1370 cm$^{-1}$ for oxidized (ferric) proteins. The shift of $\nu_4$ is thought to reflect the electron population in the porphyrin $\pi^*$ orbitals. Increasing the electron population weakens the bonds, resulting in a decrease in vibrational wavenumber. The electron population in the $\pi^*$ orbitals is thought to be increased by back donation of the electrons from the Fe atom's d orbitals. Because back donation is greater for Fe(II) than

Fe(III), the oxidation state marker bands are lower in wavenumber compared to those of Fe(II). The reliability of the correlation of $\nu_4$ with oxidation state has been questioned.[73,97] Spaulding *et al.*[73] investigated a wide range of metalloporphyrins and found $\nu_4$ to be relatively invariant despite the large anticipated differences in charge density migration from the various metals to the conjugated porphyrin ring. The result implied that the charge distribution around the central iron atom is not a fundamental factor in the positioning of this band. This hypothesis also appears to break down for excitation at 633 nm, where the position of this band is observed at 1365 cm$^{-1}$ for both oxygenated and deoxygenated cells. At 514 and 488-nm excitation wavelengths, $\nu_4$ appears in the

**Table 1.** Band positions, assignments, symmetry terms, and local coordinates for oxygenated and deoxygenated cells recorded using a range of excitation wavelengths.

| Oxy 488 nm | Deoxy 488 nm | Oxy 514 nm | Deoxy 514 nm | Oxy 568 nm | Deoxy 568 nm | Oxy 632.8 nm | Deoxy 632.8 nm | Oxy 785 nm | Deoxy 785 nm | Assignment[a] | Local coordinate |
|---|---|---|---|---|---|---|---|---|---|---|---|
|  |  |  |  |  |  |  | Amide I | 1653 (m) | 1650 (m) |  |  |
| 1638 | Absent | 1638 | Absent | 1640 | Absent | 1638 | Absent | 1639 (s) | Absent | $\nu_{10}$ | $\nu(C_aC_m)_{asym}$ |
| Absent | Absent | 1627 | Absent | Absent | Absent | 1618 | Absent | 1620 (s) | 1620 (s) | $\nu_{(C=C)}$ | $\nu(C_a{=}C_b)$ |
| 1604 | 1605 | 1604 | 1604 | Absent | 1603 | 1604 | 1608 | 1604 (s) | 1605 (s) | $\nu_{19}$ | $\nu(C_aC_m)_{asym}$ |
| 1586 | 1582 | 1585 | 1580 | 1588 | 1581 | 1581 | 1585 | 1582 (m) | 1583 (m) | $\nu_{37}$ | $\nu(C_aC_m)_{asym}$ |
| 1568 | Absent | 1557 | Absent | 1567 | Absent | 1565 | Absent | 1566 (vs) | 1563 (s) | $\nu_2$ | $\nu(C_bC_b)$ |
| Absent | 1547 | 1547 | 1546 | Absent | 1545 | 1546 | 1544 | 1547 (s) | 1548 (vs) | $\nu_{11}$ | $\nu(C_bC_b)$ |
| 1507 | Absent | Absent | Absent | 1505 | Absent | Absent | Absent | 1526 (sh) | 1526 (m) | $2\nu_{15}$ | $\nu$(pyr breathing) |
| Absent | 1473 | 1471 | 1471 | 1469 | Absent | Absent | Absent | 1448 (m) | 1446 (m) | −CH$_2$ Scissor | −CH$_2$ Scissor |
| 1432 | 1425 | 1430 | 1425 | 1431 | 1425 | 1428 | 1426 | 1430 (sh) | 1430 (sh) | $\nu_{28}$ | $\nu(C_\alpha C_m)_{sym}$ |
| 1397[c] | 1394[c] | 1397[c] | 1394[c] | 1398[c] | 1391[(c)] | 1398[c] | 1398 | 1397 (w) | 1397 (w) | $\nu_{20}$ | $\nu$(pyr quater-ring) |
| 1376 | 1357 | 1377 (m) | 1356 | 1366 | 1359 | 1367 | 1365 | 1371 (w) | 1356 (w) | $\nu_4$ | $\nu$(pyr half-ring)$_{sym}$ |
| 1344 | 1337 | 1336 | 1336 | 1341 | 1335 | 1342 | 1340 | 1337 (m) | 1336 (m) | $\nu_{41}$ | $\nu$(pyr half-ring)$_{sym}$ |
| 1308 | 1302 | 1301 | 1301 | 1305 | 1301 | 1306 | 1306 | 1322 (m) | 1325 (m) | $\nu_{21}$ | $\delta(C_mH)$ |
| 1248 | Absent | 1245 | Absent | 1248 | Absent | 1249 | Absent | 1306 (m) | 1305 (m) | $\nu_{13}$ | $\delta(C_mH)$ |
| 1227 | 1221 | 1228 | 1220 | 1223 | 1223 | 1226 | 1223 | 1248[b] | 1248[b] | $\nu_{13}$ or $\nu_{42}$ | $\delta(C_mH)$ |
| Absent | 1213 | Absent | 1212 | Absent | 1211 | Absent | 1213 | 1226 (s) | 1223 (m) | $\nu_5 + \nu_{18}$ |  |
| 1173 | 1172 | 1171 | 1172 | 1171 | 1169 | 1172 | 1171 | 1212 (m) | 1213 (s) | $\nu_{30}$ | $\nu$(Pyr half-ring)$_{asym}$ |
| 1127 | 1124 | 1134 | 1124 | 1131 | 1122 | 1129 | 1123 | 1174 (m) | 1170 (m) | $\nu_{22}$ | $\nu$(Pyr half-ring)$_{asym}$ |

**Table 1.** (*continued*).

| Oxy 488 nm | Deoxy 488 nm | Oxy 514 nm | Deoxy 514 nm | Oxy 568 nm | Deoxy 568 nm | Oxy 632.8 nm | Deoxy 632.8 nm | Oxy 785 nm | Deoxy 785 nm | Assignment[a] | Local coordinate |
|---|---|---|---|---|---|---|---|---|---|---|---|
| Absent | Absent | 1090 | 1082 | 1091 | 1084 | 1090 | 1084 | 1156 (w) | 1156 (w) | $\nu_{23}$ | $\nu(C_bC_1)_{asym}$ $B_{1g}$ |
|  |  |  |  |  |  |  |  | 1127 | 1128 | $\nu_5$ |  |
|  |  |  |  |  |  |  |  | 1082 | 1085 | $\delta(=C_bH_2)_4$ |  |
|  |  |  |  |  |  |  |  | 1074 | 1072 | $\delta(=C_bH_2)_4$ |  |
|  |  |  |  |  |  |  |  | 1047 |  | $\nu(O=O)^+$ |  |
|  |  |  |  |  |  |  |  | 1031 | 1031 | $\delta(=C_bH_2)_4$ |  |
|  |  |  |  |  |  |  |  | 1003 | 1003 | Phenylalanine |  |
| 1003 | 994 | 1001 | 993 | 995 | 998 | 996 | 996 | 993 | 995 | $\nu_{47}$ | $\nu(C_bC_1)_{asym}$ |
| 977 | 975 | 972 | 973 | 977 | 977 | 978 | 972 | 974 | 974 | $\nu_{46}$ | $\delta(Pyr\ deform)_{asym}$ and/or $\gamma(=C_bH_2)_{sym}$ |
| Absent | Absent | Absent | Absent | 827 | 827 | 827 | 827 | 827 | 827 | $\gamma_{10}$ | $\gamma(C_mH)_{sym}$ |
| Absent | Absent | Absent | Absent | 792 | 790 | 787 | 790 | 789 | 789 | $\nu_6$ | $\nu(Pyr\ breathing)$ |
| 755 | 755 | 755 | 754 | 753 | 752 | 753 | 752 | 754 | 754 | $\nu_{15}$ | $\nu(Pyr\ breathing)$ |
| 675 | 673 | 676 | 672 | 675 | 672 | 668 | 672 | 676 | 673 | $\nu_7$ | $\delta(Pyr\ deform)_{sym}$ |
|  |  |  |  |  |  |  |  | 622 | 622 | ? | $A_{1g}$ |
|  |  |  |  |  |  |  |  | 567 |  | $\nu(Fe-O_2)$ |  |
|  |  |  |  |  |  |  |  | 419 |  | $\delta(Fe-O-O)$ |  |

[a] Assignments are based mainly on labeling scheme originally devised by Abe *et al.*[28] for octaethylporphyrinato-Ni(II).

[b] Local coordinates based mainly on studies by Hu *et al.*[94] for myoglobin.

[c] Aggregation enhanced marker bands. The "a", "b", and "m" notation refer to carbon atoms shown in Figure 6.

range consistent with the Yamamoto and Palmer[79] hypothesis, while at 568 nm the band appears at 1359 cm$^{-1}$ which is at the top end of this range.

Bands observed between 1235 and 1225 cm$^{-1}$ in RR spectra of metalloporphyrins are usually assigned to C–H in-plane bending vibrations of the methine hydrogen.[12,28,94] The molecular environment of these vibrations and hence their band positions is severely affected by protein interactions and heme-stacking. An early study on metallo-OEP and hemes[12] suggested that the wavenumber of the band appearing at ~1220 cm$^{-1}$ is dependent on the strength of the

coordination between the porphyrin and metal. The decrease in wavenumber from ~1225 cm$^{-1}$ in oxy- to ~1210 cm$^{-1}$ in deoxyerythrocytes could reflect such a dependence. The low wavenumber region in the spectra of oxygenated and deoxygenated erythrocytes and the heme isolates is dominated by a number of strong bands at ca. 1169, 1122, 1091, 995, 976, 931, 823, 789, 752, and 667 cm$^{-1}$, most of which have been previously assigned.[94] While the bands in the low wavenumber region are in similar positions for oxygenated and deoxygenated cells, the relative intensities of some of these bands differ.

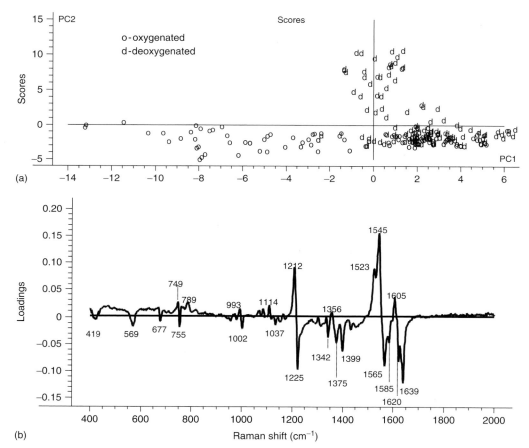

**Figure 10.** (a) PC1 versus PC2 scores plot of spectra recorded over four cycles of deoxygenation/oxygenation showing oxygenated (o) and deoxygenated (d) spectra. The separation between the groupings occurs mainly along PC1. (b) Corresponding PC1 loadings plot shows the bands important in separating the two groupings. The positive loadings are associated with the deoxygenated cells while the negative loadings are correlated with the oxygenated cells. [Reproduced from Wood *et al.*[85] © Springer, 2007.]

In a recent study, spectra from four oxygenation and deoxygenation cycles were recorded utilizing the sensitivity of the River Diagnostic Raman system (River Diagnostics B.V., The Netherlands) with 785-nm excitation over a 60 min period.[85] The low power (2–3 mW at the sample) prevented photo/thermal degradation and negated protein denaturation leading to heme aggregation. The large database consisting of 210 spectra from the four cycles was analyzed with principal component analysis (PCA). PCA reduces the dimensionality of a data set by representing the data as a linear combination of independent variables that describes the major trends. The scores are calculated projections of the Raman spectra onto the principal components (PCs), while the loading vectors report the variables (wavenumber values) that are important in explaining the observed trends in the scores plot. Figure 10 shows the PC1 versus PC2 scores plot and PC1 loadings plot for spectra recorded over the four oxygenation cycles. The scores plot shows a separation along PC1 between the oxygenated and deoxygenated cells. The PC1 loadings provide excellent detail on the important bands contributing to

the differences between oxygenated and deoxygenated states. A band at $567\,cm^{-1}$, observed in the spectra of oxygenated cells and at $569\,cm^{-1}$ in the corresponding PC1 loadings plot, was assigned to the $Fe–O_2$ stretching mode,[14] while a band appearing at $419\,cm^{-1}$ was assigned to the $Fe–O–O$ bending mode based on previous studies.[95,98] A closer inspection of this region through one oxygenation and deoxygenation cycle showed both the 567 and $419\,cm^{-1}$ bands increasing with atmospheric exposure over a period of 25 min (Figure 11).

For deoxygenated cells, the enhancement of $B_{1g}$ modes with 785-nm excitation is consistent with vibronic coupling between band III and the Soret transition. In the case of oxygenated cells, the enhancement of iron-axial out-of-plane modes and nontotally symmetric modes is consistent with excitation into the y,z polarised transition designated by Eaton *et al.*[48] as band IV $a_{1u}(\pi) \rightarrow d_{xz} + O_2(\pi_g)$ centered at 785 nm. The enhancement of nontotally symmetric $B_{1g}$ modes in oxygenated cells suggests vibronic coupling between band IV and the Soret band.

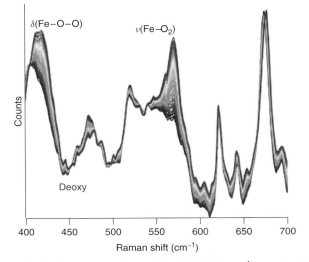

**Figure 11.** A series of spectra in the low wavenumber region ($400–700\,cm^{-1}$) recorded in 10-s intervals going from the deoxygenated state to the oxygenated state for one oxygenation/deoxygenation cycle. The spectra show the increase in the 567 and $419\,cm^{-1}$ bands assigned to the $\nu(Fe–O_2)$ and $\delta(Fe–O–O)$ ligand modes, respectively. [Reproduced from Wood *et al.*[85] © Springer, 2007.]

## 7.2 Resonance Raman spectra of ligated hemoglobin in erythrocytes

Ligand exchange can dramatically modify the heme spectral profile in RBCs. Figure 12 shows Raman spectra of oxygenated, deoxygenated, carboxy (CO)−, and metHb-erythrocytes recorded with 632.8-nm excitation at 4 °C with 10 s of laser exposure. MetRBC and deoxyRBC have the iron atom in the high-spin ferric and ferrous

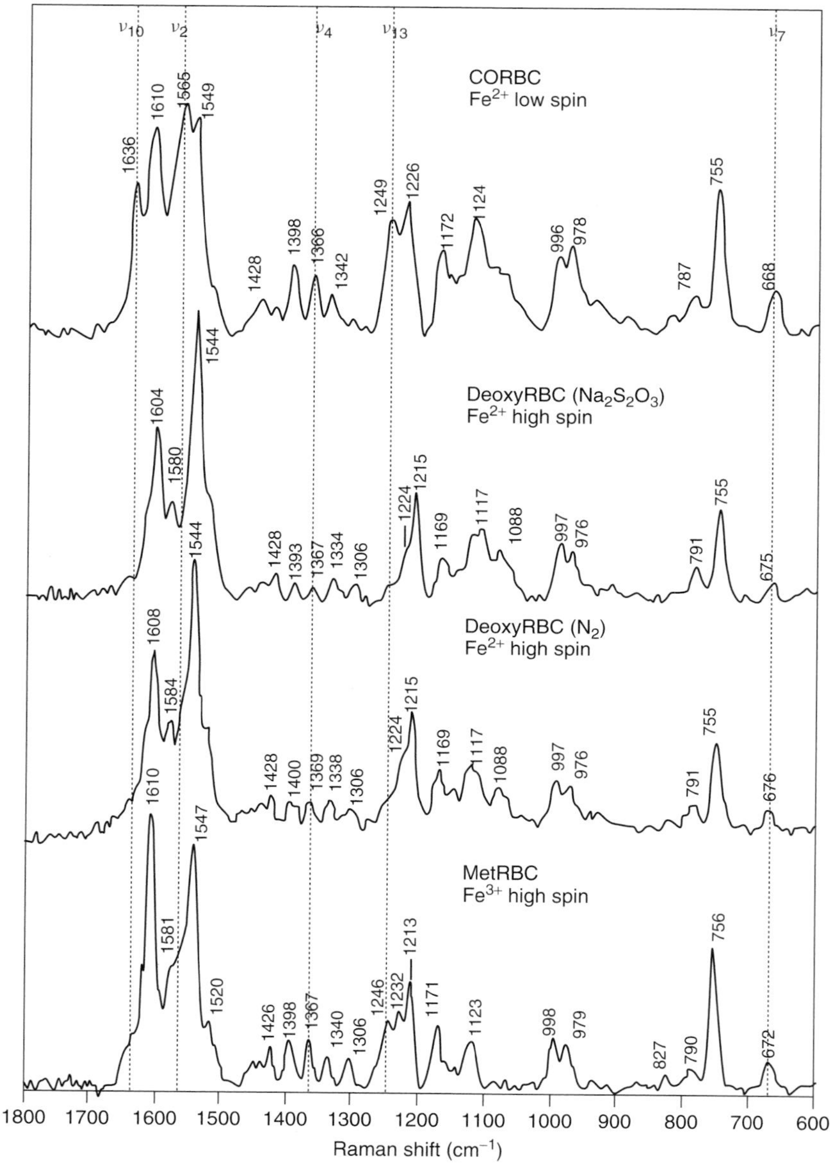

**Figure 12.** Resonance Raman spectra recorded at 632.8-nm excitation of carboxylated (CORBC), deoxygenated with nitrogen (deoxyRBC (N₂)), deoxygenated with sodium dithionite (deoxyRBC (Na₂S₂O₃)), and met (metRBC) erythrocytes. [Reproduced from Wood *et al.*[89] with permission from Elsevier, © 2001.]

state, respectively; however, metHb cannot bind oxygen. The spectra are similar in the core size marker region (1600–1500 cm$^{-1}$) showing a three band profile characteristic of ferric hemes.[89] The metHb shows an increase in the relative intensity of the 1610 cm$^{-1}$ band and the 1520 cm$^{-1}$ shoulder band. CarboxyRBCs and oxyRBCs also have a very similar profile in the 1650–1500 cm$^{-1}$ region, both exhibiting the five-band profile characteristic of low-spin complexes. The intensity ratio of the band at 1224 to that of 1248 cm$^{-1}$ is greater in carboxy RBCs compared to oxyRBCs, but otherwise the spectra are very similar. The difference in this case is most likely attributed to laser-induced effects as opposed to differences due to ligation as will be discussed.

## 7.3 Photo and thermal decomposition studies of Hb in erythrocytes

Raman imaging experiments require prolonged laser exposure, which can result in photo and/or thermal degradation of the heme molecules. To investigate these effects, we performed a number of studies where the laser exposure time, power, and temperature were varied.

## 7.4 The effect of laser exposure on single erythrocytes

To investigate the effects of laser exposure, a time series of spectra were recorded every 5 min during

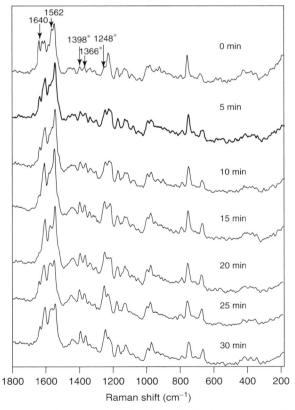

**Figure 13.** Raman spectra recorded of an oxygenated cell gradually deoxygenated (0–20 min) using nitrogen gas and then reoxygenated (25–30 min) with atmospheric $O_2$. The asterisks show the bands that appear to increase due to constant laser exposure, while the other named bands follow the oxygenation cycle of the cell. [Reproduced from Wood *et al.*[84] © SPIE, 2004.]

a 30 min oxygenation/deoxygenation experiment. The power at the sample was ~0.5 mW at 632.8-nm excitation. The cell was exposed for 10 s during each measurement and the temperature of the media held at 4 °C.

The spectra in Figure 13 show spectral changes for erythrocytes deoxygenating in the first 20 min and then reoxygenating until the 30 min mark. The $O_2$ marker band ($\nu_{10}$) at 1640 cm$^{-1}$ decreases during deoxygenation and then increases when the cells are exposed to the atmosphere. The intensity of the totally symmetric $A_{1g}$ mode at 1562 cm$^{-1}$ assigned to $\nu_2$ also appears to cycle with $O_2$ concentration. The 1248 cm$^{-1}$ band, along with the pyrrole in-phase breathing modes at 1398 and 1366 cm$^{-1}$, increase in intensity during the entire experiment, indicating a photoinduced effect caused by multiple spectral acquisitions on the same cell. In another experiment, the cells were exposed to 632.8-nm defocused laser light at 4 °C for 3 min before recording a spectrum. Spectra were then recorded at 15 and 120 min after the initial laser exposure. Figure 14 shows that the change induced by constant laser exposure is irreversible. From these studies it is clear that the 1248 cm$^{-1}$ band appears to be particular sensitive to photoinduced effects.

## 7.5 The effect of temperature on single erythrocytes

To investigate the effect of temperature, spectra were recorded every 2 °C from 4 to 52 °C. Figure 15(a)–(d) depicts baseline corrected spectra presented as interpolated matrix plots showing different spectral windows, while Figure 15(e) shows the raw spectra with no baseline correction. Due to laser exposure the actual temperature of the cell would be approximately 2° higher than that recorded by the probe, which is ~2 mm from the cell. This approximation is based on the calculations and approximations of Ramser *et al.*[91] At the higher temperatures (>42 °C), the spectra show some similar features to those presented in Figure 14, after prolonged laser exposure on the same cell. The spectra exhibit dramatic changes in band intensity as the temperature increases.[84] Bands at 1609, 1398, 1366, 1250, 1170, 1123, 998, 972, and 666 cm$^{-1}$ increase, while bands at 1636, 1565, 1621, and 1226 cm$^{-1}$ decrease in intensity. As the temperature increases so too does the signal-to-noise ratio (SNR). A plot of SNR for the 1250–1200 cm$^{-1}$ (signal) and 1800–1700 cm$^{-1}$ (noise) regions at each temperature is presented in Figure 16. The SNR values

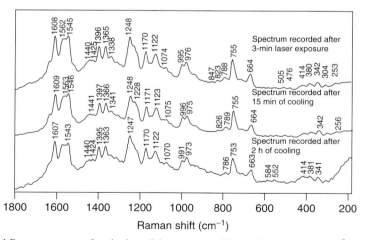

**Figure 14.** Recorded Raman spectra of a single cell in sequence. The cell was cooled to 4 °C and exposed to 100% laser power for 3 min. The cell was then allowed to cool and spectra were taken after 15 min and 2 h. Each spectrum was recorded at the center of the cell with 10 s of focused laser exposure and one accumulation. In this case, the induced effect was irreversible. [Reproduced from Wood *et al.*[84] © SPIE, 2004.]

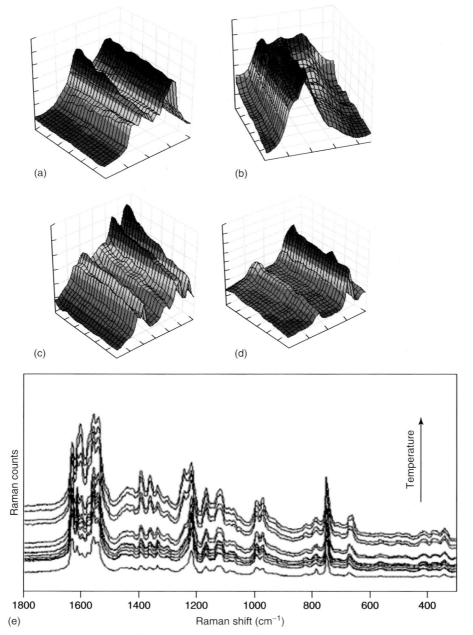

**Figure 15.** Interpolated plots showing baseline corrected Raman spectra as a function of temperature for (a) 1700–1500, (b) 1270–1190, (c) 1200–900, and (d) 800–600 cm$^{-1}$ regions, and (e) 1800–350 cm$^{-1}$ region not baseline corrected. [Reproduced from Wood *et al.*[84] © SPIE, 2004.]

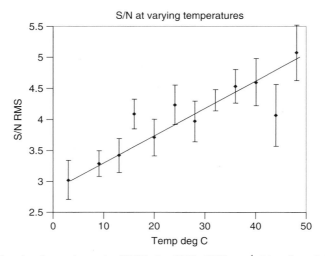

**Figure 16.** A plot of the signal-to-noise ratio (SNR) for $1250–1200\,\text{cm}^{-1}$ (signal) and $1800–1700\,\text{cm}^{-1}$ (noise) regions at specified temperatures. The SNR increases with increasing temperature indicating aggregation of heme units. [Reproduced from Wood *et al.*[84] © SPIE, 2004.]

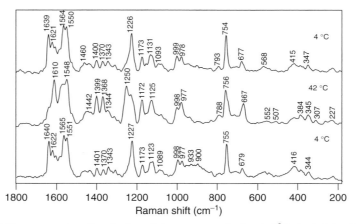

**Figure 17.** Recorded Raman spectra of a single cell in sequence, firstly at $4\,^\circ\text{C}$, heated to $42\,^\circ\text{C}$, and then cooled again to $4\,^\circ\text{C}$. Each spectrum was recorded at the center of the cell with 10 s of focused laser exposure and only one accumulation. The changes due to heat in the spectrum are shown to be reversible. [Reproduced from Wood *et al.*[84] © SPIE, 2004.]

were averaged for each temperature and the SNR root mean square (RMS) values plotted as a function of temperature. The plot shows a general linear relationship between SNR and increasing temperature.

The effects of thermal denaturation were further investigated by, (i) cooling the cells to $4\,^\circ\text{C}$, (ii) warming the suspension to $42\,^\circ\text{C}$, and

(iii) cooling back to $4\,^\circ\text{C}$, taking a spectrum at each of the three temperatures (Figure 17). The top spectrum closely matches the bottom spectrum and is characteristic of oxyHb, while the middle spectrum is a close match to spectra recorded at high temperatures and/or after continuous laser exposure, proving that the thermal denaturation effect can be reversed if the damage is not severe. This result conclusively demonstrates that the spectrum that characterizes the thermal and/or photodamaged cell with enhanced bands at 1396, 1365, and 1248 cm$^{-1}$ is not the result of photooxidation of oxyHb to metHb (Fe$^{3+}$ high-spin heme) simply because metHb cannot bind oxygen.

## 7.6 The effect of temperature on the electronic spectra of erythrocytes

Figure 18(a) shows UV–visible spectra of isolated Hb recorded from 25 to 60 °C in 5 °C intervals, and a spectrum of RBCs recorded at 25 °C. The absorbance spectra of the isolated Hb are

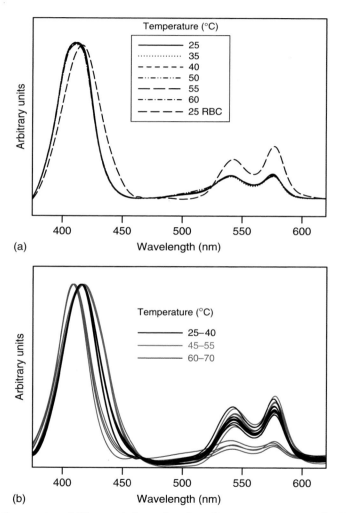

**Figure 18.** (a) Absorption spectra of Hb recorded as a function of temperature compared with a spectrum of a red blood cell (RBC) at 25 °C. Spectra were acquired in 5 °C intervals from 25 to 65 °C. (b) Corresponding absorption spectra of RBCs recorded as a function of temperature. For clarity the spectra have been grouped in ranges 25–40 (black), 45–55 (red), and 60–70 °C (blue). The spectra of the RBCs show a distinct shift in the Soret band as a function of temperature. The concentration of red blood cells is 20 μL per cm$^3$ of phosphate buffered saline. [Reproduced from Wood *et al.*[84] © SPIE, 2004.]

virtually identical for the entire temperature range. These spectra differ significantly in terms of relative band intensity, band width, and position when compared to the spectrum of the red cells recorded at 25 °C. The full-width-at-half-maximum of the Soret band ~416 nm is considerably greater in the red cells and the band is red-shifted 3 nm. The ratio of the $Q_0$ band to the Soret band in RBCs is larger than in isolated Hb. Figure 18(b) depicts UV−visible spectra of RBCs as a function of temperature. Spectra were recorded every 5 °C from 25 to 65 °C. For clarity, the spectra presented have been grouped into

three temperature intervals 25−40, 45−55, and 60−65 °C. UV−visible spectra of RBCs recorded in the 45−55 °C range clearly show broadening and a red shift, ~3 nm, of the Soret band compared to spectra recorded in the physiological tolerant temperature region ~25−40 °C. The Q bands are also slightly broader and red-shifted ~1−2 nm at the higher temperature. At >55 °C, the cells are expected to lyse, releasing Hb into the saline. The presence of free denatured Hb results in a dramatic blue shift and narrowing of the Soret band, and the Q bands are considerably weakened at these high temperatures. The spectrum of RBCs

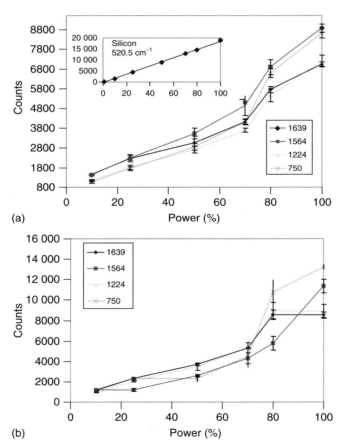

(a)

(b)

**Figure 19.** Plots showing Raman counts for a number of selected bands as a function of power and temperature. The maximum power at the sample is between 2 and 3 mW, which represents 100% in the plots. (a) Plot of Raman intensities verus power at 4 °C. Inset shows power dependence study of the 520.5 cm⁻¹ silicon band at room temperature. (b) Corresponding plot at 42 °C. [Reproduced from Wood *et al.*[84] © SPIE, 2004.]

at high temperatures is essentially the same as that of Hb in solution.

## 7.7 The effect of laser power on single erythrocytes

Figure 19 depicts the laser power dependence of the 520.5 cm$^{-1}$ band of silicon along with the power dependence of a number of bands from the heme moiety in the RBC at 4 and 42 °C. The maximum power at the sample is between 2 and 3 mW, which represents 100% in the plots. As expected, the silicon mode clearly shows a linear dependence as a function of power. However,

both the 4 and 42 °C measurements on Hb within the erythrocyte show a nonlinear dependence. Moreover, the enhancement of these vibrational modes as a function of power is greater at 42 °C than at 4 °C and appears to reach a saturation point at 80% power for some bands. Non–linear effects are attributed to excitonic interactions through π → π interactions. Such effects would be expected when hemes are in close proximity to one another. The effect would be more manifest when hemes aggregate in response to protein denaturation as would be expected at 42 °C. The UV/vis spectra support this hypothesis as the spectra of Hb at higher temperatures show a

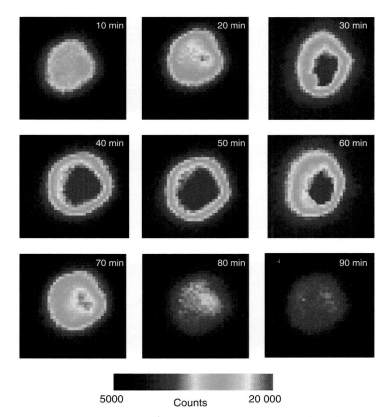

**Figure 20.** Raman images of the 1250−1210 cm$^{-1}$ region recorded on the same cell during deoxygenation (0–50 min) followed by reoxygenation (50–90 min) at 632.8 nm. Each image was acquired with 180 s of continuous defocused laser exposure. The bright red area shows high counts indicative of heme aggregation, while the blue indicates low counts where the Hb is not aggregated. The decrease in counts during the 50–90 min mark as the cell reoxygenates is possibly explained by the O$_2$ acting as a spacer molecule between aggregated hemes. [Reproduced from Wood *et al.*[84] © SPIE, 2004.]

red shifting of the Soret band, which is another indicator of excitonic interactions.

## 7.8   Raman imaging of single erythrocytes

Figure 20 shows Raman filter images of the $1248\,cm^{-1}$ band collected at 10-min intervals for 50 min from a single cell deoxygenated under $N_2$ flow at room temperature. The $1248\,cm^{-1}$ band was chosen as the imaging band because of its radical intensity change from constant exposure. On the basis of the filter band pass the actual wavenumber region imaged was found to be approximately $20\,cm^{-1}$. The $N_2$ flow was turned off after 50 min and images of the same cell were then recorded every 10 min from the 60 to the 90-min mark as the cell gradually oxygenated. Each image was recorded by defocusing the laser to illuminate the entire cell and acquiring the image with 180 s of constant laser exposure at $4\,^{\circ}$C. A Raman spectrum was recorded immediately following each image acquisition. The first image recorded after 10 min of deoxygenation shows a homogeneous green background. The intensity of

counts increases as the cell becomes more deoxygenated with continuous laser exposure. After 50 min the cell is completely deoxygenated and the image is bright red indicating a high number of Raman counts. Figure 21 shows the corresponding Raman spectra collected directly after each image was recorded. The raw spectra show a marked increase in both the baseline and the enhancement of many bands as the cell deoxygenates. At the 50-min mark the cell is reoxygenated and the subsequent images from the 60-min mark to 90 min show a decrease in the overall band intensities and baseline. However, the actual relative intensities do not change from the deoxygenated to the oxygenated state and the spectra do not resemble the oxygenated cell recorded at 632.8 nm. Small spectral changes were observed in the $755-745\,cm^{-1}$ and $1650-1500\,cm^{-1}$ regions. In particular, the band at $755\,cm^{-1}$ in the oxygenated state decreases in intensity during deoxygenation, while the band at $746\,cm^{-1}$ appears more pronounced. The band at $1636\,cm^{-1}$, which is assigned to $\nu_{10}$ and is sensitive to $O_2$ concentration, shows a slight decrease in intensity going from oxygenated to the deoxygenated state. However, after the cell was reoxygenated

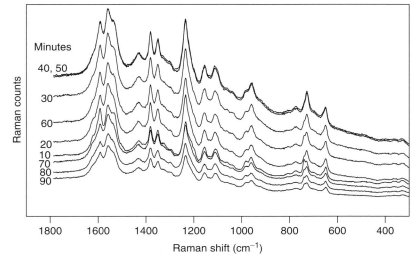

**Figure 21.**   Corresponding Raman spectra recorded after each image are presented in Figure 20. The spectra show the aggregation marker bands at 1395, 1365, and $1248\,cm^{-1}$ and the baseline increases as the cell deoxygenates (10–40 min) and decreases as the cell reoxygenates (50–90 min). [Reproduced from Wood *et al.*[84] © SPIE, 2004.]

this band did not regain its initial intensity indicating Hb dysfunction. The images presented in Figure 20 and the corresponding spectra presented in Figure 21 hence reflect spectral changes that accompany continuous laser exposure and do not reflect differences in oxygen concentration. When the oxygen flow was recommenced at the 50 min mark both the baseline and overall signal decreased. This indicated that the concentration of heme units decreased with the addition of oxygen. Thus, the oxygen may be acting as intermolecular spacer keeping the heme units apart. We hypothesize that the Hb within the cells when exposed to high power, high temperatures, and/or constant laser exposure, causes heme denaturation resulting in heme aggregation.

To both minimize the effects of laser exposure and endeavor to monitor oxygen concentration directly, we chose to record only two images of a fully deoxygenated and oxygenated cell. We used the oxygenation marker band $\nu_{10}$ at 1639 cm$^{-1}$ as the imaging band and recorded images for only 120 s at 4 °C. The image presented in Figure 22(a) shows the cell in the deoxygenated state, where the blue colors indicate very small contributions from $\nu_{10}$ at 1639 cm$^{-1}$. After oxygenation, higher intensities for $\nu_{10}$ are indicated by red hues in Figure 22(b).

## 7.9 The effect of substrate on single erythrocytes

Ramser *et al.*[91] reported effects that influenced Raman images and spectra of single functional erythrocytes using 514-nm excitation. These included changes in the cell membrane due to a surface-induced effect, possibly mitigated by the poly-L-lysine used to attach the cells, and a photoinduced fluorescence that they attributed to the conversion of oxyHb to metHb. The fluorescence effect masked any chance of obtaining reliable images of cells in the oxygenated and deoxygenated state. More recently, as described above, it was demonstrated using 633-nm excitation that exposure of red cells to prolonged laser power and/or high temperatures (>42 °C)

resulted in a general increase in SNR along with the enhancement of a number of bands.[84] Consequently, it is imperative that systematic power, laser exposure, and temperature studies are undertaken on single cells to realize the parameters required to minimize photo- and/or thermal effects that can lead to erroneous spectral interpretation.

## 7.10 Aggregation state marker bands

The increase in the overall SNR observed at high temperatures, or after prolonged laser exposure, can only be explained by an increase in heme concentration, which is a consequence of heme aggregation. The 1248 cm$^{-1}$ band, along with a number of other bands, is found to be enhanced from continuous laser exposure independent of the background contribution. When hemes are in close enough proximity to one another, they can interact through ET processes. It is hypothesized that the enhancement observed when applying 632.8-nm excitation to a single red cell results primarily from excitonic interactions between highly ordered and orientated heme groups within the cell. This enhancement becomes stronger when the heme groups within the cell aggregate in response to photo and/or thermal degradation. The enhancement of many bands, the increase in SNR, and the general increase in the background counts is a consequence of the close packing of aggregated hemes enabling energy in the form of an exciton to migrate throughout the aggregated heme network.

Exciton theory was developed by Frenkel and Wannier in the 1930s.[99–101] The exciton model is based on the quantum mechanical precept that electronic energy is distributed throughout aggregates.[102] This arises because of $\pi \rightarrow \pi$ interactions between induced transition dipole moments. The movement of electrons throughout the aggregate is via $\pi$ interactions enabling a superposition of states resulting in what is essentially an electronic band of states.[102] In a single crystal, the characterization of vibrational modes on the basis of amplitudes of individual atomic

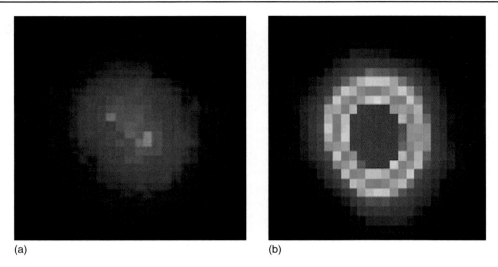

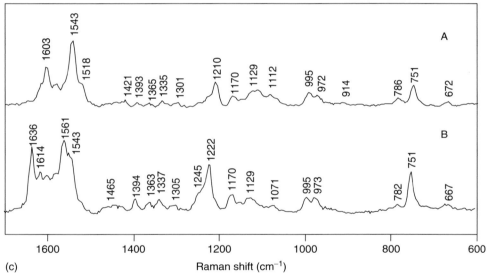

**Figure 22.** Two Raman images of the intensity of the $1639\,cm^{-1}$ oxygen marker band recorded for the same cell: (a) deoxygenated and (b) oxygenated. Each image was recorded at $4\,^{\circ}C$ with only $120\,s$ of constant laser exposure and 40 min rest between images. (c) Shows the spectra recorded after each image. [Reproduced from Wood *et al.*[84] © SPIE, 2004.]

displacements is not accurate.[103] Each excited electronic state is "in resonance" with excited states localized at other points in the lattice.[103] These running states of excitation were named "excitons" by Frenkel and represent delocalized units of excitation.[103] Electronic states in this case are linear combinations of these localized

excitations each belonging to a wave vector **K** in the reciprocal lattice.[103]

In the Wannier model the exciton is viewed as a conduction-band electron and a valence-band hole, bound together but with a considerable separation, traveling through the crystal in a state of total wave vector **K**.[103] A more generalized theory of ET accounting for all types of ET between donor and acceptor models including biological systems was developed by Förster.[104]

The transfer rate for electric dipole or long-range ET, known as Förster ET, can be written as

$$K_T = \frac{8.8 \times 10^{-25} k^2 \phi_D}{n^4 \tau_D r^6} J \qquad (10)$$

where $k$ is the orientation factor, $\phi_D$ is the donor quantum yield, $J$ represents the overlap integral function between donor emission and the acceptor absorption spectrum, $n$ is the number of molecules, $\tau_D$ is the donor lifetime, and $r$ is the distance between the donor and acceptor chromophores.[105] The ET rate has an inverse sixth power dependence on the donor–acceptor separation and the value of this has been used to determine intermolecular distances in biological systems.[105] The ET is also directly proportional to the orientation factor $k^2$, which may vary from 1 to 4 and is defined by the angles between donor and acceptor transition moments.[105] Depending on the type of chromophore and the degree of ordering, Förster distances of the order 1–10 nm are possible.[105] An early study analyzing small angle X-ray scattering (SAXS) of Hb from intact erythrocytes determined an interparticle distance of 6.2 nm,[106] indicating the Hb molecules were not distributed randomly. According to Perutz,[107] the close packing of Hb is an inevitable consequence of the high Hb concentration (22 mM). Perutz argued that if the molecules were distributed randomly, the interparticle distance would have to be 7.5 nm and a concentration of 22 mM could not be achieved. The inter-Hb molecule distance within the red cell is well within the range to enable Förster type long-range excitonic interactions between heme chromophores.

Kasha *et al.*[108] further developed exciton theory to explain the systematic spectral shifts observed in the Soret band in the absorption spectra of *meso–meso* linked porphyrin arrays. As shown above in Section 7.6 the UV–visible spectra of erythrocytes show a red shift and broadening of the Soret band and an intensity change in

the Q bands when compared with the spectrum of a suspension of red cells in saline with isolated Hb in distilled water. Upon heating the erythrocytes, the Soret band broadens and exhibits a distinct red shift. The cells lyse between 60 and 65 °C releasing denatured Hb into the surrounding solution and the Soret band is blue-shifted compared to the intact cells. Broadening and shifting of the Soret band has been interpreted to be indicative of excitonic coupling resulting from porphyrin aggregation.[109–111] It should be noted that other contributions can also affect the Soret band width, such as inhomogeneous broadening due to conformational heterogeneity, homogeneous broadening due to finite lifetime of the excited state, vibronic coupling with high-frequency nuclear vibrations, and Gaussian broadening due to coupling with a "bath" of low-frequency vibrational modes.[112] However, the fact that no broadening of the Soret band was observed in the temperature study of isolated Hb indicates that these contributions are minimal at temperatures greater than 45 °C.

Using Soret circular dichroism (SCD), Goldbeck *et al.*[113] reported that at high concentration Hb formed tetramers that had a SCD profile different from that at low concentrations of Hb, where dimers are apparently formed. These results have since been backed up by theoretical calculations by Woody[114] and support Goldbeck's *et al.*[113] hypothesis that the changes observed in the CD spectra are attributed to excitonic coupling between adjacent porphyrins. Such excitonic interactions would also occur in functional erythrocytes, where the Hb almost behaves as a liquid crystal at 22 mM concentration. Moreover, as further discussed in Section 9.2 such excitonic interactions can result in nontypical Raman scattering patterns.

## 7.11 Summary of photo and thermal decomposition studies of Hb in erythrocytes

On the basis of Raman temperature and power studies along with UV/vis measurements, it was postulated that the increase in the SNR and the shift in the Soret band observed in the electronic spectrum results from the formation of heme aggregates in RBCs in response to protein denaturation.[84] A series of bands appearing at 1398, 1366, 1248, and 972 cm$^{-1}$ were designated heme aggregation marker bands and the 1248 cm$^{-1}$ band was used to image heme aggregation in cells exposed to prolonged laser exposure[84] and/or high temperatures. By minimizing the effects of laser power and temperature, Raman images generated from the intensity of the 1638-cm$^{-1}$ band, which is particularly sensitive to oxygen concentration, were recorded in the oxygenated and deoxygenated state using only 120 s of laser exposure and 1 mW of defocused laser power (Figure 22).[84] The images show Hb in the fully oxygenated and deoxygenated state within the cell.[84] On the basis of these results and the fact that only poor quality spectra can be obtained from single crystals and highly concentrated solutions of Hb (up to 17 mM), it is postulated that the enhancement observed in the RBC at 633-nm excitation could be due to $\pi \rightarrow \pi$ excitonic interactions occurring throughout the aggregated heme network of the red cell at high temperatures and/or after prolonged laser exposure.[84]

# 8 RAMAN INVESTIGATION INTO HEME ORDERING IN THE ERYTHROCYTE

In the highly concentrated heme environment of a single erythrocyte, where the concentration of Hb approaches 22 mM, oxygen diffusion into the cell is critical for all ensuing respiration processes. As noted above, Perutz[107] argued that the close packing of Hb is an inevitable consequence of the high Hb concentration, and that the high concentration of Hb in the erythrocyte could not be achieved if the Hb molecules were distributed randomly. He suggested that the Hb within the cell must be ordered in a close-packed semicrystalline state.[107]

To investigate this hypothesis, spectra of red cells oriented horizontally and vertically to the surface of a Petri dish were recorded using both parallel and perpendicularly scattered light.[115] The results indicated an ordering of heme moieties within the cell that is not present in Hb solution. The averaged spectra from 30 cells in each orientation showed distinct differences (Figure 23), which were supported by the PCA analysis. In particular, bands at 1605, 1561, 1546, 1338, 1305, and 1227 cm$^{-1}$ are more intense in the horizontally orientated cells, while bands at 1617, 1118, and 997 cm$^{-1}$ are more intense in the vertically oriented cells. We can only hypothesize as to why the hemes are so highly oriented in the red cell. The orientation may facilitate intracellular oxygen transport in the highly concentrated Hb environment of the erythrocyte. Many models have been proposed to explain the diffusion and convection of Hb in the erythrocyte. Klug *et al.*[116] reported that an $O_2$ gradient could produce a driving force if a carrier like Hb is present that binds $O_2$ at the high $O_2$ concentration side and releases it on the low side. Zander and Schmid-Schöenbien[117] suggested that intracellular, convective oxygen transport induced by erythrocyte deformation is of greater significance than oxygen diffusion and that its relative role increases with the time of deoxygenation. Vandegriff and Olson[118] suggested a number of factors important in $O_2$ diffusion. These included intracellular Hb concentration, cell size and morphology, changes in intracellular pH, organic phosphate concentration shift, and the thickness of an "unstirred sublayer" adjacent to the cell. Bouwer *et al.*[119] determined diffusion coefficients for a range of Hb concentrations through thin liquid layers. This approach deviated from more conventional methodologies, where labeled molecules are measured through soaked porous filters. By assuming that Hb diffusion stops if Hb molecules are "just touching" and $O_2$ diffusion stops if the molecules are "closely packed" they calculated that the concentration of each would be 43 and 106 g L$^{-1}$, respectively, which is remarkably close to the reported experimental values of 46 and 100 g L$^{-1}$.[119] The

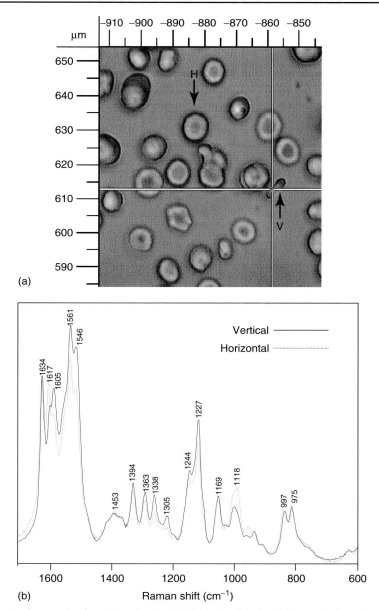

**Figure 23.** (a) Photomicrograph of red blood cells (RBCs) affixed in both the vertical and horizontal orientations relative to the Petri dish surface. (b) Average spectra from 30 cells orientated horizontally (dotted line) and another 29 cells orientated vertically (solid line) to the Petri dish surface. Each spectrum was recorded with only 10 s of constant laser exposure. [Reproduced from Wood *et al.*[115] with permission from Elsevier, © 2004.]

Raman polarization measurements provide evidence supporting an ordering of hemes within the erythrocyte. However, whether this ordering is occurring in an unstirred layer or throughout the whole cell is still in doubt. With the latest Raman imaging spectrometers that can achieve a spatial resolution of 300 nm this question may be resolved.

# 9   APPLICATIONS OF RESONANCE RAMAN SPECTROSCOPY OF ERYTHROCYTES

## 9.1   Malaria research

New drugs for the treatment of malaria are required to be effective, nontoxic, and inexpensive for use in the developing world, where the combined mortality rate is over 1 million per year.[120,121] The majority of deaths occur in children under 5 years of age and even in hospitals with good facilities child mortality from cerebral malaria is ~20%.[122] Despite the importance, only 3 of the 1223 new drugs developed during 1975–1996 were antimalarials.[123] There is an urgent need for potent, low-cost drugs, yet effective ways to screen for such drugs prior to clinical trials do not exist. Given that quinoline drugs such as chloroquine (CQ) and quinidine concentrate in millimolar levels in the food vacuole[121] a spectroscopic approach to screening such drugs in living cells would seem a viable option to the more chemically based and morphological methods, which are currently the status quo. There are three main models that endeavor to explain how quinoline-based antimalarials exert their effectiveness. In one model, the quinolines are thought to "cap" the formation of hemozoin (malarial pigment) by binding to the heme through $\pi-\pi$ stacking of the aromatic structures.[124] On the basis of isothermal titration calorimetry, a second model has been proposed, where the CQ binds between two $\mu$-oxo-dimer complexes.[125] The $\mu$-oxo-dimer

is thought to be in equilibrium with the hematin monomer and is essentially a precursor to hemozoin formation.[125] In a third model Buller *et al.*[126] made use of the crystal structure of $\beta$-hematin (a synthetic analog of hemozoin) and computed the theoretical growth morphology of hemozoin via the Hartman–Perdok approach.[127] A model of binding quinoline drugs to the fast-growing highly corrugated[128] face of hemozoin was proposed.[126] More recently, Leiserowitz and coworkers provided experimental evidence of this binding using a combination of electron diffraction (ED), electron microscopy (EM) images, powder synchrotron X-ray diffraction (PXRD), coherent grazing exit synchrotron X-ray scattering, micro-Raman spectroscopy, and IR attenuated total reflection (ATR) microimaging spectroscopy.[129]

Other antimalarials that accumulate in the food vacuole include the artemisinins, aryl-alcohol antimalarials, and the antimalarial peroxides, which are thought to exert their activity through the interaction with heme.[124]

Direct examples of drugs binding to heme in vitro are few and most rely on either measurements of crude trophozoite lysates, chemical methods under nonphysiologic conditions, or morphologic effects using microscopic techniques. Hitherto, there has been no direct molecular in vivo spectroscopic proof of the drug accumulating in the food vacuole. We hypothesized such heme–drug interactions could conceivably be monitored in the functional erythrocyte with the Raman technique, either directly by detecting Raman scattering from the drug in the food vacuole or indirectly by observing changes in the relative intensity of the hemozoin bands. In the latter case, such enhancement differences may reflect a reduction in excitonic coupling if the drug of interest indeed "caps" the formation of hemozoin. Figure 24 depicts Raman spectra of Hb and hemozoin recorded using 633-nm excitation. The hemozoin spectrum differs markedly from that of Hb and has the distinctive three band profile in the 1650–1500 cm$^{-1}$ region indicative of a high-spin complex. Other bands that distinguish hemozoin from Hb are at 1431, 1376 ($\nu_4$), 1146, 372, 344,

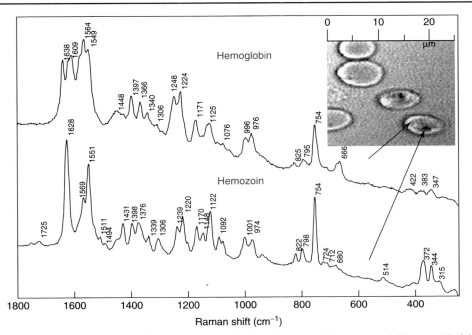

**Figure 24.** Raman spectra recorded at 633-nm excitation of a single functional *Plasmodium falciparum* (*P. falciparum*) infected erythrocyte in phosphate buffered saline (PBS) using a water immersion objective. The photomicrograph shows the functional erythrocyte with a *P. falciparum* parasite in the late trophozoite phase with hemozoin "darker area" clearly evident. Individual spectra of Hb and hemozoin at the arrow positions in this cell using laser exposure for 10 s with 2–3 mW of power were recorded. [Reproduced from Wood *et al.*[133] with permission from Elsevier, © 2003.]

and 315 cm$^{-1}$. A detailed analysis of hemozoin band assignments is available elsewhere.[130–132] Figure 25(a) shows an enhancement profile of hemozoin recorded in functional erythrocytes. It is interesting to note the increase in the relative intensity of $\nu_4$ (1376 cm$^{-1}$) when exciting into the near-IR at 780 nm. The strong intensity of $\nu_4$ relative to the other bands makes it ideal for imaging. Figure 25(b) shows a Raman image of hemozoin recorded using 780-nm excitation and filtering out the spectral region encompassing $\nu_4$. The image clearly shows the hemozoin in red surrounded by a purple haze of Hb.[133]

## 9.2 Raman "excitonic" enhancement at near-IR wavelengths

Unusual enhancement has been reported using near-IR excitation wavelengths on $\beta$-hematin,

hemozoin, hematin, hemin, and other model heme compounds.[130,133,134] Figure 26 shows the recorded RR spectra of $\beta$-hematin at a variety of excitation wavelengths, while Figure 27 shows the excitation profiles of $\beta$-hematin and hemin for the totally symmetric modes $\nu_2$ and $\nu_4$. The profile shows a strong enhancement of the electron density marker band $\nu_4$ and other totally symmetric modes including $\nu_2$ for $\beta$-hematin at the near-IR wavelengths 780–830 nm and also at the visible wavelength 564 nm. While one would expect to see the totally symmetric modes enhanced at 413 nm through resonance with the Soret band and explained by type A scattering, the enhancement at 780 and particularly 830 nm is intriguing.

To gain more insight into the mechanism responsible for the observed enhancement, UV/vis spectra were recorded during the acidification (pH 13.0–1.75) of hemin to $\beta$-hematin. The spectra

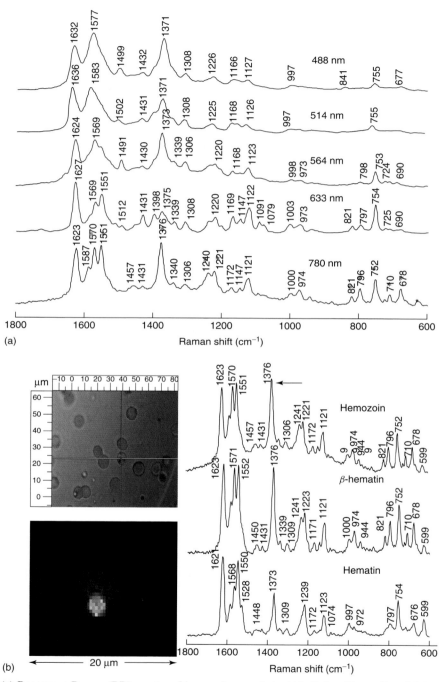

**Figure 25.** (a) Resonance Raman (RR) spectra of hemozoin recorded within living *Plasmodium falciparum* parasites at a variety of excitation wavelengths. (b) Raman image of a single functional erythrocyte infected with a living *P. falciparum* trophozoite using the 1376 cm$^{-1}$ band was recorded, which is strongly enhanced when using 780-nm excitation. [Reproduced from Wood *et al.*[133] with permission from Elsevier, © 2003.]

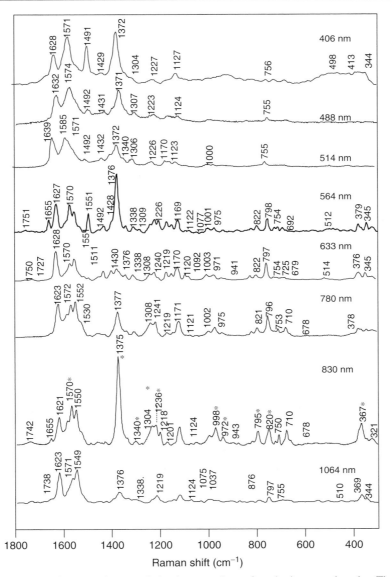

**Figure 26.** Raman spectra of $\beta$-hematin recorded using a variety of excitation wavelengths. The asterisks in the 830-nm spectrum highlight bands that appear dramatically enhanced relative to $\nu_{10}$. [Reproduced from Wood *et al.*[130] © American Chemical Society, 2004.]

presented in Figure 28 show a red shifting of the Soret and Q bands during the formation of $\beta$-hematin. Moreover, a small broad band centered at 867 nm is observed. This band was assigned to a *z*-polarized transition, designated band I, based on extended Hückel theoretical calculations on

ferric high-spin porphine complexes performed by Zerner *et al.*[51] and the MCD analysis of heme derivatives by Eaton and colleagues.[48,49,135] While no *z*-polarized transition has previously been reported for high-spin ferric hemes in the 800–1000 nm region, such *z*-polarized transitions

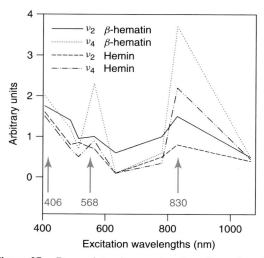

**Figure 27.** Raman intensity as a function of wavelength of $\beta$-hematin and hemin for the totally symmetric modes $\nu_2$ and $\nu_4$. The strongest enhancement of totally symmetric modes occurs in the near-IR and not in the vicinity of the Soret band.

were predicted using a semiempirical quantum chemical INDO/ROHF/CI method.[136] It was reasoned that this band might play a role in the enhancement observed at near-IR excitation

wavelengths in malaria pigment. The enhancement observed at 564 nm for $\nu_4$ and $\nu_2$ was also explained by a $z$-polarized transition assigned by Eaton and coworkers to[48] $a_{2u}(\pi) \rightarrow e_g(d_{yz})$.

It was reasoned that the enhancement observed at 830 nm could not just involve a typical type A scattering pattern because it was so much greater than that observed when using 413-nm excitation in direct resonance with the Soret band. To explain the "excitonic" enhancement, we invoked exciton theory[130] which has been introduced in Section 8.5. The excitonic coupling breaks the degeneracy of the B excited state giving rise to two Soret bands. RR spectra of porphyrin arrays in which the porphyrins are directly linked at the meso position have undergone intense study recently.[110,137] The close proximity of adjacent porphyrins (83.5 nm) and the orthogonal orientation in the arrays result in strong excitonic interactions along the axis defined by the *meso, meso*-linkage-(s), and negligible interactions along the orthogonal axis.[60] Excitonic coupling will split the electronic states into a broad band of states with different geometries, energies, and oscillator strengths. The Raman intensities for a particular wavelength

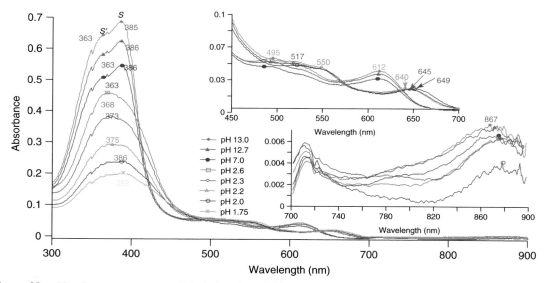

**Figure 28.** Absorbance spectra recorded during the acidification of hemin to form $\beta$-hematin. [Reproduced from Wood *et al.*[130] © American Chemical Society, 2004.]

will then reflect the extent of the excitonic coupling.

Bhuiyan *et al.*[111] recorded RR spectra of the *meso*, *meso*-linked arrays excited in the vicinity of the B state. The spectra showed a complex and unusual scattering pattern. These include (i) the enhancement of anomalously polarized vibrations with B-state excitation, (ii) the observation of only polarized (A$_{1g}$ for $D_{4h}$) and anomalously polarized modes (A$_{2g}$ for $D_{4h}$) in the RR spectrum, and (iii) the large differential enhancement of symmetric versus nontotally symmetric vibrations.[111] They postulated that these scattering characteristics were due to the effects of symmetry lowering. In this process, the asymmetric *meso* substitution pattern inherent to the *meso*, *meso*-linked arrays contributes to symmetry lowering in both the ground and excited electronic states.[111] Uniaxial excitonic interactions make an additional contribution to symmetry lowering in the excited state(s), promoting novel Franck–Condon and vibronic scattering mechanisms in the B state(s) of the arrays.[111] Collectively, the studies of the *meso*, *meso*-linked arrays provide insight into the type of RR scattering that might be anticipated for other types of systems that exhibit strong excitonic interactions among the constituent chromophores.

In other studies Akins *et al.*[102,138–145] reported that in cyanine dyes and porphyrin arrays, other enhancement effects could also be significant. One such important mechanism is aggregated enhanced Raman scattering (AERS), where bands can become enhanced through excitonic interactions between neighboring chromophores. The enhancement of vibrational modes can be explained in terms of an increase-size effect and near-resonance terms in the polarizability.[138]

It is hypothesized that the enhancement observed at 830 nm in a variety of heme molecules results from an exciton coupling mechanism that implicates the CT transition centered at ~860 nm and known as *band I*. In this scenario, the electronic symmetric component of the near-IR photon couples the excited states of the CT transitions resulting in a superposition of states increasing the contributions to the Franck–Condon integrals. The reasons for this assessment are as follows:

1. For $\beta$-hematin band I has an excited electronic configuration that corresponds to the direct product $E \times E = A_1 + A_2 + B_1 + B_2$ (under $C_{4v}$), which cannot mix with the excited state configuration of the Soret transition. Thus, band I cannot vibronically couple to the Soret band. Consequently, one would expect totally symmetric A$_1$ modes to be enhanced as predicted by the Albrecht formalism.

2. The strong enhancement of A$_1$ modes, out-of-plane modes, and some pyrrole-breathing and deformation normal modes is indicative of a $z$-polarized transition that is a characteristic of band I.

3. The stacking of hemes should result in strong excitonic interactions for $z$-polarized transitions as verified by the intensity of the A$_1$ modes in $\beta$-hematin compared to hemin at near-IR excitation wavelengths.

4. The red shifting of the Soret band, Q bands, and CT band during the acidification of hemin to form the dimer, $\beta$-hematin, is indicative of a splitting of degenerate states, which is a characteristic of excitonic coupling.

5. The nonlinear power dependence plot for the totally symmetric modes $\nu_4$ and $\nu_2$ for hemin and $\beta$-hematin is characteristic of two-photon excitonic mechanisms.

## 9.3 Resonance Raman drug studies on live cells

Drug uptake studies using Raman microscopy are inconclusive at this stage. Studies by our group using live malarial infected erythrocytes incubated with drug showed significant variation in the controls that masks to some extent the effects of drug. In particular, a band at 471 cm$^{-1}$ was observed in some controls but not others. There are no observable bands of CQ in the 780-nm spectra of live parasites inoculated with the drug. This is not so surprising given that if the CQ does in fact coat the hemozoin it would be on the order of nanometers compared to the spatial resolution of the microscope, which is 1 μm. Consequently, the CQ bands would be swamped by hemozoin bands. To add to this problem, the strongest band of CQ is located at 1370 cm$^{-1}$, which is the same position as the strongest heme band $\nu_4$. Despite these limitations, it has been possible to observe consistent spectral differences between control and CQ inoculated cells. Figure 29 presents average Raman spectra control (30 cells) and cells inoculated with 100 nM of CQ (30 cells) for 6 h. The CQ affected average spectrum shows a distinct reduction in the intensity of the $A_{1g}$ modes—1570, 1376, 796, and 678 cm$^{-1}$—and the $B_{1g}$ modes—1552 and 751 cm$^{-1}$—when compared with the controls. There is also a consistent but small shift in some of the Raman bands including $\nu_2$ (1552 cm$^{-1}$). The linewidth of $\nu_2$ broadens from 6.8 to 7.4 cm$^{-1}$ in response to the drug. These changes in position, intensity, and the shape of a number of the Raman bands in the presence of CQ can be used to infer possible changes in molecular environment

of the malaria pigment upon drug interaction. Raman studies on the synthetic precursor $\beta$-hematin inoculated with CQ during the acidification of hematin to $\beta$-hematin show inhibition of $\beta$-hematin after 6 h of incubation. Strong contributions from the hematin precursor are observed in the CQ incubated $\beta$-hematin compared to the control (Figure 30). However, once again we do not detect any bands directly assigned to the CQ moiety.

With the continued advances in speed and spatial resolution in Raman imaging techniques, it is hoped that the detection and elucidation of the quinoline binding site in hemozoin will be revealed and the technique can be utilized as a screening tool for antimalarial drug targets.

## 9.4 Sickle cell anemia

Sickle cell disease is caused by a mutant form ($\beta$6 glu → val) of hemoglobin (HbS) and affects 1 out of 600 African-American births.[146] Under hypoxic conditions, HbS polymerizes and forms

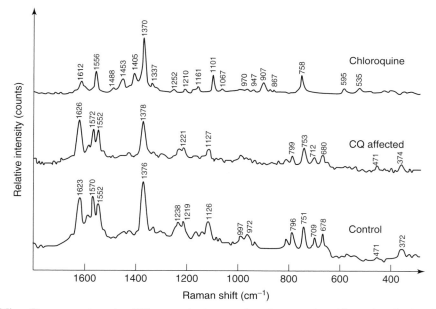

**Figure 29.** Micro-Raman spectra using 782-nm excitation wavelength comparing chloroquine disphosphate (CQ) and hemozoin (control) and CQ-affected malaria pigment in live red blood cells.

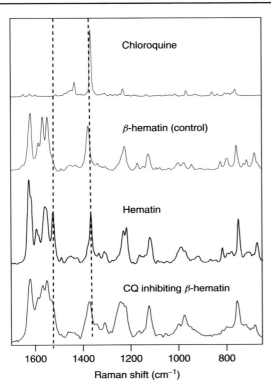

**Figure 30.** Recorded Raman spectra of chloroquine (CQ), $\beta$-hematin control after 6 h of incubation, hematin, and CQ inhibiting $\beta$-hematin after 6 h of incubation.

higher order aggregates, leading to distortion and changes in the rigidity of the erythrocyte. These rigid cells can block the microvasculature resulting in tissue ischemia, organ damage, and ultimately death.[147]

We have used the ability to monitor heme aggregation within a single erythrocyte in a preliminary study of the heme aggregation in deoxygenated sickle cells.[148] Cells were deoxygenated with sodium dithionite and 30 spectra recorded of cells from each of six patients. Only cells that showed definitive sickling were included in this study. The averaged sickle cell spectrum clearly shows the enhancement of the "aggregation marker" bands, namely 1398, 1366, 1250, and 972 cm$^{-1}$, that were identified in the thermal denaturation and photoinduced degradation studies delineated above. The 1578 cm$^{-1}$ also appears

enhanced compared to the corresponding band in a normal RBC (Figure 31a).

Figure 31(b) shows Raman images of sickle cells deoxygenated with sodium dithionite recorded using the 1250 cm$^{-1}$ aggregation marker band. The images were constructed using 120 s of laser exposure by defocusing the laser so the beam is spread over the entire cell. The red areas indicate high Raman counts while the green and blue areas show low Raman counts. The bright red area indicates high levels of heme aggregation and provides a way to visualize the aggregated hemes within the functional sickle cell.

## 9.5 Summary of applications

Raman microspectroscopic analysis of single erythrocytes offers a platform to diagnose and treat erythrocyte disorders including malaria, sickle cell disease, thalassaemia, and altitude sickness. The advantages of the technique include the following:

1. It is nondestructive;
2. Only a pinprick of blood is required for analysis;
3. It takes only 10 s to obtain a spectrum of a single cell with a high SNR;
4. No staining or fixative procedure is required thereby minimizing cost, preparative time, and structural modification to the protein matrix;
5. Drug therapy for the treatment of RBC disorders can conceivably be monitored in vivo;
6. Detailed information on the spin states and consequently the oxygen status of the cell can be inferred;
7. The physiological response of the cell to oxygen uptake can be monitored;
8. A motorized translation stage with video monitor enables a number of cells in a population to be targeted without having to seek and refocus for each cell;
9. Data can be easily stored for future statistical and multivariate analysis.

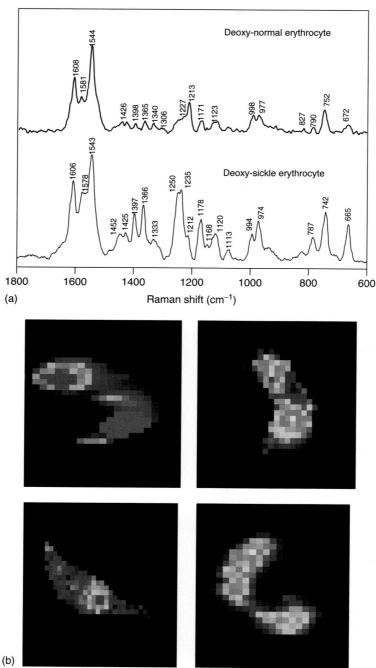

**Figure 31.** (a) Average Raman spectra of 30 deoxygenated normal erythrocytes and 30 deoxygenated sickle cells recorded using 632.8 nm excitation. (b) Raman images of deoxygenated sickle cells using the 1248 cm$^{-1}$ aggregation state marker band were recorded. The red indicates possible sites where the heme is aggregating. [Reproduced from McNaughton *et al.*[148] © SPIE, 2005.]

# ACKNOWLEDGMENTS

We would like to acknowledge Prof Leann Tilley and Ms Samantha Deed (Department of Chemistry, Latrobe University) and Dr Brian Cooke and Ms Fiona Glenister (Department of Microbiology, Monash University) for their expertise and supplying the malaria infected cells. We also would like to acknowledge Prof Steven Langford for advice on $\beta$-hematin synthesis and porphyrin chemistry. Dr Brian Tait (Australian Red Cross Blood Services) for expert advice and lab space. We would like to acknowledge all past and present postdoctoral, postgraduate, and undergraduate students who contributed to the aspects of this project including Larissa Hammer, Janelle Lim, Lara Davis, Grant Webster, Martin Duriska, Jessica Unthank, Ljiljana Puskar, Rudolf Tuckerman, and Torsten Frosch. Peter Caspers and Gerin Puppels (River Diagnostics and Erasmus University Medical Center) for use of their time, expertise and instrumentation for some of the near-IR work. Elizabeth Carter and Robert Armstrong (Sydney University) for their time, expertise and instrumentation for some of the visible work. We also thank Mr Finlay Shanks for instrumental maintenance and acknowledge the Australian Research Council for financial support. Dr Wood was financially supported by an Australian Synchrotron Research Program Fellowship and is currently supported by a Monash Synchrotron Research Fellowship.

# ABBREVIATIONS AND ACRONYMS

| | |
|---|---|
| AERS | Aggregated Enhanced Raman Scattering |
| ATR | Attenuated Total Reflection |
| CCD | Charge Coupled Device |
| CD | Circular Dichroism |
| CGD | Chronic Granulomatous Disease |
| CQ | Chloroquine |
| CT | Charge Transfer |
| CW | Continuous-wave |
| ED | Electron Diffraction |
| EM | Electron Microscopy |
| EPO | Eosinophil Peroxidase |
| ET | Energy Transfer |
| FT | Fourier Transform |
| Hb | Hemoglobin |
| IR | Infrared |
| MCD | Magnetic Circular Dichroism |
| MPO | Myeloperoxidase |
| NADPH | Nicotinamide Adenine Dinucleotide Phosphate |
| OEP | Octaethylporphyrin |
| PBS | Phosphate Buffered Saline |
| PCA | Principal Component Analysis |
| PXRD | Powder Synchrotron X-ray Diffraction |
| RBCs | Red Blood Cells |
| RMS | Root Mean Square |
| RR | Resonance Raman |
| SAXS | Small Angle X-ray Scattering |
| SCD | Soret Circular Dichroism |
| SNR | Signal-to-Noise Ratio |
| TPP | Tetraphenylporphyrin |
| UV | Ultraviolet |

# REFERENCES

1. T.G. Spiro and T.C. Strekas, *Proc. Nat. Acad. Sci. USA*, **69**, 2622–2626 (1972).

2. T.C. Strekas and T.G. Spiro, *Biochim. Biophys. Acta*, **278**, 188–192 (1972).

3. T.C. Strekas and T.G. Spiro, *Biochim. Biophys. Acta*, **263**, 830–833 (1972).

4. T.G. Spiro and T.C. Strekas, *J. Am. Chem. Soc.*, **96**, 338–345 (1973).

5. T.C. Strekas and T.G. Spiro, *J. Raman Spectrosc.*, **1**, 387–392 (1973).

6. T.G. Spiro, *Biochim. Biophys. Acta*, **416**, 169–189 (1975).

7. T.G. Spiro and J.M. Burke, *J. Am. Chem. Soc.*, **98**, 5482 (1976).

8. J.M. Burke, J.R. Kincaid, S. Peters, R.R. Gagne, J.P. Collman and T.G. Spiro, *J. Am. Chem. Soc.*, **100**, 6083–6088 (1978).

9. T.G. Spiro, in 'Infrared and Raman Spectroscopy of Biological Molecules', ed T.T., Theophanides, D. Reidel Publishing Company, Dordrecht, 267–275, Vol. 43 (1979).

10. T.G. Spiro, J.D. Stong and P. Stein, *J. Am. Chem. Soc.*, **101**, 2648–2655 (1979).

11. T.G. Spiro, 'Biological Applications of Raman Spectroscopy', Wiley, New York, (1988).

12. H. Brunner, A. Mayer and H. Sussner, *J. Mol. Biol.*, **70**, 153–156 (1972).

13. H. Brunner and H. Sussner, *Biochim et Biophys ACTA*, **310**, 20–31 (1973).

14. H. Brunner, *Naturwissenschaften*, **61**, 129–130 (1974).

15. J.A. Shelnutt, L.D. Cheung, C.C. Chang, N.-T. Yu and R.H. Felton, *J. Phys. Chem.*, **66**, 3387–3398 (1977).

16. L.D. Cheung, N.-T. Yu and R.H. Felton, *Chem. Phys. Lett.*, **55**, 527–530 (1978).

17. D.L. Rousseau, J.M. Friedman and P.F. Williams, In *Topics in Current Physics*, ed A. Weber, Springer, Berlin, **2**, 203 (1979).

18. J.A. Shelnutt, D.L. Rousseau, J.K. Dethmers and E. Margoliash, *Proc. Natl. Acad. Sci. USA.*, **76**, 3865–3869 (1979).

19. S.A. Asher, L.E. Vickery, T.M. Schuster and K. Sauer, *Biochemistry*, **16**, 5849–5856 (1977).

20. S.A. Asher, *Meth. Enzymol.*, **76**, 371–413 (1981).

21. S.A. Asher and K. Sauer, *J. Chem. Phys.* **64**, 4115–4125 (1976).

22. T. Kitagawa and K. Nagai, *FEBS Lett.*, **104**, 376–379 (1979).

23. T. Kitagawa, M. Abe and H. Ogoshi, *J. Chem. Phys.*, **69**, 4516–4525 (1978).

24. T. Kitagawa, Y. Kyogoku, T. Iizuka and M.I. Saito, *J. Am. Chem. Soc.*, **98**, 5169–5173 (1976).

25. T. Kitagawa, in 'Biological Applications of Raman Spectroscopy', ed T.G. Spiro, Wiley, New York, Vol. 3, 97–131 (1988).

26. T. Kitagawa, Y. Ozaki and K. Kyogoku, *Adv. Biophys.*, **11**, 153 (1978).

27. T. Kitagawa, H. Ogoshi, E. Watanabe and N. Yoshida, *Chem. Phys. Lett.*, **30**, 451–456 (1975).

28. M. Abe, T. Kitagawa and K. Kyogoku, *J. Chem. Phys.*, **69**, 4526–4534 (1978).

29. R.S. Armstrong, M.J. Irwin and P.E. Wright, *J. Am. Chem. Soc.*, **104**, 626–627 (1982).

30. P.M. Champion, D.W. Collins and D.B. Fitchen, *J. Am. Chem. Soc.*, **98**, 7114–7115 (1976).

31. P.M. Champion, G.M. Korenowski and A.C. Albrecht, *Solid State Commun.*, **32**, 7–12 (1979).

32. P.M. Champion and A. Albrecht, *Bull. Am. Phys. Soc.*, **24**, 319 (1979).

33. P.M. Champion, I.C. Gunsalus and G.C. Wagner, *J. Am. Chem. Soc.*, **100**, 3743–3751 (1978).

34. P.M. Champion and A.J. Sievers, *J. Chem. Phys.*, **66**, 1819–1825 (1977).

35. P.M. Champion and I.C. Gunsalus, *J. Am. Chem. Soc.*, **99**, 2000–2002 (1977).

36. D.L. Rousseau and J.M. Friedman, in 'Biological Applications of Raman Spectroscopy, Volume 3: Resonance Raman Spectra of Heme and Metalloporphyrins', ed T.G. Spiro, John Wiley & Sons, New York, Vol. 3 (1988).

37. G.J. Puppels, F.F.M. de Mul, C. Otto, J. Greve, M. Robert-Nicoud, D.J. Arndt-Jovin and T.M. Jovin, *Nature*, **347**, 301–303 (1990).

38. G.J. Puppels, G.M.J. Garritsen, G.M.J. Segers-Nolten, F.F.M. De Mul and J. Greve, *Biophys. J.*, **60**, 436–446 (1991).

39. G.J. Puppels, G.M.J. Garritsen, J.A. Kummer and J. Greve, *Cytometry*, **14**, 251–256 (1993).

40. B.L.N. Salmaso, G.J. Pupells, P.J. Caspors, R. Floris, R. Wever and J. Greve, *Biophys. J.*, **67**, 436–446 (1994).

41. C. Otto, N.M. Sijisema and J. Greve, *Eur. Biophys. J.*, **271**, 582–589 (1998).

42. N.M. Sijisema, A.G.J. Tibbe, G.M.J. Segers-Nolten, A.J. Verhoeven, R.S. Weening, J. Greve and C. Otto, *Biophys. J.*, **78**, 2606–2613 (2000).

43. R. Dickerson and I. Geis, 'The structure and action of proteins', Harper & Row, New York, (1969).

44. M.F. Perutz, *Nature*, **228**, 726 (1970).

45. J.L. Hoard, 'Hemes and Hemoproteins', Academic Press, New York, (1966).

46. G. Fermi, M.F. Perutz, B. Shannan and R. Fourme, *J. Mol. Biol.*, **175**, 154–174 (1984).

47. L. Stryer, 'Biochemistry', 3rd edition, W. H. Freeman and Company, New York, (1988).

48. W.A. Eaton, L.K. Hanson, P.J. Stephens, J.C. Sutherland and J.B.R. Dunn, *J. Am. Chem. Soc.*, **100**, 4991–5003 (1978).

49. W.A. Eaton and J. Hofrichter, in 'Hemoglobins' eds E. Antonini, L. Rossi-Bernardi and E. Chiancone, Academic Press, New York, Vol. 76 175–261 (1981).

50. J.J. Weiss, *Nature (London)*, **202**, 83–84 (1962).

51. M. Zerner, M. Gouterman and H. Kobayashi, *Theoret. Chim. ACTA*, **6**, 363–399 (1966).

52. T.G. Spiro and X.-Y. Li, in 'Biological Applications of Raman Spectroscopy', ed T.G. Spiro, John Wiley & Sons, New York, Vol. 3 1–38 (1988).

53. S. Franzen, S.E. Wallace-Williams and A.P. Shreve, *J. Am. Chem. Soc.*, **124**, 7146–7155 (2001).

54. V. Srajer and P.M. Champion, *Biochemistry*, **30**, 7390–7402 (1991).

55. J.M. Chalmers and P.R. Griffiths (eds) 'Handbook of vibrational spectroscopy', John Wiley & Sons, New Jersey, (2002).

56. D.A. Skoog and J.J. Leary, 'Principles of Instrumental Analysis', 4th edition, Saunders College Publishing, New York, (1992).

57. H.A. Kramers and W. Heisenberg, *Z. Phys.*, **31**, 681–707 (1925).

58. P.A.M. Dirac, *Proc. Roy. Soc. (London)*, **114**, 710–728 (1927).

59. V. Hizhnyakov and I. Tehver, *Phys. Stat. Sol.*, **21**, 755–768 (1967).

60. E.J. Heller, R.L. Sundberg and D. Tannor, *J. Phys. Chem.*, **86**, 1822–1833 (1982).

61. D.L. Tonks and J.B. Page, *Chem. Phys. Lett.*, **66**, 449–453 (1979).

62. D.C. Blazej and W.L. Peticolas, *J. Chem. Phys.*, **72**, 3134–3142 (1980).

63. A.B. Myers and R.A. Mathies, in 'Biological Applications of Raman Spectroscopy Volume 2: Resonance Raman Spectra of Polyenes and Aromatics', ed T.G. Spiro, John Wiley & Sons, New York, Vol. 2 1–58 (1987).

64. J.L. McHale, 'Molecular Spectroscopy', Prentice Hall, Upper Saddle River, (1999).

65. R. Kumble, T.S. Rush, M.E. Blackwood Jr, P.M. Kozlowski and T.G. Spiro, *J. Phys. Chem. B.* **102**, 7280–7286 (1998).

66. J.A. Shelnutt, *J. Chem. Phys.*, **74**, 6644–6657 (1981).

67. G. Placzek, in 'Handbuch der Radiologie', ed E. Marx, National Technical Information Service, U.S. Department of Commerce, Springfield, Vol. 2 209–374 (1934).

68. A.C. Albrecht, *J. Chem. Phys.*, **34**, 1476–1484 (1961).

69. J.A. Shelnutt, D.C. O'Shea, N.-T. Yu, L.D. Cheung and R.H. Felton, *J. Chem. Phys.*, **64**, 1156–1165 (1976).

70. M.H. Perrin, M. Gouterman and C.L. Perrin, *J. Chem. Phys.*, **50**, 4137–4150 (1969).

71. X.-l. Yan, R.-x. Dong and Q.-G. Wang, *Spectrosc. Spect. Anal.*, **24**, 576–578 (2004).

72. S. Choi and T.G. Spiro, *J. Am. Chem. Soc.*, **105**, 3683–3692 (1983).

73. L.D. Spaulding, C.C. Chang, N.-T. Yu and R.H. Felton, *J. Phys. Chem. Soc.*, **97**, 2517–2525 (1975).

74. P.V. Huong and J.-C. Pommier, *Acad. Sci. Paris, Ser. C*, **285**, 519 (1977).

75. S. Choi, T.G. Spiro, K.C. Langry, K.M. Smith, L.D. Budd and G.N. LaMar, *J. Am. Chem. Soc.*, **104**, 4345–4351 (1982).

76. J.D. Stong, R.J. Kubaska, S.I. Shupack and T.G. Spiro, *J. Raman Spec.*, **9**, 312–314 (1980).

77. G. Chottard, J.P. Baccioni, M. Baccioni, M. Lange and D. Mansuy, *Inorg. Chem.*, **20**, 1718–1722 (1981).

78. Y.A. Sarma, *Spectrochimica Acta*, **45A**, 649–652 (1989).

79. T. Yamamoto and G. Palmer, *J. Biol. Chem.*, **248**, 5211–5213 (1973).

80. C. Balagopalakrishna, P.T. Monoharan, O.O. Abugo and J.M. Rifkind, *Biochemistry*, **35**, 6393–6398 (1996).

81. L. Rimai, I. Salmeen and D.H. Petering, *Biochemistry*, **14**, 378–382 (1975).

82. P. Dhamelincourt, in 'Handbook of Vibrational Spectroscopy', eds J.M. Chalmers, P.R. Griffiths, John Wiley & Sons, Sussex, Vol. 2 1419–1428 (2002).

83. W. Rhodes and M. Chase, *Revs. Mod. Phys.*, **39**, 348–361 (1967).

84. B.R. Wood, L. Hammer, L. Davis and D. McNaughton, *J. Biomed. Opt.*, **10**, 014005 (2004).

85. B.R. Wood, P. Caspers, G. Puppels, S. Pandiancherri and D. McNaughton, *Anal. Bioanal. Chem.*, **387**, 1691–1703 (2007).

86. B.R. Wood and D. McNaughton, *Biopolymers (Biospectroscopy)*, **67**, 259–262 (2002).

87. B.R. Wood and D. McNaughton, *J. Raman Spectrosc.*, **33**, 517–523 (2002).

88. B.R. Wood and D. McNaughton, in 'New Developments in Sickle Cell Disease', ed P.D. O'Malley, NOVA, New York, 63–119 (2006).

89. B.R. Wood, B. Tait and D. McNaughton, *Biochim. Biophys. Acta*, **1539**, 58–70 (2001).

90. C. Xie, M.A. Dinno and Y.-Q. Li, *Opt. Lett.*, **27**, 249–251 (2001).

91. K. Ramser, C. Fant and M. Käll, *J. Biomed. Opt.*, **8**, 173–178 (2003).

92. K. Ramser, E.J. Bjerneld, C. Fant and M. Kall, *SPIE*, **4614**, 20–27 (2002).

93. L. Puskar, R. Tuckermann, T. Frosch, J. Popp, V. Ly, D. McNaughton and B.R. Wood, *Lab on a chip*, **7**, 1125–1131 (2007).

94. S. Hu, K.M. Smith and T.G. Spiro, *J. Am. Chem. Soc.*, **118**, 12638–12646 (1996).

95. S. Hu and J.R. Kincaid, *J. Am. Chem. Soc.*, **113**, 7189–7194 (1991).

96. S. Hoey, D.H. Brown, A.A. McConnell, W.E. Smith, M. Marabani and R.D. Sturrock, *J. Inorg. Biochem.*, **34**, 189–199 (1988).

97. A. Szabo and L.D. Barron, *J. Am. Chem. Soc.*, **97**, 660–662 (1975).

98. S. Jeyarajah, L.M. Proniewicz, H. Bronder and J.R. Kincaid, *J. Biol. Chem.*, **269**, 31047–31050 (1994).

99. J. Frenkel, *Phys. Rev.*, **37**, 17–44 (1931).

100. J. Frenkel, *Phys. Rev.*, **37**, 1276–1293 (1931).

101. G.H. Wannier, *Phys. Rev.*, **52**, 191–197 (1937).

102. D.L. Akins, S. Özcelik, H.-R. Zhu and C. Guo, *J. Phys. Chem.*, **101**, 3251–3259 (1997).

103. R.S. Knox, 'Theory of Excitons', Academic Press, New York, (1963).

104. T. Förster, *Ann. Phys. (Leipzig)*, **6**, 55–75 (1948).

105. J. Eisinger and R.E. Dale, in 'Excited States of Biological Molecules', ed J.B. Birks, John Wiley & Sons, London, 579–590 (1974).

106. D.G. Dervichian, G. Fournet and A. Guinier, *C.R. Acad. Sci. Paris*, **224**, 1848–1850 (1947).

107. M.F. Perutz, *Nature*, **161**, 204–206 (1948).

108. M. Kasha, H.R. Rawls and M.A. El-Bayoumi, *A. Pure Appl. Chem.*, **11**, 371–392 (1965).

109. D.-M. Chen, Y.-H. Zhang, T.-J. He and F.-C. Liu, *Spectrochimica Acta*, **58**, 2291–2297 (2002).

110. Y.H. Kim, D.H. Jeong, D. Kim, S.E. Jeoung, H.S. Cho, S.K. Kim, N. Aratani and A. Osuka, *J. Am. Chem. Soc.*, **123**, 76–86 (2001).

111. A.A. Bhuiyan, J. Seth, N. Yoshida, A. Osuka and D.F. Bocian, *J. Phys. Chem. B*, **104**, 10757–10764 (2000).

112. M. Leone, A. Cupane, E. Vitrano and L. Cordone, *Biophys. Chem.*, **42**, 111–115 (1992).

113. R.A. Goldbeck, L. Sagle, D.B. Kim-Shapiro, V. Flores, D.S. Kliger, *Biochem. Biophys. Res. Commun.* **235**, 610–614 (1997).

114. R.W. Woody, *Chirality*, **17**, 450–455 (2005).

115. B.R. Wood, L. Hammer and D. McNaughton, *Vib. Spectrosc.*, **38**, 71–78 (2004).

116. A. Klug, F. Kreuzer and F.J.W. Roughton, *Helv. Physiol. Acta*, **14**, 121–128 (1956).

117. R. Zander and H. Schmid- Schöenbien, *Respir. Physiol.*, **19**, 279–289 (1973).

118. K.D. Vandegriff and J.S. Olson, *J. Biol. Chem.*, **259**, 12609–12618 (1984).

119. S.T. Bouwer, L. Hoofd and F. Kreuzer, *Biochim. Biophys. Acta*, **1338**, 127–136 (1997).

120. B. Greenwood and T. Mutabingwa, *Nature*, **415**, 670–672 (2002).

121. C.R. Chong and D.J. Sullivan Jr, *Biochem. Pharmacol.* **66**, 2201–2212 (2003).

122. P.A. Holding and R.W. Snow, *Am. J. Trop. Med. Hyg.*, **64** (1 suppl), 68–75 (2001).

123. P. Troullier and P.L. Olliaro, *Int. J. Infect. Dis.*, **3**, 61–63 (1998).

124. R.G. Ridley, *Nature*, **415**, 686–693 (2002).

125. A. Dorn, S.R. Vippagunta, H. Matile, C. Jaquet, J.L. Vennerstrom and R.G. Ridley, *Biochem. Pharmacol.*, **55**, 727–736 (1998).

126. R. Buller, P.M.O. Almarsson and L. Leiserowitz, *Crystal Growth Design*, **2**, 553–562 (2002).

127. P. Hartman and W.G. Perdok, *Acta Crystallogr.*, **8**, 525–529 (1955).

128. J. Adovelande and J. Schrevel, *Life Sci.*, **59**, L309–PL315 (1996).

129. I. Solomonov, I. Feldman, C. Baehtz, K. Kjaer, I.K. Robinson, G.T. Webster, D. McNaughton, B.R. Wood, I. Weissbuch and L. Leiserowitz, *J. Am. Chem. Soc.*, **129**, 2615–2627 (2006).

130. B.R. Wood, S. Langford, B.M. Cooke, J. Lim, F.K. Glenister, M. Duriska, J. Unthank and D. McNaughton, *J. Am. Chem. Soc.*, **126**, 9233–9239 (2004).

131. B.R. Wood and D. McNaughton, *Expert Rev. Proteomics Res.*, **3**, 525–544 (2006).

132. L. Puskar, R. Tuckermann, T. Frosch, J. Popp, V. Ly, D. McNaughton and B.R. Wood, *Lab Chip*, **7**, 1125–1131 (2007).

133. B.R. Wood, S.J. Langford, B.M. Cooke, J. Lim, F.K. Glenister and D. McNaughton, *FEBS Lett.*, **554**, 247–252 (2003).

134. B.R. Wood, S. Langford, B. Cooke, F. Glenister and D. McNaughton, *Resonance Raman Spectroscopy in Malaria research*, Nakhon Ratchasima, Thailand 2003; School of Chemistry, Institute of Science, Suranaree Univesity of Technology, 17–18 (2003).

135. W.A. Eaton and R.M. Hochstrasser, *J. Chem. Phys.*, **49**, 985–995 (1968).

136. D. Harris and G. Loew, *J. Am. Chem. Soc.*, **115**, 5799–5802 (1993).

137. D. Kim and A. Osuka, *J. Phys. Chem. A*, **107**, 8791–8815 (2003).

138. D.L. Akins, *J. Phys. Chem.*, **90**, 1530–1534 (1986).

139. D.L. Akins and J.W. Macklin, *J. Phys. Chem.*, **93**, 5999–6007 (1989).

140. D.L. Akins, J.W. Macklin, L.A. Parker and H.-R. Zhu, *Chem. Phys. Lett.*, **169**, 564–568 (1990).

141. D.L. Akins, J.W. Macklin and H.-R. Zhu, *J. Phys. Chem.*, **95**, 793–798 (1991).

142. D.L. Akins, S. Özcelik, H.-R. Zhu and C. Guo, *J. Phys. Chem.*, **100**, 14390–14396 (1996).

143. D.L. Akins, H.-R. Zhu and C. Guo, *J. Phys. Chem.*, **98**, 3612–3618 (1994).

144. D.L. Akins, H.-R. Zhu and C. Guo, *J. Phys. Chem.*, **100**, 5420–5425 (1996).

145. D.L. Akins, Y.H. Zhuang, H.-R. Zhu and J.Q. Liu, *J. Phys. Chem.*, **98**, 1068–1072 (1994).

146. V.M. Ingram, *Nature*, **178**, 792–794 (1956).

147. S.H. Embury, R.P. Hebbel, N. Mohandras and M.H. Steinberg, 'Sickle Cell Disease. Basi Principles and Clinical Practise', Raven Press, New York, (1994).

148. D. McNaughton, J. Lim, S. Langford, J. Collie and B.R. Wood, *Proc. SPIE – Smart Materials, Nano-, Micro-Smart Syst.* **5651**, 52–60 (2005).

# Glossary

**Acetate paper**   A transparent sheet made of acetate.

**Actin**   A mostly helical protein, which is part of a cell's cytoskeleton and plays an important role in muscle contraction.

**Adenocarcinoma**   A form of *carcinoma* or cancerous tissue found in *glandular epithelium*.

**Agar**   A jellylike polysaccharide on which most bacterial cells are cultured.

**Alveolus (dentalis)**   The tooth bearing part of the jaw bones, tooth socket.

**Amniotic fluid**   The liquid surrounding a growing fetus within the uterus.

**Anaplasia**   Reversal of differentiation of cells, a change of adult cells toward more primitive (embryonic) cell types; a characteristic of tumor cells.

**Angiogenesis**   A process describing the growth of new blood vessels. Angiogenesis is a normal process in growth, development, and wound healing. It is also a fundamental step in the growth of tumors.

**Antigen/antigens**   A substance capable of triggering a specific immune response, usually a protein.

**Apoptosis**   Preprogrammed cell death, cell suicide. All cells have a built-in mechanism, triggered by external or internal signals, that causes a cell to self-destruct in an orderly fashion.

**Atrophic mucosa**   Mucosa thinned as a result of disease, therapy, or individual variation. Such mucosa is more vulnerable to penetration by agents applied to the surface, which can include carcinogens, especially alcohol or tobacco products and *betel* quid.

***Bacillus, pl. Bacilli***   Rod-shaped, Gram-positive bacterium, which may or may not be pathogenic.

**Barrett's esophagus**   A premalignant condition of the esophagus, often caused by prolonged acid reflux disease that can lead to esophageal cancer.

**Basal cells**   In *squamous epithelium*, the lowest layer of cells. Basal cells actively divide to form daughter cells that migrate to the surface of the epithelium while maturing.

**Basement membrane**  A ca. 50-nm-thick collagenous membrane structure that serves to anchor epithelial or endothelial cells.

**Betel**  This term is often used to describe the chewing or placing of a "quid" or "paan" in the oral cavity. Betel itself is a nut (Areca catechu) from which an addictive alkaloid substance is derived. The presence of lime in the "quid" or "paan" increases the alkaloid release. Traditionally, in many cultures the "betel quid" also contains tobacco.

**Biopsy**  The removal of a small piece of tissue for pathological analysis.

**Buccal mucosa**  The part of the oral *mucosa* specifically covering the inside of the cheeks.

**Carcinoma**  A malignant tumor of epithelial cell origin.

**Carboxy-RBC**  The hemoglobin within the red blood cell (RBC) is carboxylated by passing carbon monoxide over the cells suspended in phosphate-buffered saline. When the CO is bound to the heme, the oxidation state of the Fe ion is 2+ and the spin state is $S = 1/2$.

**Catalase reaction**  Uses a tetrameric heme enzyme that breaks down hydrogen peroxide.

**Cervical intraepithelial neoplasia (CIN)**  A premalignant condition found in the *epithelial* cells of the uterine cervix. See also *Three-tier CIN classification.*

**Choroid plexus**  Area on the ventricles of the brain where cerebrospinal fluid is produced.

**Chromatin**  A complex between DNA and histone proteins. In a mammalian cell, DNA is never found as an uncomplexed molecule, but is always tightly wrapped around histones. During transcription (copying of the genetic information from DNA to RNA) or replication (creating a new strand of DNA with complimentary information), short sections of the DNA will unwind for enzymes to copy the original information.

**CIN**  See *Cervical intraepithelial neoplasia.*

**Columnar epithelium (glandular epithelium)**  A form of *epithelium* that lines many internal organs (colon, upper esophagus, endocervix, etc.). It consists of a single layer of cells that are longer than they are wide. The nucleus is close to the base of the cell.

**Cocci**  A form of bacteria with an oval shape.

**Comorbidity**  The presence of other diseases or disorders that might make cancer treatment more hazardous and some therapeutic options impossible to offer.

**Confusion matrix**  A visualization tool typically used in supervised learning. Each column of the matrix represents the instances in a predicted class, while each row represents the instances in an actual class. One benefit of a confusion matrix is that it is easy to see if the system is confusing two classes (i.e., commonly mislabeling one as another).

**Cystitis**   An inflammation of the bladder.

**cyt b558**   Cytochrome *b*558, a key component of *phagocytic* NADPH oxidase.

**Cytoplasm**   The gelatinous body of the cell, consisting of the *cytosol* and cellular organelles.

**Cytosol**   The liquid phase inside a cell, consisting mostly of an aqueous solution of proteins and other cellular molecules.

**Cytoscopy (or cystoscopy)**   Visual inspection of the bladder lining by a flexible or rigid optical probe inserted through the urethra.

**Deoxy-RBC**   The red blood cells (RBC) are deoxygenated by passing $N_2$ over the suspension or alternatively immersing in 0.01 % solution of sodium dithionite. The Fe ion has an oxidation state of $+2$ and a spin state $S = 5/2$.

**Dura mater**   The tough and inflexible outermost of the three layers of the *meninges*.

**Dysplasia/dysplastic cells**   Abnormal maturation of cells, or the presence of immature cells in tissue at locations where mature cells are normally found.

**Ectocervical (exocervical) tissue**   The *squamous epithelium* covering the outer part of the cervix and protrudes into the vagina.

**Edema**   An increase of the interstitial fluid in any organ causing swelling.

**Encephalopathy**   A disease of the brain.

**Endometrium**   An *epithelial* lining of the uterus.

**Endoplasmic reticulum (ER)**   Membrane-rich structures adjacent to the cell nucleus. *Ribosomes* are located on the ER.

**Eosinophilic**   1. Stained by eosin, a stain routinely used in pathology.
        2. Relating to *eosinophils*. A collection of eosinophils at a greater than usual level in a tissue on histology would be referred to as an eosinophilic infiltration.

**Eosinophils**   "Acid loving" motile *phagocytic* granular *leukocytes*, which have a nucleus with two lobes connected by a thread of chromatin. The cytoplasm contains coarse, round granules of uniform size. Eosinophils kill the larvae of parasites that invade tissues.

**Eosinophil peroxidase**   EPO is a heme enzyme that is cytotoxic to bacteria.

**Epidemiology**   The study of disease by means of its incidence or prevalence at population level.

**Epigenetic**   Heritable, but potentially reversible differences in gene expression that occur without alterations in DNA sequence. This includes events such as gene methylation and phosphorylation.

**Epithelium/epithelial tissue**   Tissue composed of layers of cells. Epithelial tissue lines the inside of body cavities, as well as the outside of the body. See also *squamous and glandular epithelium*.

**Erythrocytes**   *Red blood cells (RBC)*. Cells whose sole function is to carry oxygen from the lungs to tissue. Oxygen is bound to hemoglobin, the major constituent of erythrocytes.

**Estrus/estrous**   In mammals, a cyclic event in which females are sexually receptive. Estrus may or may not be preceded by ovulation (release of an ovum).

**Ethmoid sinus**   One of the paranasal air sinuses. It forms a network of thin-walled bony cavities around the area at the bridge of the nose extending backward between the two orbital (eye) sockets.

**Eukaryote/eukaryotic cell**   Highly organized cells of plants, fungi, and animals that contain membrane-enclosed organelles such as the nucleus.

**Exfoliated cells**   Cells that have been removed from the body by gentle scraping or swiping, or by fine needle aspiration.

**Exophytic**   Describes a pattern of growth that extends outward. It is usually applied to primary cancers, which appear as raised lumps above the surface of the surrounding mucosa.

**Extracapsular spread**   Spread of cancer deposits beyond the natural boundary (capsule) of the lymph node.

**Fibroblast**   Cells that maintain the protein matrix between cells (extracellular matrix).

**Fibrosis**   The formation of excess connective tissue.

**Food vacuole**   The region within *Plasmodium* parasite where toxic iron protoporphyrin IX is oxidized into iron protoporphyrin IX hydroxide and then sequestered into chemically inert *hemozoin* (malaria pigment).

**Franck–Condon principle**   The Franck–Condon principle explains the intensity of vibronic transitions. The principle states that during an electronic transition, a change from one vibrational energy level to another will be more likely to occur if the two vibrational wave functions overlap significantly.

**Gammopathy**   A primary disturbance in immunoglobulin synthesis.

**Genotype**   Characteristics of microorganisms determined by genes at specific loci, e.g., the presence of the gene for a specific enzyme. Genotypic techniques use genetic information as measures.

**Genus, pl. genera**   The second most specific taxonomic (see *taxa*) level including closely related species.

**Gingiva(e)**   The mucosa attached to the tooth-bearing segment of the jaws.

**Glandular epithelium**   See *columnar epithelium.*

**Gleason score**   A score commonly used for grading prostate cancer. In this system, a sample is assigned a score from 1 to 10, by combining the two most important patterns of cells as Gleason grades, to give a total score (e.g., grade 3 + grade 4 = score 7). Scores in biopsies generally range between 4 and 10, with 6 and 7 being the most common.

**Glial cells (neuroglia)**   The cells in the brain that are not neurons. Glial cells are responsible for support, nutrition, and some signal transmission.

**Glioma**   Tumors of *glial* cells.

**Gliosis**   A benign proliferation of *glial* cells as a reaction of the central nervous system to neurological disorders.

**Glottis**   The area of the larynx including and immediately surrounding the vocal cord.

**Golgi apparatus**   An organelle found in *eukaryotic cells*, whose function is to process proteins and lipids synthesized in a cell.

**Gram stain**   The test for determining the Gram type of the cell wall. After treatment with gentian violet and Gram's solution, gram-negative organisms can be decolorized with alcohol, whereas gram-positive organisms retain the gentian violet.

**Gram-negative**   A group of bacteria with few layers of peptidoglycan in their cell wall, which give a negative result in the *Gram stain.*

**Gram-positive**   A group of bacteria with many layers of peptidoglycan in their cell wall, which give a positive result in the *Gram stain.*

**Hematogenous**   Mediated by the bloodstream.

**Hematin**   A toxic degradation product of hemoglobin, produced by the malaria parasite.

**Hemozoin**   An insoluble, inert form of *hematin* produced by the malaria parasite to counteract the toxic effects of hematin.

**Hematoxylin and eosin (H&E)**   A staining method commonly used in pathology to distinguish normal from cancerous tissue *biopsies*. The basic dye hematoxylin stains *basophilic* structures (DNA, RNA) with blue–purple hue. The acidic dye eosin stains *eosinophilic* structures (protein) bright pink. Most of the cytoplasm is eosinophilic.

**Heterogeneity factor**   $H$, is a measure of the similarity of sets of spectra. $H$ is calculated as follows:

$$H(r, i) = \frac{[n(p) + n(i)] \cdot D(p, i) + [n(i) + n(q)] \cdot D(q, i) - n(i) \cdot D(q, i)}{n + n(i)}$$

where $n$ is the number of spectra merged in the $i$th cluster; the clusters $p$ and $q$ are merged to the new cluster $r$; $D(p, i)$ is the spectral distance between the $p$th and $i$th clusters, while $D(q, i)$ is the spectral distance between the $q$th and $i$th clusters. $H$ is usually applied in conjunction with Ward's algorithm by which homogeneous groups are found by merging the two groups that show the smallest growth in $H$. Instead of determining the spectral distance, $D$, Ward's algorithm determines the growth of the heterogeneity. Since the value of $H(r, i)$ depends on the number of spectra, $H$ may adopt values larger than 2000.

**HPV**   Human papilloma virus, a DNA virus that affects the *mucous* membrane. A number of these are transmitted through sexual activity, affect the cervix, and a small percentage lead to precancerous lesions. Some papilloma viruses cause warts. Recent evidence supports the presence of HPV in some cancers of the head and neck, especially in the *oropharynx*.

**Hyperkeratosis (adj.: hyperkeratotic)**   A condition where excess keratin is produced. Keratin is a fibrous, structural protein.

**Hypopharynx**   The area of the pharynx behind the larynx.

**(ImH)$_2$ Fe(II)PP**   Bis-imidazole ferrous protoporphyrin IX.

**Infiltrating ductal carcinoma**   A carcinoma of the milk ducts of the breast that has invaded surrounding tissues.

**Intermediate cells**   In stratified *squamous* tissue, the layer of cells above the *parabasal* and below the *superficial layer*.

**Kappa**   A statistical term that measures the level of agreement. The higher the kappa, the more reliable is the association between the two variables measured.

**Leptomeninges**   A collective term used to refer to the pia mater and arachnoid mater, which are the inner two layers of the *meninges*.

**Leukocyte**   The white blood cell produced in the bone marrow. These cells, along with *lymphocytes*, are part of the body's immune system.

**Locoregional**  Pertaining to the primary cancer site or the area of lymphatic drainage specifically related to it. In the case of head and neck cancers, the lymphatic drainage is to the *lymph nodes* of the neck. Failure of treatment is usually divided into locoregional and distant, as problems with the former are likely to be improved by local therapies, such as surgery or radiotherapy, whereas distant failure contraindicates surgery.

**Lymphadenopathy**  Literally, the pathology affecting the *lymph nodes*. In clinical practice it is used to describe lymph node–related swelling.

**Lymph nodes**  Components of the lymphatic immune system of the body. Lymph nodes contain and produce *lymphocytes*, and filter foreign particles, such as cancer cells that may be carried through the body via the lymph.

**Lymphocytes**  A type of white blood cell (see *leukocyte*) found in the lymphatic system.

**Lysis (vb: lyse)**  The breakage of a cell's membrane, for example, by osmotic pressure changes.

**Maxillary sinus**  The largest of the paranasal air sinuses. It lies above the upper teeth within the substance of the cheek bone, lateral to the nasal cavity and extending up to the orbits (eye sockets).

**Melanoma**  Highly malignant tumor of the melanocytes, the cells in the skin that form the pigment melanin.

**Meninx, pl. meninges**  A membrane (or membranes) that envelopes the central nervous system.

**Mesenchymal (stem) cells**  The multipotent stem cells that can differentiate into connective, muscle and bone tissue cells, and several other cell types.

**Meso-meso–linked porphyrin array**  Porphyrin units linked together through the meso-carbon atoms of adjacent porphyrins.

**Metaplasia**  A reversible replacement of mature cells of one type by mature cells of another type.

**Metastatis**  The spread of a malignant tumor from one organ to another.

**MetRBC**  The hemoglobin in the *RBC* is oxidized by immersing the cells in 0.1% sodium nitrate. The Fe ion has an oxidation state of $3+$ and a spin state of $S = 5/2$. In this state the Fe ion cannot bind oxygen.

**Mitosis**  The final step in a cell's division cycle. Mitosis occurs after a cell has successfully duplicated its DNA. During mitosis, two identical new cell nuclei are formed.

**Mucin**   The glycosylated proteins secreted by *glandular* cells of the mucosal membrane.

**Mucosa**   An epithelial lining of various body cavities. Not all mucous membranes secrete mucus.

**Mucoepidermoid**   A descriptive term for a form of salivary cancer used by pathologists and clinicians. Derived from the presence of mucus secreting and epithelial cells found in this cancer.

**Murine**   Pertaining to a mouse, such as a murine model of disease.

**Myelination**   The formation of a myelin sheath around a nerve fiber. Myelin is an electrically isolating phospholipid layer.

**Myeloperoxidase**   MPO is a peroxidase enzyme found in *neutrophilic* granulocytes. It is a lysosomal protein stored in azurophilic granules of the neutrophil. MPO has a heme pigment, which gives it a green color in secretions rich in neutrophils, including pus and mucus.

**Myofibroblasts**   Cells that are, in terms of their differentiation, between *fibroblasts* and smooth muscle cells.

**NADPH**   Reduced nicotinamide adenine dinucleotide phosphate.

**Nasopharynx**   The area of the pharynx lying immediately behind the nasal cavity. It extends from the skull base to the *oropharynx*.

**Necrosis**   An accidental or traumatic cell death, caused by infection, inflammation, poison, or injury. The process is less orderly than *apoptosis*.

**Necrotic ulcer slough**   Ulcers are typically surfaced by a layer of dead (necrotic) tissue. This appears as a pale area on the surface of the ulcer and comprises dead cells, bacteria, and tissue elements.

**Neogenesis**   A new formation.

**Neoplasia (neoplasm, adj.: neoplastic)**   An abnormal and uncontrolled cell growth.

**Neurological precursor cells**   Stem cells with the ability to develop into neurological cells.

**Neurons**   Electrically excitable cells in the nervous system, also known as nerve cells.

**Neutrophils**   The most abundant type of *leukocyte* involved in the immune response. Their name is derived from the staining characteristics on *hematoxylin and eosin* (H&E) histological preparations.

**Neurons**   The cells in the nervous system that process and transmit information.

**Nuclear atypia**   Abnormal shape, size and staining patterns of the nucleus.

**Nuclear-to-cytoplasm ratio**   An estimate of the size of the nucleus of a cell compared to the size of the cytoplasm.

**Nucleolus**   A suborganelle of the cell nucleus responsible for assembly of *ribosomes*.

**Oligodendrocytes**   The cells that coat axons in the central nervous system with their cell membrane, called myelin.

**Oncotic**   Caused or marked by swelling.

**Oropharynx**   The area of the pharynx lying behind the oral cavity. It begins immediately anterior to (in front of) the tonsils.

**Oral mucosa**   The mucosal lining of the oral cavity.

**Organelles**   Subcellular organization in *eukaryotic* cells (e.g., the nucleus, *endoplasmic reticulum*, *Golgi apparatus*).

**Oxy-RBC**   The hemoglobin in the *RBC* is oxygenated by subjecting the suspension to atmospheric oxygen for 30 min. The Fe ion has an oxidation state of $+3$ and a spin state of $S = 0$.

**Parabasal cells**   In stratified *squamous* tissue, the layer of cells between *basal* and *intermediate* layers (stages).

**Parotidectomy**   The operation to remove part or all of the parotid salivary gland. The most usual is a superficial parotidectomy that removes the part of the gland lying external to the facial nerve (the nerve that moves the muscles of facial expression). Total parotidectomy describes the removal of salivary tissue both superficial and deep to the nerve.

**Pathogen**   A chemical or biological agent that causes disease.

**PCR**   An abbreviation of polymerase chain reaction. PCR is used in molecular biology to copy genes, e.g., for detection purposes.

**Pericytes**   The cells associated with the walls of small blood vessels. They are important in blood–brain barrier stability.

**Phage**   A synonym for bacteriophage; viruses that have a specific affinity for and infect bacteria.

**Phage type** The pattern of susceptibility to different bacteriophages used for subspecies discrimination.

**Phagocytic (cells)** The cells, such as neutrophils, eosinophils, basophils, etc., that are able to remove pathogens and foreign objects by engulfing them in membrane structures and rendering them harmless.

**Phenotype** Observable physiological properties of microorganisms determined by the expression of genes, e.g., the production of a specific enzyme. Phenotypic techniques use physiological tests such as the utilization of certain sugars, as measures. Organisms of the same genotype may have different phenotypes due to differences in the expression of genes.

**Pial surface** Pertaining to the pia mater, the delicate and highly vascular membrane investing the brain and spinal cord.

**Pinealocytes** The main cells of the pineal gland, which produce and secrete the hormone melatonin.

**Pleomorphic adenoma** Pleomorphic literally describes the variation on histological appearance. Adenoma refers to a tumor of glandular origin. These tumors, which are the most common type affecting the parotid gland (85%), were in the past thought to derive from both epithelial and connective tissue elements. It is now known that they are entirely of epithelial derivation but the name persists. They are considered benign but locally invasive.

**Pleomorphism** In cytology, the variability in the size and shape of cells and/or their nuclei.

**Prion protein** Small protein that can exist in a noninfectious state, and an infectious isoform responsible for *spongiform encephalopathy*.

**Prokaryote/prokaryotic cell** Cells without a nucleus, such as bacteria, in which the genetic material is contained in a circular chromosome, rather than in a cell nucleus, as in the case of *eukaryotes*.

**Proteases** A protein (enzyme) that degrades other proteins (proteolysis).

**Pyknosis** A reduction in the size of a cell or cell nucleus, often associated with terminal differentiation of a cell.

**RBC** Red blood cell, or *erythrocyte*.

**Rete ridges** On histological section a wavy pattern is seen at the epithelial/connective tissue junction. The part of the epithelium that extends downward is described as a rete ridge. They vary in size and shape by anatomical site and disease process.

**Ribosome** Small particles (20 nm) inside the *cytosol* of cells. Ribosomes consist of ribosomal RNA (r-RNA) and ribosomal proteins. Messenger RNA (m-RNA), which is produced in the cell's nucleus by transcription, binds to ribosomes where the amino acid sequence is read from the m-RNA and translated into a protein primary sequence.

**Sarcoma** Cancer of connective tissue, bone, and cartilage.

**Schwann cells** Provide *myelination* to axons in the peripheral nervous system similar to *oligodendrocytes*.

**Schwannoma** A benign encapsulated neoplasm derived from the Schwann cells that invest neural fibers.

**Scrapie** A form of *spongiform encephalopathy* in sheep and goats. Similar to Creutzfeldt–Jakob disease in humans.

**Sero** Of serum; e.g., seropositive denotes that the individual's serum carried a specific marker or antibody.

**Serology** A method for determining the serovar of a microorganism applying antibody sera.

**Serovar** A property of organisms determined by the combination of O- and H-antigens. O-antigens are located at the cell surface and determine the serogroup. H-antigens are flagella antigens. The serovar is usually determined by agglutination tests with antibodies and serves as an additional information to the species identity, e.g., for epidemiological studies.

**SIL** See *squamous intraepithelial lesion*.

**Species** A biological name designating the basic unit of biological classification.

**Spongiform** The formation of tiny holes (like in a sponge) in the brain cortex. "Spongiform *encephalopathy*" describes a class of diseases of the central nervous system caused by prions (misfolded proteins).

**Squamous epithelium** A type of epithelium consisting of flat, scale-like cells called *squamous cells*. Stratified squamous epithelium consists of four layers of cells of different degree of maturation: *basal*, *parabasal*, *intermediate*, and *superficial cells*.

**Squamous intraepithelial lesion** An abnormal growth of *squamous cells*, see also *Two-tier Bethesda system* and *Three-tier CIN classification*.

**Squamous nuclear atypia** The *squamous* cells with atypical structure such as enlargement or excessive pigmentation.

**Staging**   A formal process by which the extent of the cancer is graded, taking into account the primary cancer, the lymph nodes, and any evidence of distant spread (usually seen in the lungs).

**Strain**   A genetic variant or subtype of a bacterium within a species. This means that a species is defined by a number of stable and fundamental properties with a whole spectrum of different "strains" or "individuals", slightly differing in their genetic, phenotypic appearance, and behaviour, belonging to such a species.

**Stroma**   The connective tissue in an organ.

**Superficial cells**   In stratified *squamous epithelium*, the outermost and most mature layer of cells.

**Taxon (pl. taxa) or taxonomic unit**   The name designating a specific group of organisms. A taxon is assigned a rank and can be placed at a particular level in a systematic hierarchy reflecting evolutionary relationships. A distinction is to be made between taxa/taxonomy and classification/systematics. The former refers to biological names and the rules of naming. The latter refers to rank ordering of taxa according to presumptive evolutionary (phylogenetic) relationships. A scheme of ranks in hierarchical order for bacteria is, e.g., *genus*, *species*, and *strain*.

**Taxonomy**   The practice and science of classification in biology, which is also used in microbiology. Taxonomic schemes are composed of taxonomic units known as taxa (singular taxon) of any kind of organisms that are arranged in a hierarchical structure, typically related by subtype–supertype relationships.

**Thinprep**$^{TM}$   A methodology using a method and solutions to remove confounding cell types and debris from a Pap smear.

**Three-tier CIN classification**   In cervical pathology, a classification system that defines how far *neoplastic* cells have spread into the *squamous tissue*: CIN I (about 1/3), CIN II (about 2/3), and CIN III: total thickness up to the surface of the squamous tissue. CIN: cervical intraepithelial neoplasia.

**Two-tier Bethesda system**   In cervical cytology, a system for classification of cervical disease as low-grade (LGSIL) and high-grade cervical intraepithelial lesion (HGSIL).

**Triage**   A system to sort patients (or other items) by their needs or severity in case of insufficient resources to treat all of them.

**Trypsination**   The removal of adherent cultured cells from the culture flasks by digestion of surface proteins using trypsin.

**Ulcerative colitis**   An inflammatory bowel disease, affecting mainly the large bowel (colon).

**Urothelium**   Epithelial tissue lining the inside of the urinary bladder.

**Villoglandular adenocarcinoma**   A rare variant of a cervical adenocarcinoma with a favorable outcome.

**Vimentin**   A protein that makes up the cytoskeleton of a cell.

# Appendix: Infrared and Raman Spectra of Selected Cellular Components

In this appendix, the infrared and Raman spectra of the following biochemical components of cells and tissues are shown:

1. Nucleic acids: DNA (Figure A1)
   RNA (Figure A2)
2. Proteins: Albumin (mostly $\alpha$-helical) (Figure A3)
   $\gamma$-Globulin (mostly $\beta$-sheet) (Figure A4)
   Collagen (triple helical) (Figure A5)
3. Lipids (Figure A6) and Phospholipids (Figure A7)
4. Carbohydrate: Glycogen (Figure A8)

In each panel, the top trace represents the infrared spectrum, which is represented as absorption spectra with amplitudes between 0 and 0.5 AU. The middle trace represents the second derivative of the corresponding infrared spectrum, after multiplication by $(-1)$. The bottom trace represents the Raman spectra, which are presented in arbitrary intensity units.

All spectra were acquired microscopically from thin films cast on $CaF_2$ windows after dissolving the samples in appropriate solvents, and air drying. Infrared data were collected at $4\,cm^{-1}$ spectral resolution from sample spots measuring about $40\,\mu m$ on edge, using a PerkinElmer (Shelton, CT, USA) Spotlight 300 microspectrometer. 64 or 128 interferograms were coadded to improve the signal-to-noise ratio. Raman data were collected using a WiTEC (Ulm, Germany) Confocal Raman Microspectrometer CRM 200 using either 514.5 or 632.8 nm excitation (ca. 15 mW power at the sample). The spectral resolution of the Raman spectra, dispersed by a 30-cm focal length monochromator incorporating a $600\,mm^{-1}$ grating, varies between ca. 3 and $5\,cm^{-1}$. Data acquisition time for each spectrum was between 3 and 10 s.

The editors wish to thank Ms Tatyana Chernenko and Dr Christian Matthäus from the Laboratory for Spectral Diagnosis, Department of Chemistry and Chemical Biology, Northeastern University in Boston for collecting and presenting these reference data. These spectra will be available online as an EXCEL spreadsheet.

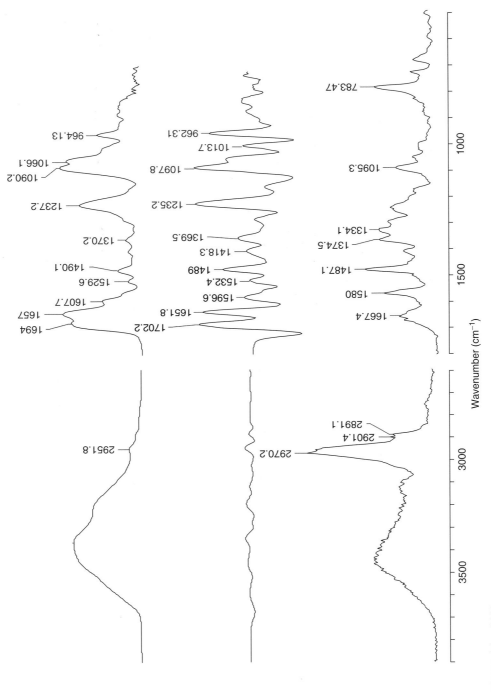

**Figure A1.** DNA.

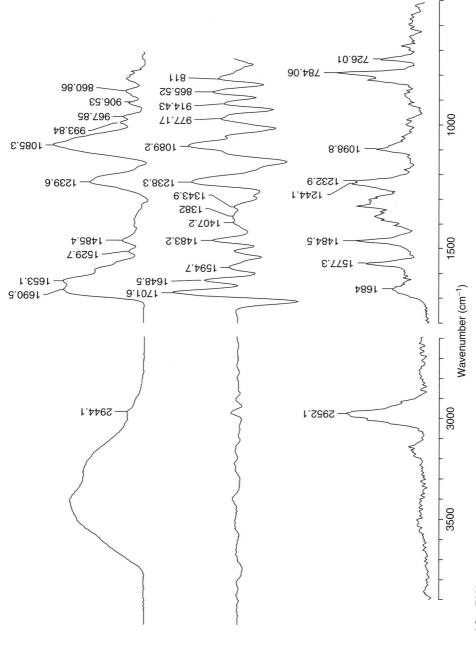

**Figure A2.**    RNA.

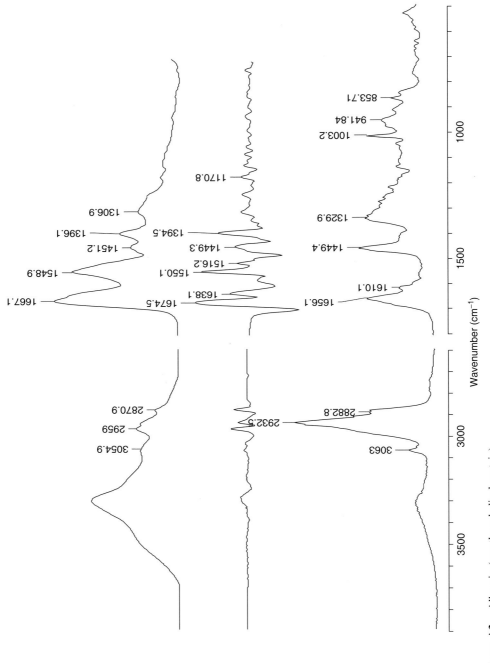

**Figure A3.**   Albumin (mostly α-helical protein).

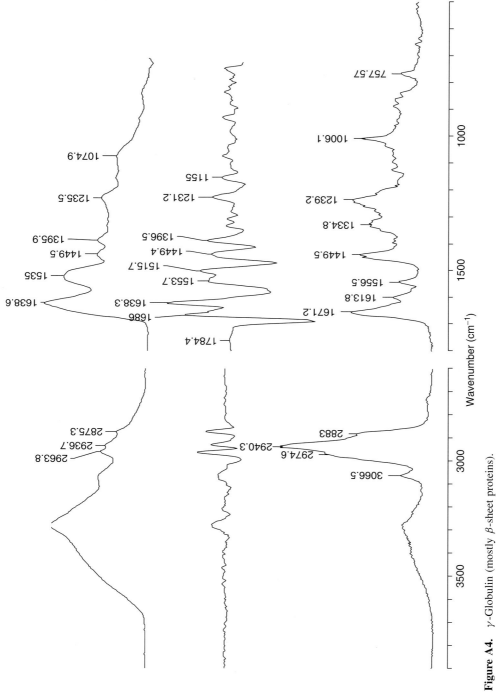

**Figure A4.** γ-Globulin (mostly β-sheet proteins).

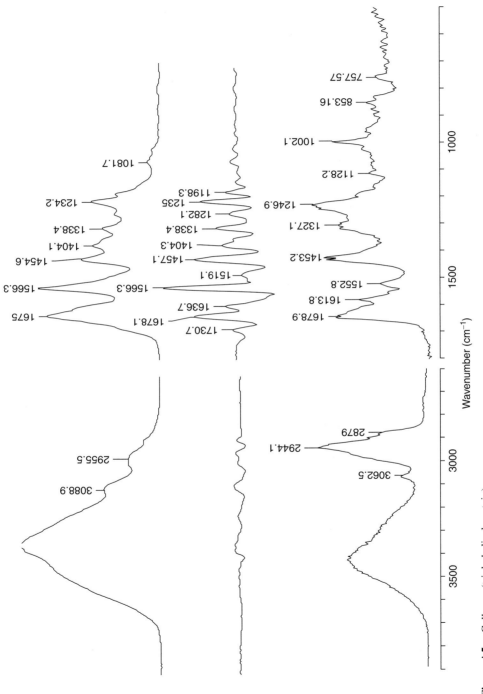

**Figure A5.** Collagen (triple helical protein).

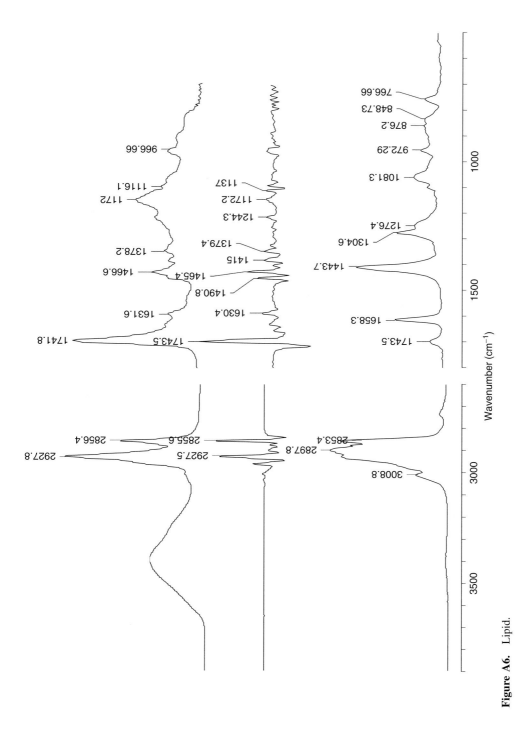

**Figure A6.** Lipid.

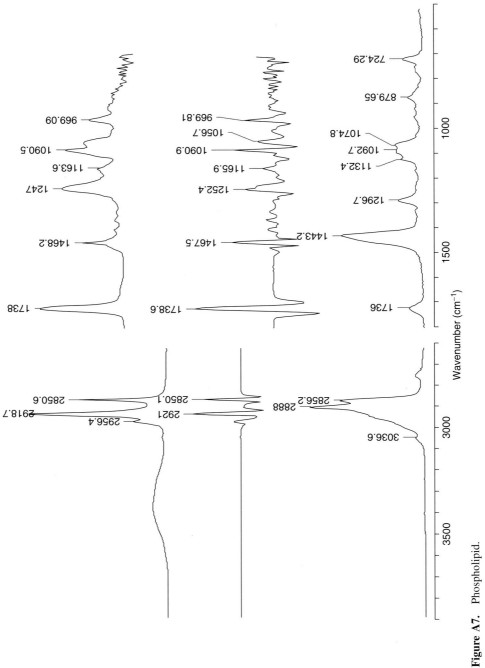

**Figure A7.** Phospholipid.

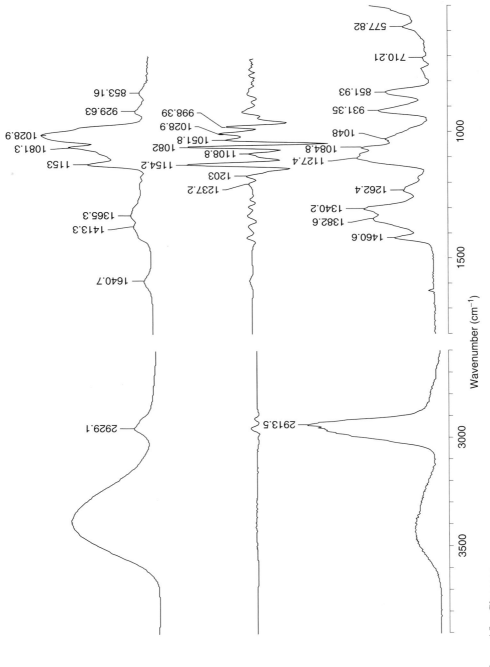

**Figure A8.** Glycogen.

# Index